Discrete mathematics

Ramanujan Mathematical Society Lectures Notes Series

Vol. 1: Number theory

Vol. 2: The Riemann zeta function and related themes

Vol. 3: Formal language aspects of natural computing

Vol. 4: Commutative algebra and combinatorics

Vol. 5: Convexity in discrete structures

Vol. 6: Number theory and discrete geometry

Vol. 7: Discrete mathematics

Vol. 8: Lectures on operator theory

Vol. 9: Essays on geometric group theory

Vol. 10: Techmüller theory and moduli problems

Ramanujan Mathematical Society

Lecture Notes Series

Volume 7

Discrete mathematics

Proceedings of the international conference on discrete mathematics, Indian Institute of Science, Bangalore, December 2006

Volume editors

R. Balakrishnan
C. E. Veni Madhavan

International Press
www.intlpress.com

Ramanujan Mathematical Society
Lecture Notes Series, Volume 7
Discrete mathematics

Volume Editors:
R. Balakrishnan
C. E. Veni Madhavan

Published in 2010 by International Press, Somerville, Massachusetts, U.S.A. under license from the Ramanujan Mathematical Society.

This work is also published in the Republic of India, exclusively by the Ramanujan Mathematical Society.

ISBN 978-1-57146-192-6

Typeset using the LaTeX system.
Printed in the United States of America.

Preface

Ramanujan Mathematical Society, which was founded in 1985, completed in 2005, twenty years of its successful activities in fostering research and education in mathematics. To mark this completion, the society organized ICDM-2006 (International Conference on Discrete Mathematics, 2006) jointly with the Institute Mathematics Initiative of the Indian Institute of Science, Bangalore during December 15–18, 2006. There were 170 registered participants in this conference.

This volume which forms the Proceedings of ICDM-2006 contains the refereed versions of the lecture notes of the invited talks as also the papers accepted from out of the lecture notes and papers submitted for the Proceedings.

We thank the members of the various committees constituted for the smooth and successful conduct of the conference. Our special thanks are due to C.R. Pradeep who helped us a great deal in looking after the conference arrangements at Bangalore. We also thank the Department of Science and Technology, Government of India, Cryptology Research Society of India and the Indo-US Science and Technology Forum, New Delhi for their financial support for the Conference. Our thanks are also due to the several referees for coming forward to spare their precious time.

We thank the Ramanujan Mathematical Society for undertaking to publish this Proceedings in their RMS-LNSM (Ramanujan Mathematical Society Lecture Notes Series in Mathematics). Finally, we thank Dr. M. Kalidoss, who did a wonderful job in bringing out both the book of lecture notes (despite severe time constraints) at the time of ICDM 2006 and the present book.

It is expected that this Proceedings which contains surveys and papers presented by several leading workers in discrete mathematics would undoubtedly prove to be a good source book for researchers in discrete mathematics.

R. Balakrishnan,
Srinivasa Ramanujan Centre,
SASTRA University,
Kumbakonam

and

C.E. Veni Madhavan,
Indian Institute of Science,
Bangalore

Volume Editors

Programme Committee

Arumugam, S., AKCE, Krishnan Koil, Tamil Nadu, India

Asha Rao, Royal Melbourne Institute of Technology, Australia

Balakrishnan, R. **(Co-Chair)**, Srinivasa Ramanujan Centre, Kumbakonam, Tamil Nadu, India

Jaikumar Radhakrishnan, Tata Institute of Fundamental Research, Mumbai, India

Kavitha, T., Indian Institute of Science, Bangalore, India

Paulraja, P., Annamalai University, Tamil Nadu, India

Pradeep, C.R., Indian Institute of Science, Bangalore, India

Raspaud, Andre, University of Bordeaux, France

Renu Laskar, Clemson University, Clemson, USA

Sampath Kumar, E., University of Mysore, Mysore, India

Sunilchandran, L., Indian Institute of Science, Bangalore, India

Veni Madhavan, C.E., **(Co-Chair)**, Indian Institute of Science, Bangalore, India

Venkatesh Raman, Institute of Mathematical Sciences, Chennai, India

Vijayakumar, A., Cochin University of Science and Technology, Kochi, India

Local Organizing Committee

Jayaram, S.R., Bangalore

Pradeep, C.R., (Convener), Indian Institute of Science, Bangalore

Ramesh Babu, D.R., PESIT, Bangalore

Rangarajan, G., Indian Institute of Science, Bangalore

Shyam Kamath, National Institute of Technology, Suratkal

Sooryanarayanan, B., Ambedkar Institute of Technology, Bangalore

Sunilchandran, L., Indian Institute of Science, Bangalore

Veni Madhavan, C.E., Indian Institute of Science, Bangalore

Venkatesh, K., Alliance Business Academy, Bangalore

INTERNATIONAL CONFERENCE ON DISCRETE MATHEMATICS

DECEMBER 15-18, 2006

VENUE : MRC Auditorium, IISc. Bangalore.

Contents

Plenary Talks

Proc. of ICDM 2006, Pages 1–8

List-colorings and the Number of Colorings of a Graph

CARSTEN THOMASSEN
Technical University of Denmark.
Email: C.Thomassen@mat.dtu.dk

Abstract. We survey some recent results and problems on the chromatic polynomial and list colorings of graphs.

1 Introduction: The Chromatic Number of a Graph

The *chromatic number* $\chi(G)$ is the smallest number k such that the vertices of G can be colored in colors $1, 2, \ldots, k$ in such a way that neighbors always receive distinct colors.

Historically, the main motivation for studying the chromatic number was the famous 4-Color-Problem, now a theorem. A *planar graph* is a graph that can be drawn in the plane such that the vertices are points and every edge xy is a simple arc joining x and y, and no edges intersect except at a common end. The 4-Color-Theorem can be formulated as follows:

Theorem 1.1. *$\chi(G) \leq 4$ for every planar graph G.*

The 4-Color-Problem goes back to the first half of the 19th century and remained open for more than a century. An affirmative answer was provided by Appel and Haken in 1976 [3, 4]. The proof is very complicated, and its completeness has been questioned. A simpler (but still complicated) proof has been found by Robertson, Seymour, Sanders and Thomas [17]. Both proofs involve a computer search, and it still remains a challenge to find a proof that is not computer based.

More recently, part of the motivation for the chromatic number has come from real life problems, as many scheduling problems may be formulated in terms of the chromatic number. Consider for example a school with teachers $t_1, t_2, \ldots, t_m$ and classes $c_1, c_2, \ldots, c_p$. Assume that teacher t_i teaches class c_j $m_{i,j}$ times during a week. Assume that the school has plenty of classrooms, but no teacher can teach two different classes at the same time and no class can be taught by two teachers at the same time. What is the smallest number of periods needed during a week? We form a new graph G whose vertices are the teachers and the classes and we add $m_{i,j}$ edges between t_i and c_j for $i = 1, 2, \ldots, m$, $j = 1, 2, \ldots, p$. The answer to the above question is then precisely the smallest number of colors needed to color the line graph of G, see the last section. This optimization problem is very well understood. The answer is simply the maximum degree (valency) of the graph G.

However, many similar examples occur, and finding the chromatic number in general is difficult. More precisely, it is NP-complete, see [11].

2 The Chromatic Number and Gallery Problems

The chromatic number shows up in many other contexts. Consider for example a closed polygonal (not necessarily convex) curve in the plane with n corners. We think of each line segment as a wall in a gallery viewed from above. In that gallery we select a number of points (in each of which we place a guard) so that every wall can be seen from one of the points. What is the number of guards needed in the worst case? It is easy to give examples where $n/3$ guards are needed. Chvátal proved that $n/3$ are always sufficient, see [1]. The following beautiful argument is due to Fisk, see also [1]: We think of the gallery as a cycle with n vertices drawn in the plane. We add successively noncrossing straight edges in the interior of the cycle until the interior is triangulated. Then we color the vertices of the resulting graph G. Clearly, we need at least three colors for that. On the other hand, it is easy to color G with only three colors: We color first a triangle with colors $1, 2, 3$. In the general step we choose a triangle in which precisely two vertices are colored, and then we give the third vertex the third color. It is easy to see that this results in a 3-coloring of G. The smallest color class has at most $n/3$ vertices, and we may place a guard in each vertex of that color class.

There are other intriguing gallery problems, some of which are unsolved, see e.g. [1].

3 The Chromatic Number and Degeneracy

A graph is *k-degenerate* if each subgraph has a vertex of degree at most k. Clearly, every k-degenerate graph is $(k+1)$-colorable. As every planar graph is 5-degenerate, it follows that every planar graph is 6-colorable. As there exist planar triangulations of minimum degree 5, for example the graph of the icosahedron, the 5-color theorem for planar graphs does not follow immediately from degeneracy. However, there is a way to get around this. Borodin [8] was perhaps the first to suggest to improve this upper bound on the chromatic number by partitioning the vertex set into two sets which induce graphs with better degeneracy properties. Specifically he conjectured that every planar graph (which is 5-degenerate by Euler's formula) has a vertex partition whose parts induce a 1-degenerate and a 2-degenerate graph, respectively, and another vertex partition whose parts induce a 0-degenerate and a 3-degenerate graph, respectively. These conjectures were proved in [22] and [25], respectively. Note that both of these results imply the 5-color theorem for planar graphs. This is a complicated way of proving the 5-color theorem but it raises the question if other coloring results follow by this method. Unfortunately, the 4-color theorem does not. It is easy to see that a planar triangulation of minimum degree 5 cannot be partitioned into a 0-degenerate and a 2-degenerate graph. Also, there exist planar triangulations which cannot be decomposed into two 1-degenerate graphs, namely the dual graphs of the counterexamples to Tait's conjecture mentioned in the next section.

A classical result of Grötzsch says that every planar graph of girth at least 4 can be colored in 3 colors only. A simple list color proof is presented in [26]. Recently, Kawarabaiashi and the present author [14] proved that Grötzsch's theorem can be proved in the same spirit as the two afore-mentioned conjectures of Borodin.

Theorem 3.1. *The vertex set of a planar graph graph G of girth 5 can be partitioned into two sets A, B such that A is an independent set and B induces a forest.*

4 The Chromatic Polynomial

The chromatic polynomial $P(G,t)$ was introduced by Birkhoff [5] in 1912. If G is a graph and t is a nonnegative integer, then $P(G,t)$ denotes the number of colorings of $V(G)$ such that all colors are one of the integers $1, 2, \ldots, t$ and neighbors receive different colors. Clearly, $P(K_n, t) = t(t-1)\ldots(t-n+1)$. If x, y are nonadjacent vertices in a graph G, and G' is obtained from G by identifying x and y, then clearly, $P(G,t) = P(G \cup \{xy\}, t) + P(G', t)$. So it follows by induction on the number of missing edges in G that $P(G,t)$ is a polynomial of degree n, where n is the number of vertices of G. By the same argument, a number of other properties of the chromatic polynomial can be derived. For example, the leading coefficient is 1, and the coefficients alternate in sign. However, one encounters quickly open questions. For example, an old conjecture of Read says that the sequence of absolute values of the coefficients is unimodal, that is, it first increases, then decreases, see [16].

The chromatic polynomial was introduced in order to understand the chromatic number better, specifically to solve the 4-Color-Problem. The 4-Color-Theorem is equivalent to the statement that $P(G,4)$ is positive for every planar triangulation. Tutte came close, see [16]:

Theorem 4.1. $P(G, (5+\sqrt{5})/2) = P(G, 3.618\ldots)$ *is positive for every planar triangulation.*

The reason that the chromatic polynomial has fascinated discrete mathematicians for years is not its potential to find the chromatic number. Ironically, there seems to be no family of graphs, perhaps even not one single graph, for which the easiest or most natural way to determine the chromatic number goes via the chromatic polynomial. It would be highly interesting to construct such a graph. The main reason for studying it is probably that it shows up in surprising contexts. One striking example is the following result of R. Stanley: If G is a graph, then an *acyclic orientation* of G is obtained by numbering it vertices $1, 2, \ldots, n$ and then orient every edge from the small end to the large end. Clearly, K_n has $n!$ acyclic orientations, and every other graph G on n vertices has fewer. To calculate the exact number is known to be hard in general. Stanley proved that it equals $|P(G,-1)|$, that is the absolute value of the chromatic polynomial evaluated at -1.

In recent years the chromatic polynomial has also been studied by theoretical physicists, see e.g Shrock and S-H Tsai [18]. In a letter of April 26, 1999, Shrock has informed me that $P(G,q)$ is equal to the zero-temperature limit of the partition function of a certain spin model called the *q-state Potts antiferromagnet* on the graph G (which is often a subgraph of a regular lattice). Also, the roots of the chromatic polynomials are studied by physicists. Thus Sokal [19] has proved that the collection of complex roots of all chromatic polynomials is a dense set in the complex plane except the disc with center 1 and radius 1.

We close this section with some comments on real roots. We say that t is a *chromatic root* of G if $P(G,t) = 0$. Clearly, 0 is a chromatic root, and so is 1 unless G has no edges. It is well known and easy to see that there are no other roots less than 1. An old (still unsettled) conjecture by Birkhoff and Lewis [6] says that a planar graph has no chromatic root less than or equal to 4. More recently, Jackson [12] proved the following fascinating result:

Theorem 4.2. *All chromatic roots distinct from* $0, 1$ *are greater than* $32/27$.

Jackson also proved that 32/27 cannot be replaced by any larger number. This was strengthened in [21] where it is proved that the set of chromatic roots consists of 0 and 1 and a dense subset of the interval from 32/27 to infinity. Inspired by these results, I found in [24] a new connection between colorings and the problem of finding a *Hamiltonian cycle* that is, a cycle containing all vertices of a graph. This problem is a special case of the *Travelling Salesman Problem*, that is, the problem of minimizing the length of a closed walk in a graph visiting every vertex at least once. The first link between colorings and hamiltonian cycles was perhaps the observation made by Tait in 1880 (see [7, page 160]) that, if a planar graph has a hamiltonian cycle, then the corresponding map is 4-colorable. The proof is easy: The Hamiltonian cycle divides the map into two parts, the interior and the exterior. The countries in each part can be colored in two colors, so that a total of only four colors are used. (Tait then made the conjecture that every 3-connected, cubic, planar graph has a Hamiltonian cycle, but that was disproved by Tutte, see [7].)

A stronger link is provided by the fact that both the problem of deciding if a graph is 3-colorable and that of deciding if a graph has a hamiltonian cycle are NP-complete, and thus these two problems are, in some sense, equivalent from a computational point of view. In [24], I proved:

Theorem 4.3. *If the chromatic polynomial of a graph has a noninteger root less than or equal to*
$t_0 = \frac{2}{3}+\frac{1}{3}\sqrt[3]{26+6\sqrt{33}}+\frac{1}{3}\sqrt[3]{26-6\sqrt{33}}= 1.29559\ldots$, *then the graph has no hamiltonian path.*

This result is best possible in the sense that it becomes false if t_0 is replaced by any larger number. Since $1.295\ldots$ is only slightly larger than $32/27 = 1.185\ldots$, this result is not easy to apply. However, it shows that, if we know how many colorings (with any number of colors) that a graph has, then in some instances we may conclude that the graph has no Hamiltonian path and hence also no Hamiltonian cycle. For Hamiltonian cycles a slightly better result may be possible: I conjecture that no graph with a Hamiltonian cycle has a chromatic root distinct from $0, 1$ and less than 2.

5 List Color Techniques

If G is a graph and, for every vertex v, $L(v)$ denotes a set of so-called *available colors* at v, then an *L-coloring* of G is a coloring of the vertices such that neighbors get different colors and every vertex gets an available color. A graph is *k-list-colorable* if it has an L-coloring whenever each vertex has k available colors. The *list-chromatic number* $\chi_l(G)$ is the smallest number k such that G is k-list-colorable. If all lists of available colors are equal to the set $\{1, 2, \ldots, k\}$, then list coloring reduces to k-coloring. Hence $\chi(G) \leq \chi_l(G)$. While the problem "Is $\chi(G) \leq k$?" clearly belongs to NP it is not known if the problem "Is $\chi_l(G) \leq k$?" belongs to NP. So it is perhaps not surprising that list coloring gives rise to many difficult questions.

Unlike the chromatic polynomial, list coloring has turned out to be very useful to attack more traditional coloring questions. One striking example is the solution of the following conjecture of Erdős which turned out to be equivalent to an old conjecture of Schur, see [9]: Let C be a cycle with $3m$ vertices. Add m pairwise disjoint triangles so that we obtain a graph H in which each vertex has degree (valency) 4. As H has triangles, $\chi(H) \geq 3$. Erdős conjectured that $\chi(H) = 3$ for each such graph H which is called a *cycle plus triangles graph.* The affirmative answer by Fleischner and Stiebitz [9] is a striking application of a list color theorem by Alon and Tarsi [2]. If D is a directed graph, that is, every edge has an orientation, then D is called *Eulerian* if, for every vertex, the indegree equals the outdegree. Every directed graph with no edge is considered to be Eulerian. Alon and Tarsi proved the following remarkable result:

Theorem 5.1. *If D is a directed graph such that the number of Eulerian subgraphs with vertex set $V(D)$ and an even number of edges is distinct from the number of Eulerian subgraphs with vertex set $V(D)$ and an odd number of edges, then the list chromatic number of D is at most the maximum outdegree plus* 1.

The two numbers of Eulerian subgraphs in Theorem 5.1 are not easy to deal with in general. However, Fleischner and Stiebitz [9] showed that the cycle plus triangles graph H satisfies the assumption of the theorem of Alon and Tarsi when the initial cycle and each of the triangles is made into a directed cycle. Hence

Theorem 5.2. *If H is a cycle plus triangles graph, then* $\chi(H) = \chi_l(H) = 3$.

Consider again the class-teacher graph G in Section 3. The minimum number of periods equals the smallest number of colors needed to color the edges of G such that two edges incident with the same vertex always receive different colors. This number is called the *chromatic index* of G and is denoted $\chi'(G)$. (It is the usual chromatic number of the so-called *line graph* of G.) As we pointed out, $\chi'(G)$ equals the maximum degree of G. The proof is easy, see e.g. [7]. This depends heavily on the fact that G is bipartite (that is, 2-colorable). For graphs G that are not necessarily bipartite, $\chi'(G)$ equals either the maximum degree of G or the maximum degree of G plus 1, by Vizing's theorem, see again [7]. The *list chromatic index* $\chi_l'(G)$ is defined in the obvious way, and it reflects to a higher degree than $\chi'(G)$ the real life scheduling problems. The *complete regular bipartite graph* $K_{m,m}$ is the bipartite graph with m vertices in each part and all possible edges between the parts (representing m classes and m teachers, each of whom teaches each class precisely once.) The conjecture that $\chi_l'(K_{m,m}) = m$ became known as *The Dinitz problem* and was open for 15 years, see [1]. Galvin [10] then proved much more:

Theorem 5.3. $\chi_l'(G) = \chi'(G)$ *for every bipartite graph G.*

One of the most prominent open problems on list coloring is to generalize this result to non-bipartite graphs.

Vizing conjectured in 1975 that every planar graph has list chromatic number at most 5. This conjecture was rediscovered by Erdős, Rubin and Taylor in 1980, who also conjectured that there exist planar graphs that are not 4-list-colorable. The former conjecture was verified by the present author, and the latter by Margit Voigt, see [1]. We shall here present the proof from [20].

Theorem 5.4. $\chi_l(G) \leq 5$ *for every planar graph G.*

Proof. It is sufficient to prove the theorem for maximal planar graphs, that is, graphs that triangulate the plane. We consider a slightly larger class of planar graphs: A *near-triangulation* is a planar graph containing a cycle $C : v_1 v_2 \dots v_m v_1$ called *the outer cycle* such that all other vertices and edges are in the interior of C and triangulate the interior of C. Assume that every vertex in the interior of C has at least 5 available colors, every vertex on C has at least 3 available colors, and v_1, v_2

are already colored. We now prove, by induction on the number of vertices, that the coloring of v_1, v_2 can be extended to a list coloring of the whole near-triangulation. Consider the vertex v_3. This vertex has three available colors. Choose two of them, say i, j distinct from the color of v_2 . Now delete v_3, and delete also i, j from the list of available colors from all neighbors of v_3 in the interior of C. These vertices used to have five available colors, and now they may have only three. But, they are now on the outer cycle so we may apply the induction hypothesis to the resulting graph. Finally, we may also color v_3 by i or j since we only have to avoid the color of v_4.

This results in a list coloring unless v_3 has a neighbor v_p on C which is distinct from v_2, v_4. In that case we argue differently. We first apply the induction hypothesis to the cycle $v_3v_2v_1 \dots v_pv_3$ and its interior. In particular, v_p, v_3 become colored. Then we apply the induction hypothesis to the cycle $v_3v_4v_5 \dots v_pv_3$ and its interior where now v_p, v_3 play the role of v_1, v_2. This completes the proof. □

The proof of Theorem 5.4 does not use Euler's formula (as do most coloring results on planar graphs). It is perhaps the simplest known proof of the 5-color theorem. More important, the list-color method of Theorem 5.4 has several applications as we point out below. A graph is *k-critical* if it has chromatic number k but every proper subgraph has smaller chromatic number.

In [23] the following was proved:

Theorem 5.5. *For each surface S, there are only finitely many 6-critical graphs on S.*

In other words, if we wish to color a graph on S in 5 colors only, then it may not be possible. (On the torus we may need 7 colors.) However, there is only a finite number of (minimal) obstructions that prevent us from coloring a graph on S in 5 colors. The presence of one of those (as a subgraph in the graph under consideration) can be checked in polynomial time.

The simple list color proof of Grötzsch's theorem in [26] is the first step in the proof of the following theorem in [27]:

Theorem 5.6. *For each surface S, there are only finitely many 4-critical graphs without triangles and quadrilaterals on S.*

Theorem 5.6 implies that the chromatic number of graphs without triangles and quadrilaterals on S can be found in polynomial time.

Theorem 5.5, 5.6 are, in a sense, best possible. Theorem 5.5 becomes false if 6 is replaced by 5. And Theorem 5.6 becomes false if quadrilaterals are allowed. Nevertheless it would be interesting to decide if there is a general 4-color theorem in the sense that 4-colororability can be checked in polynomial time on each surface.

In Theorem 5.4 a path with 2 vertices is precolored. If we precolor a path with 3 vertices, the statement becomes false. However, the exceptions are characterized completely in [29], and this is used to prove the following extension of Theorem 5.4.

Theorem 5.7. *If G is a planar graph with n vertices and, for every vertex v, $L(v)$ denotes a set of 5 available colors, then G has at least $2^{n/9}$ distinct L-colorings.*

The list-color technique can also be used to extend Grötzsch's theorem, as shown in [28]:

Theorem 5.8. *If G is a planar graph with n vertices and no cycle of length less than 5, and, for every vertex v, $L(v)$ denotes a set of 3 available colors, then G has at least $2^{n/10000}$ distinct L-colorings.*

We do not know if Theorem 5.8 extends to all planar triangle-free graphs. However, Theorem 5.8 is used to show that these have superpolynomially many 3-colorings.

Jensen and Toft [13] gave a comprehensive collection of beautiful open coloring problems. Many of these have natural list-color counterparts. The simple list color proofs of Grötzsch' theorem and Theorem 5.4 above raise the question if the 4-color theorem has a relatively short list color proof. As pointed out above, not every planar graph is 4-list-colorable. Not even if all lists are subsets of $\{1,2,3,4,5\}$ as Margit Voigt has shown. But, there might be a version where all the lists in the interior are (precisely) $\{1,2,3,4\}$, and the lists on the outer cycle are subsets of $\{1,2,3,4\}$. In order to make the induction work, some of them should be allowed to be proper subsets. However, even mild extensions of the 4-color theorem lead to false statements.

Bjarne Toft has pointed out that all known planar graphs that are not 4-list-colorable have small connectivity, and he has asked if every 4-connected planar graph is 4-list-colorable. Even for 5-connected planar graphs, this would generalize the 4-Color-Theorem.

References

[1] M. Aigner and G. Ziegler, *Proofs from THE BOOK*, Springer-Verlag, Berlin (1998).

[2] N. Alon and M. Tarsi, Colorings and orientations of graphs, *Combinatorica*, **12** (1992), pp. 125–134.

[3] K. Appel and W. Haken, Every planar graph is four colorable. Part I. Discharging, *Illinois J. Math.*, **21** (1977), pp. 429–490.

[4] K. Appel, W. Haken and J. Koch, Every planar graph is four colorable. Part II. Reducibility, *Illinois J. Math.*, **21** (1977), pp. 491–567.

[5] G.D. Birkhoff, A determinant formula for the number of ways of coloring a map, *Ann. of Math.*, **12** (1912), pp. 42–46.

[6] G.D. Birkhoff and D.C. Lewis, Chromatic polynomials, *Trans. Amer. Math. Soc.*, **60** (1946), pp. 355–451.

[7] J.A. Bondy and U.S.R. Murty, *Graph Theory with Applications*, Macmillan, London (1976).

[8] O.V. Borodin, On decomposition of graphs into degenerate subgraphs, *Discret Analiz.*, **28** (1976), pp. 3–11.

[9] H. Fleischner and M. Stiebitz, A solution to a coloring problem of P. Erdős, *Discrete Math.*, **101** (1992), pp. 39–48.

[10] F. Galvin, The list chromatic index of a bipartite multigraph, *Journal of Combinatorial Theory, Series B*, **63** (1995), pp. 153–158.

[11] M.R. Garey and D.S. Johnson, *Computers and Intractability, A guide to the Theory of NP-completeness*, Freeman, San Francisco (1979).

[12] B. Jackson, A zero-free interval for chromatic polynomials of graphs, *Combinatorics, Probability and Computing*, **2** (1993), pp. 325–336.

[13] T. Jensen and B. Toft, *Graph Coloring Problems*, John Wiley, New York (1995).

[14] K. Kawarabaiashi and C. Thomassen, Decomposing a planar graph of girth 5 into an independent set and a forest, To appear.

[15] G. Mohar and C. Thomassen, *Graphs on Surfaces*, Johns Hopkins University Press, Baltimore (2000).

[16] R.C. Read and W.T. Tutte, Chromatic polynomials, in *Selected Topics in Graph Theory* (eds. L.W. Beineke and R.J. Wilson), Academic Press, New York (1988), pp. 15–42.

[17] N. Robertson, D. Sanders, P. Seymour and R. Thomas, The four-color theorem, *Journal of Combinatorial Theory, Series B*, **70** (1997), pp. 2–44.

[18] R. Shrock and S-H. Tsai, Ground state entropy of Potts antiferromagnets on homeomorphic families of strip graphs, *Physica A*, **259** (1998), pp. 315–348.

[19] A. Sokal, Chromatic roots are dense in the whole complex plane, *Combinatorics, Probability and Computing*, **13** (2004), pp. 221–261.

[20] C. Thomassen, Every planar graph is 5-choosable, *Journal of Combinatorial Theory, Series B*, **62** (1994), pp. 180–181.

[21] C. Thomassen, The zero-free intervals for chromatic polynomials of graphs, *Combinatorics, Probability and Computing*, **6** (1997), pp. 497–506.

[22] C. Thomassen, Decomposing a planar graph into degenerate graphs, *Journal of Combinatorial Theory, Series B*, **65** (1995), pp. 305–314.

[23] C. Thomassen, Color-critical graphs on a fixed surface, *Journal of Combinatorial Theory, Series B*, **70** (1997), pp. 67–100.

[24] C. Thomassen, Chromatic roots and hamiltonian paths, *Journal of Combinatorial Theory, Series B*, **80** (2000), pp. 218–224.

[25] C. Thomassen, Decomposing a planar graph into an independent set and a 3-degenerate graph, *Journal of Combinatorial Theory, Series B*, **83** (2001), pp. 262–271.

[26] C. Thomassen, A short list color proof of Grötzsch's theorem, *Journal of Combinatorial Theory, Series B*, **88** (2003), pp. 189–192.

[27] C. Thomassen, The chromatic number of a graph of girth 5 on a fixed surface, *Journal of Combinatorial Theory, Series B*, **87** (2003), pp. 38–71.

[28] C. Thomassen, Many 3-colorings of triangle-free planar graphs, *Journal of Combinatorial Theory, Series B*, To appear.

[29] C. Thomassen, Exponentially many 5-list-colorings of planar graphs, To appear.

Proc. of ICDM 2006, Pages 9–24

Homomorphisms, Dualities and CSP

Jaroslav Nešetřil*
Department of Applied Mathematics and ITI, MFF, Charles University, CZ 11800 Praha 1, Malostranské nám. 25

Abstract. In this paper we survey the recent results on graph homomorphisms perhaps for the first time in the broad range of their relationship to wide range applications in computer science, physics and other branches of mathematics. We illustrate this development in each area by few results. This survey is prepared to accompany a lecture of the author at ICDM in Bangalore.

1 Introduction

Graph theory receives its mathematical motivation from the two main areas of mathematics: algebra and geometry (topology) and it is fair to say that the graph notions stood at the birth of algebraic topology. Consequently various operations and comparisons for graphs stress either its algebraic part (e.g. various products) or geometrical part (e.g. contraction, subdivision). It is only natural that the key place in the modern graph theory is played by (fortunate) mixtures of both approaches as exhibited best by the various modifications of the notion of graph minor. However from the algebraic point of view perhaps the most natural graph notion is the following notion of a homomorphism:
Given two graphs G and H a *homomorphism* f of G to H is any mapping $f : V(G) \to V(H)$ which satisfies the following condition :

$$[x, y] \in E(G) \text{ implies } [f(x), f(y)] \in E(H).$$

This condition should be understood as follows: on both sides of the implication one considers the same type of edges: undirected ($\{u, v\}$, often denoted by just uv) or directed ((u, v), often denoted as well by just uv). It is important that this definition is flexible enough to induce analogous definitions of the homomorphisms for hypergraphs (set systems) and relational systems (with a given signature; that will be specified later).

Homomorphisms arise naturally in various and very diverse situations

- in extremal combinatorics (and particularly in problems related to colorings, partitions and decompositions of graphs and hypergraphs);
- in statistical physics (as a model for partition functions);
- in probability (as a model of random processes, for example random walk);
- in logic (any satisfiability assignment of a formula may be viewed as a homomorphism);

*The author would like to express thanks for support by the project 1M0021620808 of the Ministry of Education of the Czech Republic.

- in Artificial Intelligence (as a model and criterion of satisfaction leading to Constraint Satisfaction Problems);
- finite model theory (as a natural way of compare and classify models);
- theory of algorithms (as an example and reduction tool);
- in complexity theory (and more recently in logic, descriptive complexity in particular);
- in algebraic combinatorics (providing the vital link to algebraic topology);
- in category theory (as a motivating example, a thoroughly studied particular case).

This paper cannot even touch upon all these topics. This is too ambitious even for a monograph. To a certain extent this has been a plan of recently published book in [22]. But the progress has been fast and we want to complement this by giving some highlights of this development. The interested reader can consult also surveys and papers [3, 30, 32, 48, 49, 64].

This paper is a (much) extended version of the talk given by the author at Bangalore conference *International Conference on Discrete Mathematics*, December 15-18, 2006. The purpose of this text is to illustrate the rich conceptual framework of the contemporary study of homomorphisms in various mathematical as well as non mathematical context of various related applications.

We rely on standard texts such as [2], [41] (for graph theory), [22] (for graphs and homomorphisms), [35, 41] (for general combinatorics). However the field of combinatorial structures related to homomorphisms (which this author likes to call *combinatorics of mappings*, see forthcoming [47]) is currently developing very fast and it is the purpose of this short survey to cover this recent development.

2 Existence and CSP

Given graphs G, H we can either count the number of homomorphisms from G to H or just to decide whether such a homomorphism exists. Neither of these questions subsumes the other. In fact both parts of the theory point to different directions: the counting relates to probability and properties of random structures in general, to partition models of physical phenomena; on the other side the existence problems relate to computational complexity of decision models (such as Constraint Satisfaction Problem (CSP), logic and descriptive complexity and dualities. Some of this will be covered in this and next sections. For counting part see e.g. [3, 49].

We consider here the following decision problem:

2.1 H-coloring Problem

Consider the following decision problem (for a fixed graph H):

Instance: A graph G.

Question: Does there exist a homomorphism $G \longrightarrow H$.

This problem covers many concrete problems which were and are studied (see [22]):

(i) For $H = K_k$ (the complete graph with k vertices) we get a k-coloring problem;

(ii) For graphs $H = K_{k/d}$ we get *circular chromatic numbers*' see e.g. [70];

(iii) For H Kneser graphs $K\binom{k}{d}$ we get so called *multicoloring*; [22].

Further examples include so called T-colorings, see e.g.[63, 70], which in turn are related to the recently popular *channel assignment problem.*

Perhaps the most extensively studied aspect of H-coloring problems is its complexity. This is interesting and generally still unresolved. The situation is well understood for complete graphs: For any fixed $k \geq 3$ the K_k-coloring problem (which is equivalent to the deciding whether $\chi(G) \leq k$) is NP-complete. On the other hand K_1- and K_2-coloring problems are easy. Thus, in the undirected case, we may always assume that the graph H is not bipartite.

Some other problems are easy to discuss. For example, if $H = C_5$ then we can consider the following (*arrow replacement*) construction:

For a given graph G let G^{**} be the graph which we obtain from G by replacing of every edge of G by a path of length 3 (these paths are supposed to be internally disjoint). Another way to say this is to consider a subdivision of G where each edge is subdivided by exactly two new vertices. It is now easy to prove that for any undirected graph G the following two statements are equivalent:

(i) $G \longrightarrow K_5$;

(ii) $G^{**} \longrightarrow C_5$.

This example is not isolated (the similar trick may be used e.g. for any odd cycle). Using analogous, but more involved, edge-, vertex- and other replacement constructions (called *indicators, subindicators*, and *edge - subindicators*) the following has been proved in [23]:

Theorem 2.1. *For a graph H the following two statements are equivalent:*

(1) *H is non-bipartite;*

(2) *H-coloring problem is NP-complete.*

We have to stress that the analogy of Theorem 5.3 for oriented graphs fails to be true for one can construct easily an orientation $\vec{H}$ of bipartite graph H such that $\vec{H}$ - coloring problem is NP-complete. Even more so, one can construct a balanced oriented graph $\vec{H}$ with the property that $\vec{H}$-coloring problem is NP-complete (an oriented graph is called *balanced* if every cycle has the same number of forward and backward arcs).

One can go even further and (perhaps a bit surprisingly) omit all cycles. Namely, one has the following [24]:

Theorem 2.2. *There exist an oriented tree T (i.e. T is an orientation of an undirected tree) such that T-coloring problem is NP-complete.*

Presently the smallest such tree T has 45 vertices.

3 Constrained Satisfaction Problems (CSP)

Every part of mathematics has some typical features which presents both its advantages and its limitations. One of such feature for the study of homomorphisms is the fact that its problems are usually easy to generalize and formulate. The H-coloring problem (explained in the previous section) is a good example of this. One can formulate it more generally for finite structures.

3.1 Relational Structures

A relational structure of a given type generalizes the notion of a relation and of a graph to more relations and to higher (non-binary) arities. The concept was isolated in the thirties by logicians (e.g. Löwenheim, Skolem) who developed logical "static" theory. As we shall see this influenced terminology even today as we find useful to speak about models (of our chosen relational language). In the sixties new impulses came from the study of algebraic categories and the resulting "dynamic" studies called for a more explicit approach, see e.g [62]. We shall adopt here a later notation (with a touch of logical vocabulary).

A type Δ is a sequence $(\delta_i; i \in I)$ of positive integers. A *relational system* $\mathbf{A}$ of *type* Δ is a pair $(X, (R_i; i \in I))$ where X is a set and $R_i \subseteq X^{\delta_i}$; that is R_i is a δ_i-nary relation on X. In this paper we shall always assume that X and I are finite sets (thus we consider finite relational systems only).

The type $\Delta = (\delta_i; i \in I)$ will be fixed throughout this paper. Note that for the type $\Delta = (2)$ relational systems of type Δ correspond to directed graphs, the case $\Delta = (2, 2)$ corresponds to directed graphs with blue-green colored edges (or rather arcs).

Relational systems (of type or signature Δ) will be denoted by capital letters **A, B, C,....** A relational system of type Δ is also called a Δ-system (or a model). If $\mathbf{A} = (X, (R_i; i \in I))$ we also denote the base set X as A and the relation R_i by $R_i(\mathbf{A})$. Let $\mathbf{A} = (X, (R_i; i \in I))$ and $\mathbf{B} = (Y, (S_i; i \in I))$ be Δ-systems. A mapping $f : X \to Y$ is called a *homomorphism* if for each $i \in I$ the following holds: $(x_1, \ldots, x_{\delta_i}) \in R_i$ implies $(f(x_1), \ldots, f(x_{\delta_i})) \in S_i$.

In other words a homomorphism f is any mapping $f : A \to B$ which satisfies $f(R_i(\mathbf{A})) \subseteq R_i(\mathbf{B})$ for each $i \in I$. (Here we extended the definition of f by putting $f(x_1, \ldots, x_t) = (f(x_1), \ldots, f(x_t))$.)

For Δ-systems $\mathbf{A}$ and $\mathbf{B}$ we write $\mathbf{A} \to \mathbf{B}$ if there exists a homomorphism from $\mathbf{A}$ to $\mathbf{B}$. Hence the symbol $\to$ denotes a relation that is defined on the class of all Δ-systems. This relation is clearly reflexive and transitive, thus induces a quasi-ordering of all Δ-systems. As is usual with quasi-orderings, it is convenient to reduce it to a partial order on classes of equivalent objects: Two Δ-systems $\mathbf{A}$ and $\mathbf{B}$ are called *homomorphically equivalent* if we have both $\mathbf{A} \to \mathbf{B}$ and $\mathbf{B} \to \mathbf{A}$; we then write $\mathbf{A} \sim \mathbf{B}$. For every $\mathbf{A}$ there exists up to an isomorphism a unique $\mathbf{A}'$ such that $\mathbf{A} \sim \mathbf{A}'$ and $\mathbf{A}'$ has smallest size $|A'|$. Such $\mathbf{A}'$ is called the *core* of $\mathbf{A}$.

The relation $\to$ induces an order on the classes of homomorphically equivalent Δ-system, which we call the *homomorphism order.* (So this is a partial order when restricted to non-isomorphic core structures.)

The homomorphism order will be denoted by $\mathcal{C}_\Delta$ (as it is also called coloring order). We denote by $\mathbf{Rel}(\Delta)$ the class of all finite relational structures of type Δ and all homomorphisms between them. This category plays a special role in the model theory and theory of categories [62]. It is also central in the branch of Artificial Intelligence (AI) dealing with *Constraint Satisfaction Problems* [11]. The expressive power of homomorphisms between relational structures leads to the following:

Theorem 3.1 ([14]). *Every Constraint Satisfaction Problem can be expressed as a membership problem for a class* $\mathbf{CSP}(\mathbf{B})$ *of relational structures (of a certain type* Δ*) defined as follows:*

$$\mathbf{CSP}(\mathbf{B}) = \{\mathbf{A}; \mathbf{A} \longrightarrow \mathbf{B}\}.$$

Recall, that the *membership problem* for a class $\mathcal{K}$ is the following problem: Given a structure $\mathbf{A}$ does $\mathbf{A}$ belong to $\mathcal{K}$? Is $\mathbf{A} \in \mathcal{K}$? For brevity we call the membership problem for class $\mathbf{CSP}(\mathbf{B})$ simply the $\mathbf{B}-$*coloring problem*, or $\mathbf{CSP}(\mathbf{B})$ *problem*. The structure $\mathbf{B}$ is usually called *template* of $\mathbf{CSP}(\mathbf{B})$.

(Generalized) coloring problems cover a wide spectrum of (applications rich) problems. This attracted recently a very active research on the boundary of complexity theory, combinatorics, logic and universal algebra. Only some of it will be reviewed in this paper.

However the complexity status of the $\mathbf{CSP}(\mathbf{B})$ problem is solved only for special and rather restricted situations. The following are principal results:

1. undirected graph coloring (i.e. Theorem 5.3), see [23];
2. the characterization of complexity of $\mathbf{B}-$coloring problems for structures $\mathbf{B}$ which are *binary* (i.e. for which $|B| = 2$), see [65];
3. the characterization of complexity of $\mathbf{B}-$coloring problems for structures $\mathbf{B}$ which are *ternary* (i.e. for which $|B| = 3$), see [7].

The last two results may seem to be easy, or limited, but reader should realize that while the size of the $|B|$ may be small (such as 2 or 3) the relational system can in fact be very large as the arities δ_i of relations $R_i(\mathbf{B})$ may be arbitrarily large. Whole book [11] is devoted to the case $|B| = 2$.

Nevertheless in all known instances one proves that the $\mathbf{CSP}(\mathbf{B})$ problem is either polynomial (the class of polynomial problems is denoted by P) or NP-complete. This is remarkable as such *dichotomy* generally does not hold. Of course, there is a possibility that the classes P and NP coincide (this constitutes the famous P-NP problem; one of the *millennium problems*). But if these classes are distinct (i.e. if P $\subset$ NP) then there are infinitely intermediate classes (by a celebrated result of Ladner [39]). This (and other more theoretical evidence) prompted Feder and Vardi [14] to formulate the following by now well known problem:

3.2 Dichotomy Conjecture

Every $\mathbf{CSP}(\mathbf{B})$ problem is either P or NP-complete.

Although this is still open, a lot of work has been done. Let us finish this section by formulating two related results.

Oriented Graphs Suffice

At the first glance the complexity of the $\mathbf{CSP}(\mathbf{B})$ problem lies in the great variety of possible relational structures. Already in [14] it has been realized that it is not so.

Theorem 3.2. *The dichotomy conjecture follows from the dichotomy of the complexity of H-coloring problem where H is an oriented graph.*

This is interesting as this positions Theorem 5.3 in the new light and shows a surprising difference between colorings (partitions) of undirected and directed graphs which was not before realized. See [14] for the original proof; see also [22].

Dichotomy is Asymptotically Almost Surely True

Relational structures and homomorphisms express various decision and counting combinatorial problems such as coloring, satisfiability, and linear algebra problems. Many of them can be reduced to special cases of a general Constraint Satisfaction Problem **CSP(B)**. A number of such problems have been studied and have known complexity, e.g., when we deal with undirected graphs or the problem is restricted to small sets A (see [7, 23, 65]). However at this moment we are far from understanding the behavior of **CSP(B)** problem even for binary relations (i.e., when relational systems of type $\Delta = (2)$). It seems that the Dichotomy Conjecture holds in a stronger sense:

Dichotomy Conjecture*

Most **CSP(B)** problems are NP-complete problems with a few exceptions which are polynomial.

For example for undirected graphs the **CSP(B)**-problem is always hard problem with exactly 3 exceptions: **B** is homomorphically equivalent either to the loop graph, or to the single vertex graph (with no edges) or the symmetric edge, [23]. Results for other solved cases have a similar character supporting the modified Dichotomy Conjecture* (see [7, 65]). One can confirm this feeling by proving that both Dichotomy and Dichotomy* Conjectures are equivalent. In order to make this statement precise, let $\mathcal{R}(n,k)$ denote a random k-ary relation defined on a set $[n] = \{1, 2, \ldots, n\}$, for which the probability that $(a_1, \ldots, a_k) \in \mathcal{R}(n,k)$ is equal to $1/2$ independently for each $(a_1, \ldots, a_k)$, where $1 \le a_r \le n$ for $r = 1, \ldots, k$. Assume further that not all a_i's are equal: for $a \in [n]$, we assume $(a, a, \ldots, a) \notin \mathcal{R}(n,k)$. Let $([n], (\mathcal{R}(\delta_i, n)_{i \in I}))$ denote the random relational system of type $\Delta = (\delta_i)_{i \in I}$. In this situation we show that the probability that $([n], (\mathcal{R}(\delta_i, n)_{i \in I}))$−coloring is NP-complete tends to one as either n, or $\max_i \delta_i$ tends to infinity.

Note that the **B**−coloring problem for relational system **B** is NP-complete, provided it is NP-complete for the system $(B, R_{i_0}(\mathbf{B}))$ of type (δ_{i_0}) for some $i_0 \in I$. (The converse implication, in general, does not hold). Thus, we prove our result for 'simple' relational systems which consist of just one k-ary relation.

Theorem 3.3 ([40]). *For a fixed $k \ge 2$,*

$$\lim_{n \to \infty} \Pr\big(([n], \mathcal{R}(n,k)) \textit{ is NP complete }\big) = 1,$$

while for a given $n \ge 2$,

$$\lim_{k \to \infty} \Pr\big(([n], \mathcal{R}(n,k)) \textit{ is NP complete }\big) = 1.$$

The proof uses properties of random graphs together with an algebraic approach to the dichotomy conjecture (based on the analysis of clones of polymorphisms) which was pioneered by [8, 28].

4 Dualities

From the combinatorial point of view there is a standard way how to approach (and sometimes to solve) a monotone property P: one investigates those structures without the property P which are *critical*, (or *minimal*) without P. One proceeds as follows: denote by $\mathcal{F}$ the class of all critical structures and define the class Forb($\mathcal{F}$) of all structures which do not "contain" any $\mathbf{F} \in \mathcal{F}$. The class Forb($\mathcal{F}$) is the class of all structures not containing any of the critical substructures and thus it is easy to see that Forb($\mathcal{F}$) coincides with the class of structures with the property P. Of course in most cases the class $\mathcal{F}$ is infinite yet a structural result about it may shed some light on property P. For example this is the case with 3-colorability of graphs where 4-critical graphs were (and are) studied thoroughly (historically mostly in relationship to Four Color Conjecture).

When the family $\mathcal{F}$ of obstructions is finite (or algorithmically "well behaved"), then such theorems clearly provide an example of *good characterizations* (in the sense of Edmonds). Any instance of such good characterization is called a *homomorphism duality*. This concept was introduced in [53] and applied to various graph theoretical good characterizations. The simplest homomorphism dualities are those where the family of obstructions consists of only singletons (i.e. single structures). In the other words such homomorphism dualities are described by a pair $\mathbf{A}$, $\mathbf{B}$ of structures as follows:

(Singleton) Homomorphism Duality Scheme

A structure $\mathbf{C}$ admits a homomorphism into $\mathbf{B}$ if and only if $\mathbf{A}$ does not admit a homomorphism into $\mathbf{C}$.

Despite the fact that singleton homomorphism dualities are scarce for both undirected and directed graphs, for more general structures (such as oriented matroids with suitable version of strong maps) the (singleton) homomorphism duality may capture general theorems such as Farkas Lemma (see [25]). In [53] are described all singleton homomorphism dualities for undirected graphs. As a culmination of several partial results all homomorphism dualities for general relational structures were finally described in [58]. This is not the end and more recently the homomorphism dualities emerged as an important phenomena in a new context. This will be briefly described below.

4.1 Generalized CSP Classes

For a finite set of structures $\mathcal{D}$ in $\mathbf{Rel}(\Delta)$ we denote by $\mathbf{CSP}(\mathcal{D})$ the class of all structures $\mathbf{A} \in \mathbf{Rel}(\Delta)$ satisfying $\mathbf{A} \longrightarrow \mathbf{D}$ for some $\mathbf{D} \in \mathcal{D}$. Thus $\mathbf{CSP}(\mathcal{D})$ is the union of classes $\mathbf{CSP}(\mathbf{D})$ for all $\mathbf{D} \in \mathcal{D}$. This definition (of *generalized CSP class* is sometimes more convenient and in fact generalized CSP classes are polynomially equivalent to the classes described syntactically as MMSNP ([14]; the equivalence is non trivial and follows from [31]).

These classes are sometimes called *color classes* and denoted by

$$\longrightarrow \mathcal{D} = \mathbf{CSP}(\mathcal{D}).$$

4.2 Forb Classes

Let $\mathcal{F}$ be a finite set of structures of $\mathbf{Rel}(\Delta)$. Denote by $\mathbf{Forb}(\mathcal{F})$ the class of all structures $\mathbf{A} \in \mathbf{Rel}(\boldsymbol{\Delta})$ which do not permit a homomorphism $\mathbf{F} \longrightarrow \mathbf{A}$ for any $\mathbf{F} \in \mathcal{F}$. Formally:

$$\mathbf{Forb}(\mathcal{F}) = \{\mathbf{A} \mid \text{there is no } f : \mathbf{F} \to \mathbf{A} \text{ with} \mathbf{F} \in \mathcal{F}\}$$

Let us remark that these classes are sometimes denoted by

$$\mathbf{Forb}(\mathcal{F}) = \mathcal{F} \not\longrightarrow .$$

4.3 Finite Duality

Finite duality is the equation of two classes: of the class $\mathbf{Forb}(\mathcal{F})$ and of the class $\mathbf{CSP}(\mathcal{D})$ for a particular choice of *forbidden set* $\mathcal{F}$ and *dual set* $\mathcal{D}$. Formally:

$$\mathbf{Forb}(\mathcal{F}) = \mathbf{CSP}(\mathcal{D}).$$

We also say that $\mathcal{D}$ *has finite duality.* Finite dualities were defined in [53]. They are being intensively studied from the logical point of view, and also in the optimization (mostly CSP) context.

We say that a class $\mathcal{K} \subset \mathbf{Rel}(\Delta)$ is *First Order definable* if there exists a first order formula ϕ (i.e. quantification is allowed only over elements) such that the class $\mathcal{K}$ is just the class of all structures $\mathbf{A} \in \mathbf{Rel}(\Delta)$ where ϕ is valid. Formally:

$$\mathcal{K} = \{\mathbf{A}; \mathbf{A} \models \phi\}.$$

It has been recently showed [1, 64] that if the class $\mathbf{CSP}(\mathcal{D})$ is first order definable then it has finite duality. (This is a consequence of the solution of an important *homomorphism preservation conjecture* solved in [64].) On the other hand the finite dualities in categories $\mathbf{Rel}(\Delta)$ were characterized in [15] as an extension of [58]. By combining these results we obtain:

Theorem 4.1. *For a finite set $\mathcal{D}$ relational structures in* $\mathbf{Rel}(\Delta)$ *are the following statements equivalent:*

(i) *The class* $\mathbf{CSP}(\mathcal{D})$ *is first order definable;*

(ii) *$\mathcal{D}$ has finite duality; explicitly, there exists a finite set $\mathcal{F}$ such that* $\mathbf{Forb}(\mathcal{F}) = \mathbf{CSP}(\mathcal{D})$*;*

(iii) $\mathbf{Forb}(\mathcal{F}) = \mathbf{CSP}(\mathcal{D})$ *for a finite set $\mathcal{A}$ of finite forests.*

We did not define what is a forest in a structure(see [15, 58]). For the sake of completeness let us say that a *forest* is a structure not containing any cycle. And a *cycle* in a structure $\mathbf{A}$ is either a sequence of distinct points and distinct tuples $x_0, r_1, x_1, \ldots, r_t, x_t = x_0$ where each tuple r_i belongs to one of the relations $R(\mathbf{A})$ and each x_i is a coordinate of r_i and r_{i+1}, or, in the degenerated case $t = 1$ a relational tuple with at least one multiple coordinate. The *length* of the cycle is the integer t in the first case and 1 in the second case. Finally the *girth* of a structure $\mathbf{A}$ is the shortest length of a cycle in $\mathbf{A}$ (if it exists; otherwise it is a forest).

In a sharp contrast with that, there are no finite dualities for (general) finite algebras. It has been recently shown [33] that there are no such dualities at all. Namely, one has

Theorem 4.2. *For every finite set $\mathcal{A}$ of finite algebras of a given type $(\delta_i)_{i\in I}$ and every finite algebra* $\mathbf{B}$ *there exists a finite algebra* $\mathbf{A}$ *such that* $\mathbf{A} \in \mathbf{Forb}(\mathcal{A})$ *and* $\mathbf{A} \notin \mathbf{CSP}(\mathbf{B})$.

(This concerns the standard homomorphisms $f : (X, (\alpha_i)_{i\in T}) \to (X', (\alpha'_i)_{i\in T})$ satisfying

$$x = \alpha_i(x_1, \ldots, x_{n_i}) \Rightarrow f(x) = \alpha'_i(f(x_1), \ldots, f(x_{n_i}))). \,) \qquad (*)$$

5 Restricted Dualities

5.1 Special Classes

In this section we deal with graphs only. We motivate this section by the following two examples.

Example 1. Celebrated Grötzsch's theorem (see e.g. [2]) says that every planar graph is 3-colorable. In the language of homomorphisms this says that for every triangle free planar graph G there is a homomorphism of G into K_3.

Using the partial order terminology (for the homomorphism order $\mathcal{C}_\Delta$) the Grötzsch's theorem says that K_3 is an upper bound (in the homomorphism order) for the class $\mathcal{P}_3$ of all planar triangle free graphs. As obviously $K_3 \notin \mathcal{P}_3$ a natural question (first formulated in [48]) suggests: Is there yet a smaller bound? The answer, which may be viewed as a strengthening of Grötzsch's theorem, is positive: there exists a triangle free 3-colorable graph H such that $G \to H$ for every graph $G \in \mathcal{P}_3$. Explicitly:

$$K_3 \not\longrightarrow G \Leftrightarrow G \longrightarrow H$$

for every planar graph G. Because of this we call such theorem *restricted duality*. A restricted duality asserts the duality but only for structures in a restricted class of graphs. The (non-trivial) existence of graph H above has been proved in [50] (in a stronger version for proper minor closed classes). The case of planar graphs and triangle is interesting in its own as it is related to the Seymour conjecture and its partial solution [18], see [44]; it seems that a proper setting of this case is in the context of *TT-continuous mappings*, see [56]. Restricted duality results have been generalized since to other classes of graphs and to other forbidden subgraphs. In fact for every "forbidden" finite set of connected graphs we have a duality restricted to any proper subclass $\mathcal{K}$ of all graphs which is minor closed, see [50]. This then implies that Grötzsch's theorem can be strengthened by a sequence of ever stronger bounds and that the supremum of the class of all triangle free planar graphs does not exist.

Example 2. Let us consider all sub-cubic graphs (i.e. graph with maximum degree ≤ 3). By Brooks theorem (see e.g. [2]) all these graphs are 3-colorable with the single connected exception K_4. What about the class of all sub-cubic *triangle free* graphs? Does there exists a triangle free 3-colorable bound? The positive answer to this question is given in [20]. In fact for every finite set $\mathcal{F} = \{F_1, F_2, \dots, F_t\}$ of connected graphs there exists a graph H with the following properties:

- H is 3-chromatic;
- $G \longrightarrow H$ for every subcubic graph $G \in \mathbf{Forb}(\mathcal{F})$.

It is interesting to note that while sub-cubic graphs have restricted dualities (and, more generally, this also holds for the classes of bounded degree graphs) for the classes of degenerated graphs a similar statement is not true (in fact, with a few trivial exceptions, it is never true).

Where lies the boundary for validity of restricted dualities? We clarify this after introducing the formal definition.

Definition 5.1. *A class $\mathcal{K}$ of graphs has a* all restricted dualities *if, for any finite set of connected graphs $\mathcal{F} = \{F_1, F_2, \dots, F_t\}$, there exists a finite graph $D_{\mathcal{F}}^{\mathcal{K}}$ such that $F_i \not\longrightarrow D_{\mathcal{F}}^{\mathcal{K}}$ for $i = 1, \dots, t$ and such that for all $G \in \mathcal{K}$ holds. Explicitly:*

$$(F_i \not\longrightarrow G), i = 1, 2, \dots, t, \iff (G \longrightarrow D_{\mathcal{F}}^{\mathcal{K}}).$$

It is easy to see that using the homomorphism order we can reformulate this definition as follows: A class $\mathcal{K}$ has restricted dualities if for any finite set of connected graphs $\mathcal{F} = \{F_1, F_2, \ldots, F_t\}$ the class $\mathbf{Forb}(\mathcal{F}) \cap \mathcal{K}$ is bounded in the class $\mathbf{Forb}(\mathcal{F})$.

The main result of [51] can be then stated as follows:

Theorem 5.2. *Any class of graphs with bounded expansion has all restricted dualities.*

Of course we have to yet define bounded expansion (and we do so in the next section). But let us just note that both proper minor closed classes and bounded degree graphs form classes of bounded expansion. Consequently this result generalizes both Examples 1 and 2. In fact the seeming incomparability of bounded degree graphs and minor closed classes led the authors of [51] to the definition of bounded expansion classes.

Theorem 5.3 (Bounded expansion classes). *Recall that the* maximum average degree *$mad(G)$ of a graph G is the maximum over all subgraphs H of G of the average degree of H, that is $mad(G) = \max_{H \subseteq G} \frac{2|E(H)|}{|V(H)|}$. The* distance *$d(x, y)$ between two vertices x and y of a graph is the minimum length of a path linking x and y, or ∞ if x and y do not belong to same connected component.*
We introduce several notations:

- The *radius* $\rho(G)$ of a connected graph G is:

$$\rho(G) = \min_{r \in V(G)} \max_{x \in V(G)} \mathrm{d}(r, x)$$

- A *center* of G is a vertex r such that $\max_{x \in V(G)} \mathrm{d}(r, x) = \rho(G)$.

Definition 5.4. *Let G be a graph. A* ball *of G is a subset of vertices inducing a connected subgraph. The set of all the families of pairwise disjoint balls of G is noted $\mathfrak{B}(G)$.*

Let $\mathcal{P} = \{V_1, \ldots, V_p\}$ be a family of pairwise disjoint balls of G.

- The *radius* $\rho(\mathcal{P})$ of $\mathcal{P}$ is $\rho(\mathcal{P}) = \max_{X \in \mathcal{P}} \rho(G[X])$
- The *quotient* $G/\mathcal{P}$ of G by $\mathcal{P}$ is a graph with vertex set $\{1, \ldots, p\}$ and edge set $E(G/\mathcal{P}) = \{\{i, j\} : (V_i \times V_j) \cap E(G) \neq \emptyset \text{ or } V_i \cap V_j \neq \emptyset\}$.

We introduce several invariants that generalize the one of maximum average degree:

Definition 5.5. *The* greatest reduced average density (grad)*of G with rank r is*

$$\nabla_r(G) = \max_{\substack{\mathcal{P} \in \mathfrak{B}(G) \\ \rho(\mathcal{P}) \leq r}} \frac{|E(G/\mathcal{P})|}{|\mathcal{P}|}$$

The following is our key definition:

Definition 5.6. *A class of graphs $\mathcal{K}$ has* bounded expansion *if there exists a function $f : \mathbb{N} \to \mathbb{N}$ such that for every graph $G \in \mathcal{K}$ and every r holds*

$$\nabla_r(G) \leq f(r). \tag{1}$$

f is called the expansion function. Proper minor closed classes have expansion function bounded by a constant, regular graphs by the exponential function, geometric graphs such as d-dimensional meshes have polynomial expansion. Expansion function can grow arbitrary fast.

Finally note that bounded expansion classes have many applications. Some of them are included in [52].

5.2 Lifts and Shadows

We return to the general relational structures. Duals of structures are a fascinating subject (see e.g. [34, 57, 59]). In closing these sections on dualities we want to briefly stress the following aspect which is on the first glance surprising.

Consider again the general duality scheme: structure $\mathbf{C}$ admits a homomorphism to $\mathbf{D} \in \mathcal{D}$ if and only if $\mathbf{F}$ does not admit a homomorphism into $\mathbf{C}$ for any $\mathbf{F} \in \mathcal{F}$; formally:

$$\mathbf{Forb}(\mathcal{F}) = \mathbf{CSP}(\mathcal{D}).$$

One is somehow tempted to think that the left side of the definition is somewhat more restrictive, that the finitely many obstacles make the problem easy if not trivial. After all, while the class $\mathbf{CSP}(\mathbf{D})$ may be complicated and have membership problem NP-complete even for a simple graph $\mathbf{D}$ (such as K_3) the left side is always polynomially decidable (for every finite set $\mathcal{F}$). But in a way this is a misleading argument. The expressing power of the classes $\mathbf{Forb}(\mathcal{F})$ for finite sets $\mathcal{F}$ is very large. This follows from a recent work [32] which we now briefly describe. We start with an example.

Think of 3-coloring of graph $G = (V, E)$. This is a well known hard problem and there is a multiple evidence for this: concrete instances of the problem are difficult to solve (if you want a non-trivial example consider Kneser graphs; [42]), there is an abundance of minimal graphs which are not 3-colorable (these are called 4-critical graphs, see e.g. [29]) and in the full generality (and even for important "small" subclasses such as 4-regular graphs or planar graphs) the problem is a canonical NP-complete problem.

Yet the problem has an easy formulation. A 3-coloring is simple to formulate even at the kindergarten level. This is in a sharp contrast with the usual definition of the class NP by means of polynomially bounded non-deterministic computations. Fagin [13] gave a concise description of the class NP by means of logic: NP languages are just languages accepted by an Existential Second Order (ESO) formula of the form

$$\exists P \Psi(S, P),$$

where S is the set of input relations, P is a set of existential relations, the proof for the membership in the class, and Ψ is a first-order formula without existential quantifiers. This definition of NP inspired a sequence of related investigations and these *descriptive complexity* results established that most major complexity classes can be characterized in terms of logical definability of finite structures. Particularly this led Feder and Vardi [14] to their seminal reduction of *Constraint Satisfaction Problems* to so called MMSNP (*Monotone Monadic Strict Nondeterministic Polynomial*) problems which also nicely link MMSNP to the class NP in computational sense. Inspired by these results we would like to ask an even simpler question:

Can one express the computational power of the class NP by combinatorial means?

It may seem to be surprising that the classes of relational structures defined by ESO formulas (i.e. the whole class NP) are polynomially equivalent to canonical *lifts* of structures which are defined by a finite set of forbidden substructures.

Shortly, finitely many forbidden lifts determine any language in NP.

Let us define lifts and shadows more formally: We will work with two (fixed) signatures, Δ and $\Delta \cup \Delta'$ (the signatures Δ and Δ' are always supposed to be disjoint). For convenience we denote structures in $Rel(\Delta)$ by $\mathbf{A}, \mathbf{B}$ etc. and structures in $Rel(\Delta \cup \Delta')$ by $\mathbf{A}', \mathbf{B}'$ etc. For convenience we shall denote $Rel(\Delta \cup \Delta')$ by $Rel(\Delta, \Delta')$. The classes $Rel(\Delta)$ and $Rel(\Delta, \Delta')$ will be considered as categories endowed with all homomorphisms. The interplay of categories $Rel(\Delta, \Delta')$ and $Rel(\Delta)$ is here the central theme. Towards this end we define the following notion: Let $\Phi : Rel(\Delta, \Delta') \to Rel(\Delta)$ denote the natural *forgetful functor* that "forgets" relations in Δ'. Explicitly, for a structure $\mathbf{A'} \in Rel(\Delta, \Delta')$ we denote by $\Phi(\mathbf{A}')$ the corresponding structure $\mathbf{A} \in Rel(\Delta)$ defined by $X(\mathbf{A'}) = X(\mathbf{A})$, $R(\mathbf{A'}) = R(\mathbf{A})$ for every $R \in \Delta$ (for homomorphisms we have $\Phi(f) = f$).

These object-transformations call for a special terminology: For $\mathbf{A}' \in Rel(\Delta, \Delta')$ we call $\Phi(\mathbf{A}') = \mathbf{A}$ the *shadow* of $\mathbf{A}'$. Any $\mathbf{A}'$ with $\Phi(\mathbf{A}') = \mathbf{A}$ is called a *lift* of $\mathbf{A}$. The analogous terminology is used for subclasses $\mathcal{K}$ of $Rel(\Delta, \Delta')$ and $Rel(\Delta)$.

The following combinatorial characterization of NP was recently proved in [32]:

Theorem 5.7. *For every language $L \in NP$ there exist relational types Δ, Δ' and a finite set $\mathcal{F}'$ of structures in $Rel(\Delta, \Delta')$ such that L is computationally equivalent to $\Phi(Forb(\mathcal{F}'))$. Moreover, we may assume that the relations in Δ' are at most binary.*

References

[1] A. Atserias, On digraph coloring problems and treewidths duality, in *Proc. 20th IEEE Symposium on Logic in Computer Science (LICS)* (2005), pp. 106–115.

[2] B. Bollobás, *Modern Graph Theory*, Springer, (2002).

[3] Ch. Borgs, J. Chayes, L. Lovász, V.T. Sós and K. Vestergombi, Counting Graph Homomorphisms, in *Topics in Discrete Mathematics, Dedicated to Jarik Nešetřil on the Occassion of his 60th Birthday* (eds. M. Klazar, J. Kratochvíl, M. Loebl, J. Matoušek, R. Thomas, P. Valtr), Springer, (2006), pp. 315–372.

[4] Ch. Borgs, J. Chayes, L. Lovász, V.T. Sós, B. Szegedy and K. Vestergombi, in *Graph limits and parameter testing, STOC'06, ACM* (2006), pp. 261–270.

[5] G. Birkhoff, Generalized arithmetics, *Duke Math. J.*, **9** (1942), pp. 283–302.

[6] G. Brightwell and P. Winkler, Graph homomorphisms and phase transitions, *J. Comb Th. B*, **77** (1999), pp. 415–435.

[7] A. Bulatov, A dichotomy theorem for constraints on a three element set, in *FOCS'02* (2002), 649–658.

[8] A. Bulatov, P.G. Jeavons and A. Krokhin, The complexity of maximal constraint languages, in *STOC'01* (2001), pp. 667–674.

[9] M. Chudnovsky, N. Robertson, P. Seymour and R. Thomas, The strong perfect graph theorem, *Annals of Mathematics*, **164 (1)** (2006), pp. 51–229.

[10] F. Chung, R.L. Graham and R.M. Wilson, Quasi-random graphs, *Combinatorica*, **9** (1989), pp. 345–362.

[11] N. Creignou, S. Khanna and M. Sudan, *Complexity classification of Boolean constraint satisfaction problems*, SIAM, (2001).

[12] A. Dochtermann, *Hom complexes and homotopy theory in the category of graphs*, arXiv:math.CO/0605275, (2006).

[13] R. Fagin, Generalized first-order spectra and polynomial-time recognizable sets, Complexity of Computation ed. R. Karp, in *SIAM-AMS Proceedings* **7** (1974), pp. 43–73.

[14] T. Feder and M. Vardi, The computational structure of monotone monadic SNP and constraint satisfaction: A study through Datalog and group theory, *SIAM J. Comput.*, **28(1)** (1999), pp. 57–104.

[15] J. Foniok, J. Nešetřil and C. Tardif, Generalized dualities and maximal finite antichains in the homomorphism order of relational structures, *European J. Comb.*, To appear.

[16] M. Freedman, L. Lovász and L. Schrijver, Reflection positivity, rank connectivity, and homomorphism of graphs, To appear.

[17] T. Feder, P. Hell, S. Klein and R. Motwani, Complexity of graph partition problems, in *31st Annual ACM STOC* (1999), pp. 464–472.

[18] B. Guenin, Edge coloring plane regular multigraphs, manuscript

[19] R.L. Graham, J. Spencer and B.L. Rothschild, *Ramsey Theory*, Wiley, New York (1980).

[20] R. Häggkvist and P. Hell, Universality of A-mote graphs, *Europ. J. Combinatorics*, **14** (1993), pp. 23–27.

[21] Z. Hedrlín, On universal partly ordered sets and classes, *J. Algebra*, **11** (1969), pp. 503–509.

[22] P. Hell and J. Nešetřil, Graphs and Homomorphisms, Oxford University Press, Oxford (2004).

[23] P. Hell and J. Nešetřil, Complexity of H-coloring, *J. Comb. Th. B*, **48** (1990), pp. 92–110.

[24] P. Hell, J. Nešetřil and X. Zhu, Duality and polynomial testing of tree homomorphisms, *Trans. Amer. math. Soc.*, **348** (1996), pp. 1281–1297.

[25] W. Hochstätter and J. Nešetřil, Linear Programming Duality and Morphisms, *Comment. Math. Univ. Carolinae*, **40** (1999(3)), pp. 557–592.

[26] J. Hubička and J. Nešetřil, Finite paths are universal, *Order*, **21(3)** (2004), pp. 181–200. *Order*, **22** (2005), pp. 21–40.

[27] J. Hubička and J. Nešetřil, Universal partial order represented by means of oriented trees and other simple graphs, *European J. Comb.*, **26** (2005), pp. 765–778. (2005), pp. 21–40.

[28] P.G. Jeavons, On the algebraic structure of combinatorial problems, *Theor. Comp. Sci.*, **200** (1998), pp. 185–204.

[29] T. Jensen and B. Toft, *Graph Coloring Problems*, Wiley, (1995).

[30] D.N. Kozlov, Chromatic numbers, morphism complexes, and Stiefel-Whitney characteristic classes, in *Ceometric combinatorics, AMS* (2006).

[31] G. Kun, Constraints, MMSNP and expander structures, manuscript, 2006.

[32] G. Kun and J. Nešetřil, Forbidden Lifts NP and CSP for combinatorists, submitted.

[33] G. Kun and J. Nešetřil, Density and Dualities for Algebras, submitted.

[34] B. Larose, C. Lotte and C. Tardif, A characterisation of first-order constraint satisfaction problems,, LICS 2006..

[35] L. Lovász, The rank of connection matrices and the dimension of graph algebras, *European J. Comb.*, To appear.

[36] L. Lovász, Operations with Structures, *Acta. Math. Acad. Sci. Hung.*, **18** (1967), pp. 321–329.

[37] L. Lovász and B. Szegedy, Szemerédi's Lemma for the Analyst, To appear.

[38] L. Lovász and B. Szegedy, Limits of dense graph sequences, To appear.

[39] R.E. Ladner, On the structure of Polynomial Time Reducibility, *Journal of the ACM*, **22(1)** (1975), pp. 155–171.

[40] T. Luczak and J. Nešetřil, A probabilistic approach to the dichotomy problem, *SIAM J. Comput.*, **36** (2006), pp. 835–843.

[41] J. Matoušek and J. Nešetřil, *Invitation to Discrete Mathematics*, Oxford University Press, (1998).

[42] J. Matoušek, *Using Borsuk-Ulam Theorem*, Springer Verlag, Berlin (2003).

[43] R. McKenzie, The zig-zag property and exponential cancellation of ordered sets, To appear.

[44] R. Naserasr, Homomorphisms and edge-coloring of planar graphs, *J. Comb. Th. B*, **97** (2007), pp. 394-400.

[45] J. Nešetřil, Ramsey Theory, *Handbook of Combinatorics*, (eds. R.L. Graham, M. Grötschel and L. Lovász), Elsevier, pp. 1331–1403. (1995).

[46] J. Hubička and J. Nešetřil, Universal partial order represented by means of trees and other simple graphs, *European J. Comb.*, **26(5)** (2005), pp. 765–778.

[47] J. Nešetřil, *Combinatorics of Mappings*, Birkhäser, Verlag To appear.

[48] J. Nešetřil, Aspects of structural combinatorics, *Taiwanese J. Math.*, **3(4)** (1999), pp. 381–424.

[49] J. Nešetřil, *Homomorphisms of Structures (Concepts and Highlights)*, Cargese School Physics and Computer Science, To appear.

[50] J. Nešetřil and P. Ossona de Mendez, Tree depth, subgraph coloring and homomorphism bounds,, *European Journal of Combinatorics*, **27** (2006), pp. 1022–1041.

[51] J. Nešetřil and P. Ossona de Mendez, Grad and Classes with Bounded Expansion III. Restricted Dualities, submitted

[52] J. Nešetřil and P. Ossona de Mendez, Linear Time Low tree-width Partitions and Algorithmic Consequences, in *STOC'06*, ACM (2006), pp. 391–400.

[53] J. Nešetřil and A. Pultr, On classes of relations and graphs determined by subobjects and factorobjects, *Discrete Math.*, **22** (1978), pp. 287–300.

[54] J. Nešetřil, A. Pultr and C. Tardif, Gaps and dualities in Heyting categories, submitted.

[55] J. Nešetřil and V. Rödl, Chromatically optimal rigid graphs, *J. Comb. Th. B*, **46** (1989), pp. 133–141.

[56] J. Nešetřil and R. Šámal, Tension continuous maps-their structure and applications, in *ITI Series 2005-242* (2005).

[57] J. Nešetřil and I. Švejdarová, Diameter of duals are linear, in *KAM-DIMATIA Series 2005-729* (2005).

[58] J. Nešetřil and C. Tardif, Duality theorems for finite structures characterizing gaps and good characterizations, *J. Comb. Th. B*, **80** (2000), pp. 80–97.

[59] J. Nešetřil and C Tardif, Short answers to exponentially long questions: Extremal aspects of homomorphism duality, in *KAM-DIMATIA Series 2004-714 SIAM J. Disc. Math.*, To appear.

[60] J. Nešetřil and X. Zhu, Path Homomorphisms, *Proc. Camb. Phil. Soc.*, **120** (1996), pp. 207–220.

[61] A. Pultr, The right adjoints into the categories of relational systems, *Lecture Notes in Mathematics*, **137** (1970), pp. 100–113.

[62] A. Pultr and V. Trnková, *Combinatorial, Algebraic and Topological Representations of Groups, Semigroups and Categories*, North Holland, (1980).

[63] F.S. Roberts, T-colorings of Graphs Recent Results and Open Problems, *Discrete Math.*, **93** (1991), pp. 229–245.

[64] B. Rossman, Existential positive types and preservation under homomorphisms, in *20th IEEE Symposium on Logic in Computer Science LICS*(2005), pp. 467–476.

[65] T. Schaefer, The complexity of satisfiability problems, in *STOC'78* (1978), pp. 216–226.

[66] C. Schulz, Graph colourings, spaces of edges, and spaces of circuits, Prepint.

[67] E. Szemerédi, On sets of integers containing no k elements in arithmetic progression, *Acta Arithmetica*, **27** (1975), pp. 299–345.

[68] A. Thomasson, Pseudorandom graphs, in *Random Graphs '85* North Holland, (1987), pp. 307–331.

[69] E. Welzl, Color Families are Dense, *J. Theoret. Comp. Sci.*, **17** (1982), pp. 29–41.

[70] X. Zhu, Circular chromatic number: a survey,in *Combinaotorics, Graph Theory, Algorithms and Applications* (eds. M. Fiedler, J. Kratochvíl and J. Nešetřil), *Discrete Math.*, **229(1–3)** (2001), pp. 371–410.

Proc. of ICDM 2006, Pages 25–37

Introduction to the Minimum Cost Homomorphism Problem for Directed and Undirected Graphs

GREGORY GUTIN*
Department of Computer Science, Royal Holloway University of London, Egham, Surrey TW20 OEX, UK
Email: gutin@cs.rhul.ac.uk
Also at
Department of Computer Science, University of Haifa, Israel
EUN JUNG KIM
Department of Industrial Engineering, KAIST 373-1 Kusong-dong, Yusong-ku, Taejon, 305-791, Republic of Korea
Email: masquenada@kaist.ac.kr

Abstract. For digraphs D and H, a mapping $f : V(D) \rightarrow V(H)$ is a homomorphism of D to H if $uv \in A(D)$ implies $f(u)f(v) \in A(H)$. If, moreover, each vertex $u \in V(D)$ is associated with costs $c_i(u), i \in V(H)$, then the cost of the homomorphism f is $\sum_{u \in V(D)} c_{f(u)}(u)$. For each fixed digraph H, we have the *minimum cost homomorphism problem for* H. The problem is to decide, for an input graph D with costs $c_i(u)$, $u \in V(D), i \in V(H)$, whether there exists a homomorphism of D to H and, if one exists, to find one of minimum cost. Minimum cost homomorphism problems encompass (or are related to) many well studied optimization problems. We overview results and proof methods that have already been obtained for the problem and initiate the study of the problem for digraphs with possible loops.

1 Introduction, Terminology and Notation

The purpose of this paper is twofold: (a) To give an introduction to a new area, the minimum cost homomorphism problem for directed and undirected graphs; (b) To initiate the study of the problem for digraphs with possible loops and to prove new results for such digraphs. Our proofs of the new results illustrate proof methods known from the literature.

In most of the literature on digraphs, only digraphs without loops and multiple arcs are studied. This is justified by the fact that multiple arcs and especially loops play no role when we investigate path and cycle structure of digraphs and several other central topics. However, digraphs with possible loops appear naturally in many applications such as digraphs of relations, automata, Markov chains, etc. In homomorphism problems for undirected graphs, graphs with loops are often investigated [8, 14, 17] (multiple edges are of no interest for such problems). Nevertheless, loops are usually not taken into consideration when digraph homomorphisms are studied [14, 17] (one notable exception is [6]). We think the situation

*Corresponding author

should be changed and we initiate research of the minimum cost homomorphism problem for digraphs with possible loops.

In this paper, we consider only directed (undirected) graphs that do not have multiple arcs (edges), but may have loops. If a directed (undirected) graph G has no loops, we call G *loopless*. If a directed (undirected) graph G has a loop at every vertex, we call G *reflexive*. When we wish to stress that a family of digraphs contains digraphs with loops we will speak of digraphs *with possible loops (w.p.l.)* For an undirected graph H, $V(H)$ and $E(H)$ denote its vertex and edge sets, respectively. For a digraph H, $V(H)$ and $A(H)$ denote its vertex and arc sets, respectively.

For directed (undirected) graphs G and H, a mapping $f : V(G) \rightarrow V(H)$ is a *homomorphism of G to H* if uv is an arc (edge) implies that $f(u)f(v)$ is an arc (edge). A homomorphism f of G to H is also called an *H-coloring* of G, and $f(x)$ is called the *color* of the vertex x in G. We denote the set of all homomorphisms from G to H by $HOM(G, H)$. Recent treatments of homomorphisms in directed and undirected graphs can be found in [14, 17]. Let H be a fixed directed or undirected graph. The *homomorphism problem* for H asks whether a directed or undirected input graph G admits a homomorphism to H. The *list homomorphism problem* for H asks whether a directed or undirected input graph G with lists (sets) $L_u \subseteq V(H), u \in V(G)$ admits a homomorphism f to H in which $f(u) \in L_u$ for each $u \in V(G)$.

Notice that if H has a loop ii, the homomorphism problem for H is trivial ($f(x) = i$ for each $x \in V(G)$ is a homomorphism), but the list homomorphism problem for H is far from trivial even for reflexive undirected graphs [5].

Suppose G and H are directed (or undirected) graphs, and $c_i(u)$, $u \in V(G)$, $i \in V(H)$ are nonnegative *costs*. The *cost of a homomorphism* f of G to H is $\sum_{u \in V(G)} c_{f(u)}(u)$. If H is fixed, the *minimum cost homomorphism problem*, MinHOM(H), for H is the following optimization problem. Given an input graph G, together with costs $c_i(u)$, $u \in V(G)$, $i \in V(H)$, we wish to find a minimum cost homomorphism of G to H, or state that none exists.

The minimum cost homomorphism problem was introduced in [11], where it was motivated by a real-world problem in defence logistics. We believe it offers a practical and natural model for optimization of weighted homomorphisms. The problem's special cases include the list homomorphism problem [14, 17] and the general optimum cost chromatic partition problem, which has been intensively studied [13, 18, 19], and has a number of applications, [22, 24].

By a *directed path (cycle)* we mean a simple directed path (cycle) (i.e., with no self-crossing). We assume that a directed cycle has at least two vertices. In particular, a loop is not a cycle. A directed cycle with k vertices is called a *directed k-cycle* and denoted by $\vec{C}_k$. A digraph H is *cyclic* if H has a cycle and *acyclic* if H has no cycle.

Let H be a digraph. The *converse* of H is the digraph obtained from H by replacing every arc xy with the arc yx. If xy is an arc of a digraph H, we will say that x *dominates* y, y *is dominated by* x and denote it by $x \rightarrow y$. For a pair X, Y of vertex sets of a digraph H, we define $X \times Y = \{xy : x \in X, y \in Y\}$; $X \rightarrow Y$ means that $x \rightarrow y$ for each $x \in X, y \in Y$.

An *empty digraph* is a digraph with no arcs. A loopless digraph D is a *tournament* (*semicomplete digraph*) if there is exactly one arc (at least one arc) between every pair of vertices. We will consider *tournaments with possible loops (w.p.l.)*, i.e., digraphs obtained from tournaments by appending some number of loops (possibly zero loops). A *k-partite tournaments* (*semicomplete k-partite digraph*) is a digraph obtained from a complete k-partite graph by replacing every edge xy with one of the two arcs xy, yx (with at least one of the arcs xy, yx). We will also consider k-partite tournaments w.l.p. An acyclic semicomplete digraph on p vertices with no loops is denoted by TT_p and called a *transitive tournament*. The vertices of a transitive tournament TT_p can be labeled $1, 2, \dots, p$ such that $ij \in A(TT_p)$ if and only if $1 \leq i < j \leq p$. By TT_p^- $(p \geq 2)$, we denote TT_p without the arc $1p$.

Let H be a loopless digraph with vertices $x_1, x_2, \dots, x_p$ and let $S_1, S_2, \dots, S_p$ be digraphs. Then the *composition* $H[S_1, S_2, \dots, S_p]$ is the digraph obtained from H by replacing x_i with S_i for each $i = 1, 2, \dots, p$. In other words,

$$V(H[S_1, S_2, \dots, S_p]) = V(S_1) \cup V(S_2) \cup \dots \cup V(S_p) \text{ and}$$
$$A(H[S_1, S_2, \dots, S_p]) = \cup\{V(S_i) \times V(S_j) : x_i x_j \in A(H), 1 \leq i \neq j \leq p\} \cup (\cup_{i=1}^{p} A(S_i)).$$

If every S_i is an empty digraph, the composition $H[S_1, S_2, \dots, S_p]$ is called an *extension* of H.

2 Polynomial Time Solvable Cases

In the three subsections of this section, we describe various approaches used in the literature to prove that MinHOM(H) is polynomial time solvable for a certain directed or undirected graph H. While the approaches described in Subsection 2.1 are elementary, the methods described in the other two subsections are more sophisticated.

2.1 Directed Cycles and Extensions

Recall that $\vec{C}_k$ denotes a directed cycle on k vertices, $k \geq 2$; let $V(\vec{C}_k) = \{1, 2, \dots, k\}$. One can check whether $HOM(D, \vec{C}_k) \neq \emptyset$ using the following algorithm $\mathcal{A}$ from Section 1.4 of [17]. First, we may assume that D is connected (i.e., its underlying undirected graph is connected) as otherwise $\mathcal{A}$ can be applied to each component of D separately. Choose a vertex x of D and assign it color 1. Assign every out-neighbor of x color 2 and each in-neighbor of x color k. For every vertex y with color i, we assign every out-neighbor of y color $i+1$ modulo k and every in-neighbor of y color $i-1$ modulo k. We have $HOM(D, \vec{C}_k) \neq \emptyset$ if and only if no vertex is assigned different colors.

To solve MinHOM(H) for $H = \vec{C}_k$, choose an initial vertex x in each component D' of D (a component of its underlying graph). Using the algorithm $\mathcal{A}$ from the previous paragraph, we can check whether each D' admits an H-coloring. If the coloring of D' exists, we compute the cost of this coloring and compute the costs of the other $k-1$ H-colorings when x is colored $2, 3, \dots, k$, respectively. Thus, we can find a minimum cost homomorphism in $HOM(D', H)$. Thus, in polynomial time, we can obtain a H-coloring of the whole digraph D of minimum cost. In other words, we have the following lemma, which was first proved in [9].

Lemma 2.1. *For $H = \vec{C}_k$, $k \geq 2$, MinHOM(H) is polynomial time solvable.*

The following simple lemma proved in [9] is useful when dealing with extensions of certain digraphs.

Lemma 2.2. *Let H be a loopless digraph. If MinHOM(H) is polynomial time solvable then, for each extension H' of H, MinHOMP(H') is also polynomial time solvable.*

Proof. Recall that we can obtain H' from H by replacing every vertex $i \in V(H)$ with an empty digraph S_i. Consider an H'-coloring h' of an input digraph D. We can reduce h' into an H-coloring of D as follows: if $h'(u) \in S_i$, then $h(u) = i$.

Let $u \in V(D)$. Assign $\min\{c_j(u) : j \in S_i\}$ to be a new cost $c_i(u)$ for each $i \in V(H)$. Observe that we can find an optimal H-coloring h of D with the new costs in polynomial time and transform h into an optimal H'-coloring of D with the original costs using the obvious inverse of the reduction described above. □

Corollary 2.3. *Let H be an extension of a directed cycle. Then MinHOM(H) is polynomial time solvable.*

2.2 Min-Max Ordering Theorem

Let H be a digraph and let $v_1, v_2, \ldots, v_p$ be an ordering of vertices of H. Let $e = v_i v_r$ and $f = v_j v_s$ be two arcs in H. The pair $v_{\min\{i,j\}} v_{\min\{s,r\}}$ ($v_{\max\{i,j\}} v_{\max\{s,r\}}$) is called the *minimum* (*maximum*) of the pair e, f. (The minimum (maximum) of two arcs is not necessarily an arc.) An ordering $v_1, v_2, \ldots, v_p$ is a *Min-Max ordering of* $V(H)$ if both minimum and maximum of every two arcs in H are in $A(H)$. Two arcs $e, f \in A(H)$ are called a *non-trivial pair* if $\{e, f\} \neq \{g', g''\}$, where g' (g'') is the minimum (maximum) of e, f. Clearly, to check that an ordering is Min-Max, it suffices to verify that the minimum and maximum of every non-trivial pair of arcs are arcs, too.

The following approach is based on some results for the *valued constraint satisfaction problem (VCSP)* [3,4]. Let Z be the set consisting of of all nonnegative integers and ∞, and let Φ be a set of functions $\phi : W^{r(\phi)} \to Z$, where $r(\phi)$ is the arity of ϕ. An instance $\mathcal{I}$ of VCSP(Φ) is a triple (V, W, C), where V is a finite set of *variables*, which are to be assigned values from the set W, and C is a set of (valued) constraints. Each element of C is a pair $c = (\sigma, \phi)$, where σ is a $|\sigma|$-tuple of variables and $\phi : W^{|\sigma|} \to Z$ is a (cost) function, $\phi \in \Phi$. An *assignment* for $\mathcal{I}$ is a mapping s from V to W. The *cost* of s is defined as follows: $c_{\mathcal{I}}(s) = \sum_{((v_1,\ldots,v_m),\phi)\in C} \phi(s(v_1), \ldots, s(v_m))$.

An *optimal solution* of $\mathcal{I}$ is an assignment s of minimum cost.

Let W be a totally ordered set. A binary function $\phi : W^2 \to Z$ is called *submodular* if, for all $x, y, u, v \in W$, we have

$$\phi(\min\{x, u\}, \min\{y, v\}) + \phi(\max\{x, u\}, \max\{y, v\}) \leq \phi(x, y) + \phi(u, v).$$

The following theorem is the main 'positive' result in [4].

Theorem 2.4. *For each Φ consisting of some unary functions and some binary submodular functions, VCSP(Φ) can be solved in time $O(|V|^3|W|^3)$.*

The following theorem was proved in [9] for loopless digraphs. In fact, the same proofs is valid for digraphs with possible loops. We give the proof from [9] for the sake of completeness. In fact, the theorem can be proved directly using a transformation from MinHOM(H) to the minimum cut problem in a special flow network similar to networks used in [4, 8, 10, 20].

Theorem 2.5. *Let H be a digraph and let an ordering $1, 2, \ldots, p$ of $V(H)$ is a Min-Max ordering, i.e., for any pair ik, js of arcs in H, we have $\min\{i, j\} \min\{k, s\} \in A(H)$ and $\max\{i, j\} \max\{k, s\} \in A(H)$. Then MinHOM(H) is polynomial time solvable.*

Proof. Let $1, 2, \ldots, p$ be a Min-Max ordering of vertices of H. The Min-Max property of the ordering ensures that the binary function ϕ, defined by $\phi(i, j) = 0$ if $ij \in A(H)$ and $\phi(i, j) = \infty$ otherwise, is submodular. We will reduce MinHOMP(H) to VCSP(Φ), where Φ satisfies the conditions of Theorem 2.4. Let $\phi_u(i) = c_i(u)$ for all $u \in V(D)$ and $i \in V(H)$. Let $V = V(D)$ and $W = V(H)$. An assignment is an arbitrary function f from $V(D)$ to $V(H)$. Let $C = C' \cup C''$, where $C' = \{(u, \phi_u) : u \in V(D)\}$ (for a fixed u, ϕ_u is a unary function from $V(H)$ to Z) and $C'' = \{((u, v), \phi_{uv}) : uv \in A(D)\}$, where each $\phi_{uv} = \phi$. Since each ϕ_{uv} is submodular, $\Phi = \{\phi_u : u \in V(D)\} \cup \{\phi_{uv} : uv \in A(D)\}$ satisfies the conditions of Theorem 2.4.

Let $\mathcal{I}$ be an instance of the above-constructed VCSP(Φ). It remains to observe that, if an assignment f is an H-coloring of D, then

$$c_{\mathcal{I}}(f) = \sum_{u \in V(D)} \phi_u(f(u)) + \sum_{uv \in A(D)} \phi_{uv}(f(u), f(v)) = \sum_{u \in V(D)} c_{f(u)}(u),$$

which is the cost of f in MinHOMP(H) (an integer), and if f is not an H-coloring, then $c_{\mathcal{I}}(f) = \infty$. Thus, by solving VCSP($\Phi$) we will determine whether $HOM(H) \neq \emptyset$, and find an optimal $h \in HOM(H)$, if $HOM(H) \neq \emptyset$. □

2.3 Applications of Min-Max Ordering Theorem

We start with the following result from [9], which has an almost trivial proof provided we apply Theorem 2.5. It seems it is not easy to show directly (without the use of Theorem 2.5) that MinHOM(TT_p^-) is polynomial time solvable.

Theorem 2.6. *If H is TT_p ($p \geq 1$) or TT_p^- ($p \geq 3$), then MinHOM(H) is polynomial time solvable.*

Proof. The first case is trivial. To show the case $H = TT_p^-$, label the vertices of TT_p^- by $1, 2, \ldots p$ such that $ij \in A(TT_p^-)$ if and only if $1 \leq i < j \leq p$, but $ij \neq 1p$. Observe that $1, 2, \ldots, p$ is a Min-Max ordering since $1p$ can be neither the minimum nor the maximum of a non-trivial pair of arcs. □

While studying homomorphisms, we can view an undirected graph G as a directed graph by replacing every edge xy of G by the pair xy, yx of arcs. This way we can define Min-Max orderings for undirected graphs and apply Theorem 2.5.

Consider two families of undirected graphs. A graph H with vertices $\{1, 2, \ldots, p\}$ is called *proper interval* if there is an inclusion-free family $\{I_1, I_2, \ldots, I_p\}$ of intervals on the real line such that $ij \in E(H)$ ($1 \leq i, j \leq p$) if and only if $I_i \cap I_j \neq \emptyset$. The definition implies that each vertex of a proper interval graph has a loop. A bipartite graph H with partite sets $P = \{1', 2', \ldots, p'\}$ and $Q = \{1'', 2'', \ldots q''\}$ is called a *proper interval bigraph* if there are two inclusion-free families $\{I_1, I_2, \ldots I_p\}$, $\{J_1, J_2, \ldots, J_q\}$ of intervals on the real line such that $i'j'' \in E(H)$ if and only if $I_i \cap J_j \neq \emptyset$.

The following result was proved in [15, 23].

Theorem 2.7. *The vertices of a reflexive graph H has a Min-Max ordering if and only if it is a proper interval graph.*

A similar result for loopless graphs contains proper interval bigraphs rather than proper interval graphs [10]. These theorems and Theorem 2.5 imply the following:

Corollary 2.8 ([10]). *Let H be a graph in which every component is either a proper interval graph or a proper interval bigraph. Then MinHOMP(H) is polynomial time solvable.*

This result is of importance due to the following 'opposite' result:

Theorem 2.9 ([10]). *Let H be a graph in which at least one component is neither a proper interval graph nor a proper interval bigraph. Then MinHOMP(H) is NP-hard.*

Notice that if a component of H has a pair of vertices one with a loop and the other without a loop, MinHOMP(H) is NP-hard due to Lemma 4.2.

2.4 k-Min-Max Ordering

A collection $V_1, V_2, \dots V_k$ of subsets of a set V is called a *k-partition* of V if $V = V_1 \cup V_2 \cup \cdots \cup V_k$, and $V_i \cap V_j = \emptyset$ provided $i \neq j$.

Let $H = (V, A)$ be a loopless digraph and let $k \geq 2$ be an integer. We say that H has a *k-Min-Max ordering* if there is a k-partition of V into subsets $V_1, V_2, \dots V_k$ and there is an ordering $v^i_1, v^i_2, \dots, v^i_{\ell(i)}$ of V_i for each i such that

(i) Every arc of H is an arc from V_i to V_{i+1} for some $i \in \{1, 2, \dots, k\}$,

(ii) $v^i_1, v^i_2, \dots, v^i_{\ell(i)} v^{i+1}_1 v^{i+1}_2, \dots, v^{i+1}_{\ell(i+1)}$ is a Min-Max ordering of the subdigraph of H induced by $V_i \cup V_{i+1}$ for each $i \in \{1, 2, \dots, k\}$,

where all indices $i + 1$ are taken modulo k.

Note that if H is a loopless strong digraph in which the greatest common divisor of all cycle lengths is k, then $V(H)$ has a k-partition, $k \geq 2$, satisfying (i) (see Theorem 10.5.1 in [1]). A simple example of a digraph having a k-Min-Max ordering, but no Min-Max ordering, is an extension of $\vec{C}_k$. To see than an extension H of $\vec{C}_k$ has no Min-Max ordering, consider an ordering $1, 2, \dots, p$ of the vertices of H, and an arc leaving 1 and an arc coming into 1. The minimum arc of the two arcs is the loop 11 not in H.

The following theorem from [10] establishes usefulness of k-Min-Max orderings. The proof of Theorem 2.10 in [10] is based on a reduction from MinHOM(H) to the minimum cut problem in a special network. The reduction is somewhat similar to the ones used in in [20] and [4].

Theorem 2.10. *If a loopless digraph H has a k-Min-Max ordering, then MinHOM(H) is polynomial time solvable.*

Consider the following simple application of Theorem 2.10.

Proposition 2.11. *Let H' be a directed k-cycle and let H be obtained from H' by appending a set S of $s \leq k$ new vertices each dominated by exactly one vertex of H' such that every vertex of H' dominates at most one vertex of S. Then MinHOM(H) is polynomial time solvable.*

The proof of Proposition 2.11 that uses Theorem 2.10 is trivial. However, it is not so obvious how to prove Proposition 2.11 without using Theorem 2.10. It is interesting to compare Proposition 2.11 with Theorem 4.5.

Theorem 2.10 allows us to prove much more difficult results such as Theorem 2.12 formulated below. We will start from a number of definitions. A bipartite graph H with vertices $x_1, x_2, x_3, x_4, y_1, y_2, y_3$ is called

a bipartite claw if its edge set $E(H) = \{x_4y_1, y_1x_1, x_4y_2, y_2x_2, x_4y_3, y_3x_3\}$;
a bipartite net if its edge set $E(H) = \{x_1y_1, y_1x_3, y_1x_4, x_3y_2, x_4y_2, y_2x_2, y_3x_4\}$;
a bipartite tent if its edge set $E(H) = \{x_1y_1, y_1x_3, y_1x_4, x_3y_2, x_4y_2, y_2x_2, y_3x_4\}$.

The graphs are depicted in Figure 2.1.

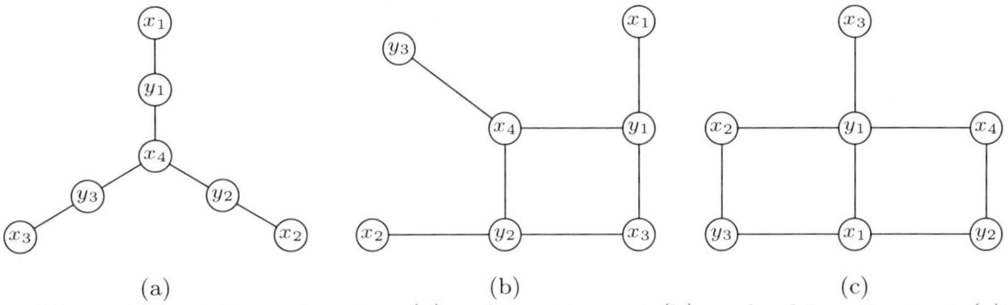

Figure 2.1: A bipartite claw (a), a bipartite net (b) and a bipartite tent (c).

Let us introduce five special digraphs for which the minimum homomorphism problem is NP-hard [10]. The digraph C_4' has vertex set $\{x_1, x_2, y_1, y_2\}$ and arc set

$$\{x_1y_1, y_1x_2, x_2y_2, y_2x_1, y_1x_1\}.$$

The digraph C_4'' has the same vertex set, but its arc set is $A(C_4') \cup \{x_2y_1\}$. The digraph H^* has vertex set $\{x_1, x_2, y_1, y_2, y_3\}$ and arc set $\{x_1y_1,\ y_1x_2,\ x_2y_2,\ y_2x_1,\ x_1y_3,\ x_2y_3\}$.

Let N_1 be a digraph with $V(N_1) = \{x_1, x_2, x_3, y_1, y_2, y_3\}$ and

$$A(N_1) = \{x_1y_1, y_1x_1, x_2y_2, y_2x_2, x_3y_3, y_3x_3, y_1x_2, y_1x_3, x_1y_2, x_1y_3, x_3y_2, x_2y_3\}.$$

Let N_2 be a digraph with $V(N_2) = \{x_1, x_2, x_3, y_1, y_2, y_3\}$ and

$$A(N_2) = \{x_1y_1, x_2y_2, y_2x_2, x_3y_3, y_3x_3, y_1x_2, y_1x_3, x_1y_2, x_1y_3, x_3y_2, x_2y_3\}.$$

Let H be a bipartite digraph with partite sets P, Q. Then $H^{\rightarrow}$ ($H^{\leftarrow}$, $H^{\leftrightarrow}$) denotes a bipartite digraph obtained from H by deleting all arcs from Q to P (a bipartite digraph obtained from H by deleting all arcs from P to Q, a bipartite digraph obtained from H by deleting all arcs not in directed 2-cycles).

A digraph H belongs to the family $\mathcal{HFORB}$ if H or its converse is isomorphic to one of the five digraphs above or $UN(H^s)$ is isomorphic to bipartite claw, bipartite net, bipartite tent or even cycle with at least 6 vertices, where $s \in \{\rightarrow, \leftarrow, \leftrightarrow\}$.

The family $\mathcal{HFORB}$ is of importance due to the following two results proved in [10].

Theorem 2.12. *If a semicomplete bipartite digraph H does not contain a digraph from $\mathcal{HFORB}$ as an induced subdigraph, then MinHOM(H) is polynomial time solvable.*

Theorem 2.13. *The problem MinHOM(H) is NP-hard if $H \in \mathcal{HFORB}$.*

3 Known Dichotomies

Corollary 2.8 and Theorem 2.9 provide a complete dichotomy of the computational complexity of MinHOM(H) when H is an undirected graph. Theorems 2.12 and 2.13

give a dichotomy when H is a semicomplete bipartite digraph. Interestingly, the following dichotomy for semicomplete k-partite digraphs ($k \geq 3$) is less complicated than the one for semicomplete bipartite digraphs in both formulation and proof.

Theorem 3.1. *Let H be a semicomplete k-partite digraph, $k \geq 3$. If H is an extension of TT_k, $\vec{C}_3$ or TT_{k+1}^-, then MinHOMP(H) is polynomial time solvable. Otherwise, MinHOMP(H) is NP-hard.*

The 'polynomial part' of this theorem follows immediately from Lemmas 2.1 and 2.2 and Theorem 2.6.

4 New Dichotomies

The following lemma is an obvious basic observation often used to obtain dichotomies.

Lemma 4.1 ([9]). *Let H' be an induced subdigraph of a digraph H. If MinHOMP(H') is NP-hard, then MinHOMP(H) is also NP-hard.*

4.1 Directed Cycles w.p.l.

NP-completeness of the independence set problem for graphs is often used to prove NP-hardness of MinHOMP(H) for certain digraphs H. The next lemma gives a simple example of this reduction.

Lemma 4.2. *If $V(H) = \{1, 2\}$ and $A(H) = \{12, 21, 22\}$, then MinHOM(H) is NP-hard.*

Proof. Let D be an input of MinHOM(H), let D be a symmetric digraph (i.e., $xy \in A(D)$ implies $yx \in A(D)$) and let G be the underlying graph of D. Set $c_1(x) = 0$ and $c_2(x) = 1$ for each vertex $x \in V(D)$. Clearly, $h(x) = 2$ for each $x \in V(D)$ defines a homomorphism of D to H. Let f be a minimum cost homomorphism of D to H. Then $f(x) = f(y) = 1$ implies that x and y are not adjacent in G. Hence $f^{-1}(1)$ is a maximum size independent set in G.

Let I is a maximum size independent set in G. Then a homomorphism g of D to H such that $g(x) = 1$ if and only if $x \in I$ is of minimum cost. Since the maximum size independent set problem is NP-hard, MinHOM(H) is NP-hard as well. □

We have seen that MinHOM($\vec{C}_2$) is polynomial time solvable. The above lemma shows that if we add just one loop to $\vec{C}_2$, we get an NP-hard case. It is easy to see that addition of two loops returns us to polynomial time solvable cases. Similar changes in the complexity are not valid for $\vec{C}_k$, $k \geq 3$, as shown in the next lemma.

Lemma 4.3. *Let a digraph H is obtained from $\vec{C}_k$, $k \geq 3$, by adding at least one loop. Then MinHOM(H) is NP-hard.*

Proof. Assume that kk is a loop.

Let G be a loopless digraph with n vertices. Construct a digraph D as follows: $V(D) = \{x_1, x_2, \ldots, x_{k-1} : x \in V(G)\}$ and $A(D) = \{x_1x_2, x_2x_3, \ldots, x_{k-2}x_{k-1} : x \in V(G)\} \cup \{x_{k-1}y_1 : xy \in A(G)\}$. For each $x \in V(G)$, set the costs for vertices of D as follows: $c_i(x_i) = 0$ for each $i = 1, 2, \ldots, k-1$, $c_j(x_i) = (k-1)n + 1$ for each $i = 1, 2, \ldots, k-1$ and $j \in \{1, 2, \ldots k-1\} - \{i\}$, and $c_k(x_i) = 1$ for each $i = 1, 2, \ldots, k-1$.

Clearly, $h(x_i) = k$ for each $x \in V(D)$ and $i = 1, 2, \ldots, k-1$ defines a homomorphism of D to H. Let f be a minimum cost homomorphism of D to H. It follows from the fact that the cost of h is $(k-1)n$ that $f(x_i) \in \{i, k\}$ for each $x \in V(D)$ and $i = 1, 2, \ldots, k-1$. Thus, for every vertex $x \in V(G)$ we have either $f(x_i) = i$ or $f(x_i) = k$ for each $i = 1, 2, \ldots, k-1$.

Let $f(x_1) = f(y_1) = 1$, where x, y are distinct vertices of G. If $xy \in A(G)$, then $x_{k-1}y_1 \in A(D)$, which is a contradiction since $f(x_{k-1}) = k-1$. Thus, x and y are non-adjacent in G. Hence, $I = \{x \in V(G) : x_1 \in f^{-1}(1)\}$ is an independent set in G. Observe that the cost of f is $(k-1)(n-|I|)$.

Conversely, if I is an independent set in G, we obtain a homomorphism g of D to H by fixing $g(x_i) = i$, $i = 1, 2, \ldots, k-1$, for every $x \in I$ and $g(x_i) = k$, $i = 1, 2, \ldots, k-1$, for every $x \in V(G) - I$. Observe that the cost of g is $(k-1)(n-|I|)$. Hence a homomorphism g of D to H is of minimum cost if and only if the corresponding independent set I is of maximum size in G. Since the maximum size independent set problem is NP-hard, MinHOM(H) is NP-hard as well. □

Our first dichotomy follows directly from Lemmas 2.1, 4.2 and 4.3 and a simple observation that MinHOM($\vec{C}_2^*$) is polynomial time solvable, where $\vec{C}_2^*$ is the digraph obtained from a directed 2-cycle by adding loop to each vertex.

Proposition 4.4. *Let H be a digraph obtained from $\vec{C}_k$, $k \geq 2$, by possibly adding loops. Then MinHOM(H) is polynomial time solvable if $H \in \{\vec{C}_2^*\} \cup \{\vec{C}_k : k \geq 2\}$. Otherwise, MinHOM(H) is NP-hard.*

4.2 Tournaments w.p.l.

A special case of the following lemma was proved in [9]. Our proof is a modification of the proof in [9].

Lemma 4.5. *Let H' be a digraph obtained from $\vec{C}_k = 12\ldots k1$, $k \geq 2$, by adding an extra vertex $k+1$ dominated by at least two vertices of the cycle and let H'' is the digraph obtained from H' by adding the loop at vertex $k+1$. Let H be H' or its converse or H'' or its converse. Then MinHOM(H) is NP-hard.*

Without loss of generality we may assume that $H = H'$ and that vertices 1 and ℓ dominate vertex $k+1$. We will reduce the maximum independent set problem to MinHOMP(H). Let G be a graph. Construct a digraph D as follows:

$$V(D) = V(G) \cup \{v_i^e : e \in E(G)\ i \in V(H)\},\ A(D) = A_1 \cup A_2, \text{ where}$$

$$A_1 = \{v_1^e v_2^e, v_2^e v_3^e, \ldots v_{k-1}^e v_k^e, v_k^e v_1^e : e \in E(G)\}$$

and

$$A_2 = \{v_1^{uv} u, v_{k+1}^{uv} u, v_\ell^{uv} v, v_{k+1}^{uv} v : uv \in E(G)\}.$$

Let all costs $c_i(t) = 1$ for $t \in V(D)$ apart from $c_{k+1}(p) = 2$ for all $p \in V(G)$.

Consider a minimum cost homomorphism f of D to H. By the choice of the costs, f assigns the maximum possible number of vertices of G (in D) a color different from $k+1$. However, if pq is an edge in G, by the definition of D, f cannot assign colors different from $k+1$ to both p and q. Indeed, if both p and q were assigned colors different from $k+1$, then the existence of arcs $v_{k+1}^{pq} p$ and $v_{k+1}^{pq} q$ would imply that they would be assigned the same color, which however is impossible by the existence of arcs $v_1^{pq}p$ and $v_\ell^{pq}q$. Observe that f may assign exactly one of the vertices p, q color $k+1$ and the other a color different from $k+1$. Also f may assign both of them color $k+1$. Thus, G has a maximum independent set with α vertices if and only if D has a minimum cost H-coloring of cost $|E(G)| \cdot |V(H)| + 2|V(G)| - \alpha$. This reduces the maximum independent set problem to MinHOMP(H). □

The proof of the following theorem is similar to the proof of the 'loopless version' of the theorem in [9]. The only significant difference is the use of Lemma 4.3.

Theorem 4.6. *Let T be a tournament w.p.l. If H is an acyclic tournament w.p.l. or $H = \vec{C}_3$, then MinHOM(H) is polynomial time solvable. Otherwise, MinHOM(H) is NP-hard.*

Proof. The polynomial time solvability part follows from Lemma 2.1 and Theorem 2.6. Now assume that H is not an acyclic tournament and $H \neq \vec{C}_3$. Suppose MinHOM(H) is not NP-hard. Since H is a cyclic tournament, it is easy to show that H contains a directed 3-cycle C. By Lemmas 4 and 4.3, C has no loops, and, thus, there is a vertex $i \in V(H)$ not on C.

Assume that the subdigraph L of H induced by $V(C) \cup \{i\}$ is strong. Assume that i belongs to a directed 3-cycle. Then, if i has a loop, MinHOM(H) is NP-hard, a contradiction. Thus, i has no loop. However, MinHOM(H) is NP-hard by Lemmas 4 and 4.5, a contradiction. □

4.3 Cyclic Multipartite Tournaments w.p.l.

Lemma 4.7. *Let H' be given by $V(H') = \{1,2,3,4\}$, $A(H') = \{12, 23, 31, 34, 41\}$. Let H'' be obtained from H' by adding the loop 44. If H is H' or H'', then MinHOMP(H) is NP-hard.*

Proof. Since the case $H = H'$ was proved in [9], we will consider only the case $H = H''$. We will reduce the maximum independent set problem to MinHOMP(H). Consider a digraph $T(u,v)$ defined as follows:

$V(T(u,v)) = \{x, y, u', u, v', v, z_1, z_2, \ldots, z_{12}\}$,
$A(T(u,v)) = \{xy, xz_1, yz_1, z_6u', u'u, z_{11}v', v'v, z_1z_2, z_2z_3, z_3z_4, \ldots, z_{11}z_{12}, z_{12}z_1\}$.

Let G be a graph with n vertices and m edges. Construct a digraph D as follows. Start with $V(D) = V(G)$ and, for each edge $uv \in E(G)$, add a distinct copy of $T(u,v)$ to D. Note that the vertices in $V(G)$ form an independent set in D and that $|V(D)| = n + 16m$.

Let $M = 2n + 16m$. Let the cost $c_i(t) = 1$ for each $t \in V(D)$ and $i = 1,2,3,4$ with the following exceptions: $c_j(p) = 2$ for each $p \in V(G)$ and $j \neq 2$, and $c_4(z_i) = 1 + M$ for each i not divided by 4.

Consider a homomorphism h of $T(u,v)$ to H defined as follows: $h(z_{3i+1}) = 1$, $h(z_{3i+2}) = 2$, $h(z_{3i+3}) = 3$ for each $i = 0,1,2,3$, $h(u') = h(u) = h(v) = h(y) = 4$ and $h(v') = h(x) = 3$. The cost of h equals M.

Let f be a minimum cost homomorphism of $T(u,v)$ to H. Observe that we must have $f(z_1) = 1$ since $x \rightarrow z_1$, $y \rightarrow z_1$ and $c_4(z_1) = 1 + M$. This and the fact that f is of cost at most M imply that $(f(z_1), f(z_2), \ldots, f(z_{12}))$ has to coincide with one of the following two sequences: $(1,2,3,1,2,3,1,2,3,1,2,3)$ or $(1,2,3,4,1,2,3,4,1,2,3,4)$.

If the first sequence is the actual one, then we have $f(z_6) = 3$, $f(u') \in \{1,4\}$, $f(u) \in \{1,2,4\}$, $f(z_{11}) = 2$, $f(v') = 3$ and $f(v) \in \{1,4\}$. If the second sequence is the actual one, then we have $f(z_6) = 2$, $f(u') = 3$, $f(u) \in \{1,4\}$, $f(z_{11}) = 3$,

$f(v') \in \{1,4\}$ and $f(v) \in \{1,2,4\}$. We can color one of u and v with color 2 and the other with color 1 or 4. However we cannot assign color 2 to both u and v in a homomorphism.

Let g be a minimum cost homomorphism of D to H. Clearly, g must aim at assigning as many vertices of $V(G)$ in D color 2. Notice that if pq is an edge in G, by the arguments above, g cannot assign color 2 to both p and q. However, g can assign color 2 to either p or q (or neither). Thus, g corresponds to a maximum independent set in G and vise versa (the vertices of a maximum independent set are assigned color 2 and all other vertices in $V(G)$ are assigned color 1 or 4). □

Theorem 4.8. *Let H be a cyclic k-partite tournament w.p.l., $k \geq 2$. If H is an extension of $\vec{C}_3$ or $\vec{C}_4$, then MinHOMP(H) is polynomial time solvable. Otherwise, MinHOMP(H) is NP-hard.*

Proof. We assume that P≠NP as otherwise this theorem is of no interest. Assume also that MinHOMP(H) is polynomial time solvable and H is not an extension of $\vec{C}_3$ and $\vec{C}_4$. Since H is a k-partite tournament, $k \geq 2$, and H has a cycle, there can be two possibilities for the length of a shortest cycle C in H: 3 or 4. Thus, we consider two corresponding cases. Observe that, by Lemma 4.3, no cycle in H has a loop. In particular, C has no loop.

Case 1. We have $C = ijli$. If H has at least four partite sets, then i, j, l together with a vertex from a partite set containing none of i, j, l induce a tournament w.p.l. By Theorem 4.6, MinHOMP(H) is NP-hard, a contradiction. So, H has three partite sets.

Let I, J and L be partite sets of H such that $i \in I$, $j \in J$ and $l \in L$. Let s be a vertex outside C and let $s \in I$. If s is dominated by j and l or dominates j and l, then MinHOMP(H) is NP-hard by Lemma 4.5, a contradiction. If $j \rightarrow s \rightarrow l$, then MinHOMP($H$) is NP-hard by Lemma 4.7, a contradiction. Thus, $l \rightarrow s \rightarrow j$. Similar arguments show that $l \rightarrow I \rightarrow j$. Similarly we can prove that $L \rightarrow I \rightarrow J \rightarrow L$, i.e., H is an extension of $\vec{C}_3$, contradiction.

Case 2. We have $C = i_1 i_2 i_3 i_4 i_1$. Since C is a shortest cycle, i_1, i_3 are belong to the same partite set, say L, and i_2, i_4 belong to the same partite set, say M. Assume first that H is not bipartite. Then there is a vertex q belonging to a partite set different from L and M. Since H has no directed 3-cycle, either q dominates $V(C)$ or $V(C)$ dominates q. In both cases, MinHOMP(H) is NP-hard by Lemma 4.5, a contradiction.

Assume that H is bipartite. By Lemma 4.5, the vertices of H can be partitioned into four sets: $I_j = \{p : i_{j-1} \rightarrow p \rightarrow i_{j+1}\}$, $j = 1, 2, 3, 4$, where all indices are taken modulo 4. Suppose that there is pair $q_1 \in I_1, q_2 \in I_2$ of vertices such that $q_2 \rightarrow q_1$. Let H' denote the subdigraph of H induced by q_1, i_2, i_3, i_4, q_2. Observe that MinHOM(H') is NP-hard by Lemma 4.5. Similar arguments imply that $I_1 \rightarrow I_2 \rightarrow I_3 \rightarrow I_4 \rightarrow I_1$, i.e., H is an extension of $\vec{C}_4$, a contradiction. □

5 Further Research

We have managed to obtain dichotomies for tournaments w.p.l. and cyclic multipartite tournaments w.p.l. The reader may ask why we have not obtained dichotomies, for example, for semicomplete digraphs w.p.l. and acyclic multipartite tournaments w.p.l. Our research indicates significant difficulties in obtaining these dichotomies.

One of them is a significantly larger number of polynomial time solvable (and NP-hard) cases. In fact, we have very recently obtained a dichotomy for semicomplete digraphs w.p.l. [21], but its proof is far too long to be included here. A dichotomy for acyclic multipartite tournaments w.p.l. remains an open problem.

The above indicates that addition of loops may well make dichotomies significantly more complicated. Thus, the problem of obtaining a dichotomy for all semicomplete k-partite digraphs, $k \geq 2$, w.p.l. appears to be a very difficult open problem. We conjecture that there is a dichotomy for the whole class of digraphs w.p.l. and it would be very interesting to verify this conjecture.

Very recently, Gupta et al. [7] verified the conjecture for reflexive digraphs.

References

[1] J. Bang-Jensen and G. Gutin, *Digraphs: Theory, Algorithms and Applications*, Springer-Verlag, London (2000).

[2] J. Bang-Jensen, P. Hell and G. MacGillivray, The complexity of colouring by semicomplete digraphs, *SIAM J. Discrete Math.*, **1** (1988), pp. 281–298.

[3] D. Cohen, M. Cooper, P. Jeavons and A. Krokhin, Soft constraints: complexity and multimorphisms, in *Proc. CP'03 Lect. Notes Comput. Sci.*, **2833** (2003), pp. 244–258.

[4] D. Cohen, M. Cooper, P. Jeavons and A. Krokhin, A maximal tractable class of soft constraints, *J. Artif. Intell. Res.*, **22** (2004), pp. 1–22.

[5] T. Feder and P. Hell, List homomorphisms to reflexive graphs, *J. Combin. Theory B.*, **72** (1998), pp. 236–250.

[6] T. Feder, P. Hell and J. Huang, List homomorphisms to reflexive digraphs, Manuscript.

[7] A.Gupta, P.Hell, M.Karimi and A. Rafiey, Minimum Cost Homomorphism to Reflexive Digraphs, *Proc. of LATIN 2008, Lecture Notes Comput. Sci.*, **4957** (2008), pp. 182–193.

[8] G. Gutin, P. Hell, A. Rafiey and A. Yeo, A Dichotomy for Minimum Cost Graph Homomorphisms, *Europ. J. Combin.*, **29** (2008), pp. 900-911.

[9] G. Gutin, A. Rafiey and A. Yeo, Minimum Cost and List Homomorphisms to Semicomplete Digraphs, *Discrete Appl. Math.*, **154** (2006), pp. 890–897.

[10] G. Gutin, A. Rafiey and A. Yeo, Minimum Cost Homomorphisms to Semicomplete Bipartite Digraphs, *Discrete Appl. Math.*, To appear.

[11] G. Gutin, A. Rafiey, A. Yeo and M. Tso, Level of repair analysis and minimum cost homomorphisms of graphs, *Discrete Appl. Math.*, **154** (2006), pp. 881–889.

[12] W. Gutjarh, *Graph coloring, PhD thesis*, Free University, Berlin (1991).

[13] M. M. Halldorsson, G. Kortsarz, and H. Shachnai, Minimizing average completion of dedicated tasks and interval graphs. Approximation, Randomization, and Combinatorial Optimization (Berkeley, Calif, 2001), *Lecture Notes in Computer Science vol. 2129*, Springer, Berlin pp. 114–126. (2001).

[14] P. Hell, Algorithmic aspects of graph homomorphisms, in *Survey in Combinatorics 2003, London Math. Soc. Lecture Note Series 307, Cambridge University Press* (2003), pp. 239–276.

[15] P. Hell and J. Huang, Certifying LexBFS recognition algorithms for proper inteval graphs and proper interval bigraphs, *SIAM J. Discrete Math.*, **18** (2005), pp. 554–570.

[16] P. Hell and J. Nešetřil, On the complexity of H-colouring, *J. Combin. Theory B*, **48** (1990), pp. 92–110.

[17] P. Hell and J. Nešetřil, *Graphs and Homomorphisms*, Oxford University Press, Oxford (2004).

[18] K. Jansen, Approximation results for the optimum cost chromatic partition problem, *J. Algorithms*, **34** (2000), pp. 54–89.

[19] T. Jiang and D.B. West, Coloring of trees with minimum sum of colors, *J. Graph Theory*, **32** (1999), pp. 354–358.

[20] S. Khanna, M. Sudan, L. Trevisan and D. Williamson, The approximability of constraint satisfaction problems, *SIAM J. Comput.*, **30** (2000), pp. 1863–1920.

[21] G. Gutin and E.J. Kim, Complexity of the Minimum Cost Homomorphism Problem for Semicomplete Digraphs with Possible Loops, Submitted.

[22] L.G. Kroon, A. Sen, H. Deng, and A. Roy, The optimal cost chromatic partition problem for trees and interval graphs, *Graph-Theoretic Concepts in Computer Science (Cadenabbia, 1996), Lecture Notes in Computer Science vol. 1197*, Springer, Berlin pp. 279–292. (1997).

[23] J. Spinrad, Efficient Graph Representations, *AMS*, (2003),

[24] K. Supowit, Finding a maximum planar subset of a set of nets in a channel, *IEEE Trans. Computer-Aided Design*, **6** (1987), pp. 93–94.

Proc. of ICDM 2006, Pages 39–41

Refinements of Ramanujan's Congruence Modulo 5

HERBERT S. WILF*
Department of Mathematics, University of Pennsylvania, Philadelphia, PA 19104-6395, USA

1 Introduction

Ramanujan, in the course of his path-breaking researches on integer partitions, conjectured a number of remarkable congruences that are satisfied by the partition function $p(n)$, which counts the integer partitions of n. He went on to prove many of these and started a line of research that remains very active today, with many new discoveries and conjectures, and unsolved problems being generated.

His first observation was that the values $p(4), p(9), p(14), p(19), \ldots$ all seemed to be divisible by 5. He proved this by a variety of methods, the most remarkable of which is perhaps his celebrated identity

$$\sum_{n\geq 0} p(5n+4)x^n = 3D5\frac{\prod_{j\geq 1}(1-x^{5j})^5}{\prod_{j\geq 1}(1-x^j)^6},$$

from which the congruences

$$p(5n+4) \equiv 0 \pmod 5 \qquad (n = 3D0, 1, 2, 3, \ldots) \tag{1}$$

follow by inspection. There are a number of proofs of this identity in the literature (e.g. [1, 5, 7]), though none of them are really simple. Afterwards he made a number of conjectures using the moduli 7, 11, and others, almost all of which have been proved since.

Here I want to discuss some refinements of the mod 5 congruence (1). By a refinement of that congruence I mean that not only is the total number of partitions of $5n+4$ divisible by 5, but we can divide up the set of all of those partitions, according to the value of some parameter, into a union of subsets, *each one of which has cardinality divisible by 5.*

2 The BG-Rank

The parameter that will be used to subdivide the set of all partitions of $5n+4$ in the above manner is the BG-rank of the partition, which was first studied in depth in [2].

If π is a partition of n we define the *BG-rank* $\beta(\pi)$, of π as follows. First draw the Ferrers diagram of π. Then fill the cells with alternating ± 1's, chessboard style, beginning with a $+1$ in the $(1,1)$ position. The sum of these entries is $\beta(\pi)$, the BG-rank of π. For example, the BG-rank of the partition $13 = 3D4+3+3+1+1+1$ is -1.

*This article has been published by the Electronic Journal of Combinatorics. It is paper #N18, vol. 13, no. 1, 2006.

+1	−1	+1	−1
−1	+1	−1	
+1	−1	+1	
−1			
+1			
−1			

Figure 2.1: A partition with BG-rank -1

3 2-Cores

Following the approach in [4], the other ingredient that will be needed in order to prove the refinements in question is the idea of the 2-core of a partition. If π is a partition of n, then draw the Ferrers diagram of π. Then remove either a vertically adjacent pair of cells or a horizontally adjacent pair of cells, subject to the condition that after removal the diagram is still the Ferrers board of some partition. Repeat this removal process, making arbitrary choices, until no further choices remain. The Ferrers diagram that remains is that of a certain partition π_c, which is the 2-core of π. The 2-core is independent of the order in which the various choices for removal of an adjacent pair were made. See [8] for more information on the cores of partitions.

Now it is easy to see that the 2-core of a partition π is always a *staircase* partition, i.e., a partition of the form $k + (k-1) + \cdots + 2 + 1$, for some k. The utility of the Bg rank in the present context resides in the fact that the BG rank of a partition and of its 2-core are the same, since at each removal of a pair of adjacent cells, the BG rank does not change.

Further, the BG rank of a staircase partition is easy to calculate. Specifically, the BG rank of the staircase of height k is $(k+1)/2$, if k is odd, and $-k/2$, if k is even. Therefore, if π is a partition of BG-rank $= 3Dj$ then its 2-core is a staircase partition of height $2j-1$, if $j > 0$, and $-2j$, if $j \leq 0$. In either case, if π is a partition whose BG-rank is j, then its 2-core is a diagram of exactly $j(2j-1)$ cells, i.e., a partition of the integer $j(2j-1)$.

4 The Results

By putting together the above methods, we give a shorter proof of the congruences that were found in [3], namely

$$
\begin{aligned}
p_j(5n) &\equiv 0 \pmod 5, = \text{if } j \equiv 1, 2 \pmod 5, && (2)\\
p_j(5n+1) &\equiv 0 \pmod 5, \text{ if } = j \equiv 0, 3, 4 \pmod 5, && (3)\\
p_j(5n+2) &\equiv 0 \pmod 5, \text{ if } = j \equiv 1, 2, 4 \pmod 5, && (4)\\
p_j(5n+3) &\equiv 0 \pmod 5, \text{ if } = j \equiv 0, 3 \pmod 5, && (5)\\
p_j(5n+4) &\equiv 0 \pmod 5, \ \forall j. && (6)
\end{aligned}
$$

In particular, the last one of these shows that for each j, the number of partitions of $5n+4$ whose BG rank is j is always a multiple of 5, which accomplishes the refinement which was spoken of at the beginning of this article.

To prove the above congruences, a few additional ingredients are needed. In particular we make strong use of the fact that

$$pp(n) \equiv 0 \pmod 5 \text{ if } n \equiv 2, 3, 4 \pmod 5, \tag{7}$$

where $pp(n)$ is the number of ordered pairs of partitions of integers whose sum is n. This result was found first in [5] by another method.

References

[1] G.E. Andrews, *The Theory of Partitions*, Addison-Wesley, (1976).

[2] A. Berkovich and F.G. Garvan, On the Andrews-Stanley refinement of Ramanujan's congruence modulo 5 and generalization, *Trans. Amer. Math. Soc.*, **358** (2006), pp. 703–726.

[3] A. Berkovich and F.G. Garvan, *The BG rank of a partition and its applications*, Preprint, The Mathematics arXiv [math/0602362] (2005).

[4] William Y.C. Chen, Kathy Q. Ji, and Herbert S. Wilf, BG ranks and 2-cores, *Electronic J. Combinat.*, **13(1)** (2006), #N18.

[5] John L. Drost, A shorter proof of the Ramanujan congruence modulo 5, *Amer. Math. Monthly*, **104(10)** (1997), pp. 963–964.

[6] G.H. Hardy and E.M. Wright, *An Introduction to The Theory of Numbers, 5th Ed.*, Oxford, (1979).

[7] Michael D. Hirschhorn, Another short proof of Ramanujan's mod 5 partition congruence, and more, *Amer. Math. Monthly*, **106(6)** (1999), pp. 580–583.

[8] G. James and A. Kerber, *The Representation Theory of the Symmetric Group*, Addison-Wesley, Reading (1981).

Invited Talks

Proc. of ICDM 2006, Pages 43–53

Almost Ramsey Type Theorems for Directed Graphs

S.B. RAO[1] AND SUSHMITA RUJ[2]
Indian Statistical Institute, Kolkata
Email: [1]raosb@isical.ac.in [2]sush_r@isical.ac.in

Abstract. The famous Ramsey Theorem [6] for graphs states that given any graph **H**, there is a positive integer $p(H)$, such that for every graph **G** of order $p \geq p(H)$, either **G** or its complement $\overline{\mathbf{G}}$ has **H** as a subgraph. This Theorem and its relatives have a wide variety of applications in diverse fields such as number theory, algebra, geometry, topology, set theory, logic, ergodic theory, information theory, theoretical computer science, to name only a few [17]. However in several practical applications such as social or communication networks and others the binary relations considered on a set used need not be symmetric. Hence one is forced to consider directed graphs instead. Unfortunately, the analog of Ramsey Theorem does not hold for directed graphs even in the simplest case when **H** happens to be a directed cycle on 3 vertices as a transitive tournament on n-vertices shows.

Here we consider a probabilistic analog of the Ramsey Theory for directed graphs (digraphs), for the case when **H** is a directed cycle on n vertices. It has been proved by Acharya et.al. [1] that if **D** is a diconnected digraph of order $p \geq 6$, then **D** or $\overline{\mathbf{D}}$ has a 3-dicycle, and hence for almost all digraphs **D**, either **D** or its complement $\overline{\mathbf{D}}$ has a 3-dicycle. Further a good characterization of all digraphs **D** such that neither **D** nor its complement $\overline{\mathbf{D}}$ has a 3-dicycle was given in [1] with applications to behavioral sciences. Probabilistic Ramsey Theory for graphs was investigated earlier by Graham, Rödl and Ruciński in a series of papers in[4] and [5].

In this paper we show that for the directed cycle C_n on n vertices, almost all directed graphs **D** have the property that either **D** or its complement (taken in the symmetric complete digraph) $\overline{\mathbf{D}}$ has a C_n where all digraphs of order p are equally likely. Further, we define for every $n \geq 3$, $f(n)$ as the minimum p, such that for every Hamiltonian digraph **D** of order $p \geq f(n)$, either **D** or $\overline{\mathbf{D}}$ has a n-dicycle. We show that $f(n)$ exists and determine the exact values for $f(n)$ for all values of $n \geq 3$. The methods used are graph theoretic, probabilistic and elementary number theoretic in nature.

---

1 Introduction

In any collection of six people either three of them mutually know each other or three of them mutually do not know each other.

This puzzle has always been considered the first non-trivial example of what we call Ramsey Theory.

We have tacitly assumed that the relation of "knowing" is symmetric, that is, if A knows B then B knows A. We do not assume transitivity; if A knows B and B knows C then A may or may not C.

Let G_1 and G_2 be two graphs. The *Ramsey number* R=$R(G_1,G_2)$ is the smallest integer R with the property that a complete graph of order at least R whose edges are partitioned into two color classes contains either a monochromatic G_1 or G_2. Similarly, the *Ramsey number* $R = R(n)$ is the smallest integer R with the property that any complete graph of at least R vertices whose edges are partitioned into two color classes contains a monochromatic complete subgraph with n vertices. Let C_n denote a cycle of length n. The generalized Ramsey number of the pair (C_n, C_k), denoted by $R(C_n, C_k)$, is the smallest positive integer R, such that any complete graph with R vertices whose edges are colored with two different colors contain either a monochromatic cycle of length n in the first color or a monochromatic cycle of length k of the second color. The Generalized Ramsey number for cycles were determined by Faudree-Schelp and Rosta and later an easier proof was provided by Karolyi and Rosta [12]. They proved that

$$C(n,n) = \begin{cases} 6 & \text{if } n = 3 \text{ or } 4 \\ 3n/2 - 1 & \text{if } n \text{ is even} \\ 2n - 1 & \text{if } n \text{ is odd} \end{cases} \tag{1}$$

The analog of the Generalized Ramsey numbers for dicycles does not hold even in the simplest case when the graph happens to be a directed cycle on 3-vertices as a transitive tournament on n-vertices. Ramsey number of digraphs was first discussed by Harary and Hell in [10]. They showed that $R(D_1, D_2)$ exists if and only if D_1 or D_2 is acyclic. Their paper also contains a number of results on tournaments and star graphs. Some results on Ramsey number for tournaments has been given in [13].

Here we consider the fact that the labeled $2^{p(p-1)}$ digraphs on p vertices are equally likely. It can be shown that the ratio of the number of labeled Hamiltonian digraphs of order p to that of all the digraphs of order p tends to 1 as p tends to infinity. With this probability scheme, it can also be shown that ratio of the number of labeled tournaments of order p to the total number of digraphs of order p goes to zero as p tends to infinity.

The paper is organized as follows. Section 2 contains some notations and states two main theorems which are proved in Section 3. Section 4 gives some applications of the Ramsey Theory.

2 Definitions and Notations

In this section we define few terms and give a notations and conventions which we follow throughout the rest of the paper. We state the two theorems which we prove in Section 3 and Section 4.

Definition. *A directed graph* $\boldsymbol{D}$ *is said to be* k*-good if* $\boldsymbol{D}$ *or* $\overline{\boldsymbol{D}}$ *has a directed* k*-cycle.*

Definition. *Given any positive integer* k ($k \geq 3$)*, define* $f(k)$ *as the smallest positive integer* n*, such that given any Hamiltonian digraph* $\boldsymbol{D}$ *on* n *or more vertices,* $\boldsymbol{D}$ *or* $\overline{\boldsymbol{D}}$ *has a* k*-dicycle.*

Definition. *A positive integer* n *is* k*-good* ($k \leq n$) *if every Hamiltonian digraph on* n*-vertices is* k*-good.*

Note that given any positive integer k ($k \geq 3$)*,* $f(k)$ *is the smallest positive integer* n'*, such that for every* $n \geq n'$*,* n *is* k*-good.*

We prove the following two theorems.

Theorem I.
1. $f(m)$ *exists for all* $m \geq 3$.
2. $f(3) = f(4) = 6$.
3. $f(m) = m$, *m even*, $m \geq 6$.
4. $f(m) = 2m - 1$, *m odd* $m \geq 5$.

Theorem II. *For any integer* $k \geq 2$, *almost all digraphs are* k*-good.*

Notations. Throughout this paper we consider $\mathbf{D}$ to be a digraph with n-vertices whose vertex set is $V = \{v_1, v_2, \cdots, v_n\}$. For brevity we consider $V = \{1, 2, \cdots, n\}$ to be the vertex set of $\mathbf{D}$.

All additions are done modulo the number of vertices in the cycle under consideration. For definitions not given and notations not explained we follow [12] and [18].

3 Proofs of the Main Theorems

The following Lemma is crucial and reduces the verification about the existence of $f(k)$ considerably.

Lemma 1. *If for some* n, *every hamiltonian digraph* $\mathbf{D}$ *on* n *and* $n+1$ *vertices is* k*-good, then so is every hamiltonian digraph on* $n' \geq n$ *vertices.*

Proof. It is enough to prove that every hamiltonian digraph on $n+2$ vertices is k-good. Suppose that there is a hamiltonian digraph $\mathbf{D}$ on $n+2$ vertices which is not k-good. Let $V = \{1, 2, \cdots, n+2\}$ be the vertex set of $\mathbf{D}$ and $(1, 2, 3, \cdots, n+2, 1)$ be a hamiltonian dicycle in $\mathbf{D}$. Then, for $1 \leq i \leq n+2$, $(i, i+3)$ is an arc in $\overline{\mathbf{D}}$. If not, then $C = (1, 2, 3, \cdots, n, 1)$ is a n-dicycle in $\mathbf{D}$ and the subdigraph D' induced on the vertices of C is a hamiltonian digraph, hence by hypothesis is k-good. Hence $\mathbf{D}$ is k-good, a contradiction. Now for $1 \leq i \leq n+2$, $(i, i+2)$ is an arc in $\overline{\mathbf{D}}$. If not, then $(1, 2, 3, \cdots, n+1, 1)$ is a $n+1$-cycle in $\mathbf{D}$, and by hypothesis and as above, the subdigraph induced on $(1, 2, \cdots, n+1, 1)$ is k-good. Hence $\mathbf{D}$ is k-good, a contradiction.
We consider the following two cases:

Case 1. n is odd, $n \geq 3$
It follows that $(1, 3, 5, \cdots, n+2, 2, 4, \cdots, n+1, 1)$ is a dicycle in $\overline{\mathbf{D}}$. Now, for $1 \leq i \leq n+2$, $(i, i+4)$ is an arc in $\mathbf{D}$, otherwise we will get a dicycle of length $n+1$ in $\overline{\mathbf{D}}$. Since $(i, i+3)$ and $(i, i+2)$ are arcs in $\overline{\mathbf{D}}$, the dicycle $(1, 4, 6, 8, \cdots, n+1, 2, 5, 7, 9, \cdots, n, 1)$ omits only vertices 3, $n+2$, hence it is a n dicycle in $\overline{\mathbf{D}}$. Therefore, the subdigraph of $\overline{\mathbf{D}}$ induced on $(1, 4, 6, 8, \cdots, n+1, 2, 5, 7, 9, \cdots, n, 1)$ is k-good. Thus, $\overline{\mathbf{D}}$ is k-good. Hence $\mathbf{D}$ is k-good, a contradiction.

Case 2. n is even, $n \geq 3$.
Already, $(1, 3, 5, \cdots, n+1, 1)$ and $(2, 4, 6, \cdots, n+2, 2)$ are dicycles in $\overline{\mathbf{D}}$. Since $(i, i+3)$ is an arc in $\overline{\mathbf{D}}$, the dicycle $C = (1, 4, 6, 8, \cdots, n, n+2, 2, 5, 7, 9, \cdots, n+1, 1)$ is a dicycle of length $n+1$ in $\overline{\mathbf{D}}$. Thus the subdigraph of $\overline{\mathbf{D}}$ induced on the vertices of C is a hamiltonian digraph of order $n+1$, hence by hypothesis is k-good. Thus $\overline{\mathbf{D}}$ is k-good. So $\mathbf{D}$ is k-good a contradiction. □

Lemma 2. $f(3) = 6$.

Proof.

Part 1. To prove 6 is 3-good.
Let **D** be a directed hamiltonian graph on 6 vertices. Suppose **D** is not 3-good. Let $V = \{1, 2, \cdots, 6\}$ be the vertex set of **D**. for every $i, 1 \leq i \leq 6$, $(i+2, i)$ be an arc in $\overline{D}$, otherwise $C = (i+2, i, i+1, i+2)$ would form a 3-dicycle in **D**. Hence, **D** is 3-good. Thus, $(3,1), (4,2), (5,3), (6,4), (1,5), (2,6)$ are arcs in $\overline{D}$.Now, $(1,5,3,1)$ is a 3-dicycle in $\overline{D}$. Hence 6 is 3-good.

Part 2. To prove 7 is 3-good.
Let **D** be a directed hamiltonian graph on 7 vertices. Suppose **D** is not 3-good. Let $V = \{1, 2, \cdots, 7\}$ be the vertex set of **D**. for every i, $1 \leq i \leq 7$, $(i+5, i)$ be an arc in $\overline{D}$, otherwise $C = (i, i+1, i+2, i+3, i+4, i+5, i)$ would form a 6-dicycle in **D**, which is 3-good, a contradiction. $C' = (1,3,5,7,2,4,6,1)$ is a 7-dicycle in $\overline{D}$. Now, for every i, $1 \leq i \leq 7$, $(i, i+4)$ be an arc in **D**. Since $(i, i+1)$ is an arc in **D**, $(i, i+1, i+5, i+2, i+6, i+3, i)$ is a 6-dicycle in **D**, which is 3-good, a contradiction. Hence 7 is 3-good. □

The following Lemma gives a lower bound on $f(k)$ when k is odd.

Lemma 3. $f(2k+1) \geq 4k+1$.

Proof. We construct a hamiltonian digraph on $4k$ vertices which is not $(2k+1)$-good. Let $D = K^*_{2k,2k}$ be the complete bipartite symmetric digraph. **D** does not have any odd cycle, however it is Hamiltonian. $\overline{D}$ is $K^*_{2k} \bigcup K^*_{2k}$. Hence $\overline{D}$ is not $(2k+1)$-good.
Hence $f(2k+1) \geq 4k+1$. □

The following two lemmas are number theoretic in nature and used in the proofs of subsequent lemmas.

Lemma 4(a). $k^t \equiv 1 \bmod (2k-1)$ *$(t \geq 3)$, if and only if* $k^{t-2} \equiv 4 \bmod (2k-1)$. $(k \geq 3)$.

Proof. Only if:

$$
\begin{aligned}
& & k^t &\equiv 1 \bmod (2k-1) & \\
&\Rightarrow & k^t &= m(2k-1)+1 & \\
&\Rightarrow & k^t &= (m-1)(2k-1)+2k & \\
&\Rightarrow & k^{t-1} &= (m-1)(2k-1)/k+2 & \\
& & k^{t-1} &\equiv 2 \bmod (2k-1) & (\text{As, } k \mid (m-1)) \\
&\Rightarrow & k^{t-1} &= m'(2k-1)+2 & \\
&\Rightarrow & k^{t-1} &= (m'-2)(2k-1)+4k & \\
&\Rightarrow & k^{t-2} &= (m'-2)(2k-1)/k+4 & \\
& & k^{t-2} &\equiv 4 \bmod (2k-1) & (\text{As, } k \mid (m'-2))
\end{aligned}
$$

The other direction is trivial. □

Lemma 4(b). *If* $k \geq 3, k^t \equiv 4 \bmod (2k-1)$ *for some positive integer* $t \leq 2k-2$.

Proof. We note that $\gcd(k^t, 2k-1) = 1$, for all positive integers t. By Euler's theorem, $k^{\phi(2k-1)} \equiv 1 \bmod (2k-1)$. Hence by Lemma 4(a), $k^{\phi(2k-1)-2} \equiv 4 \bmod (2k-1)$, since $\phi(2k-1) \geq 3$. □

Lemma 5. $f(3m+1) \leq 6m+1$, *for every* m.

Proof. Let $6m+1$ not be $3m+1$-good. Let **D** be a Hamiltonian digraph on $6m+1$ vertices which is not $3m+1$-good. Let $V = \{1, 2, \cdots, 6m+1\}$ be the vertex set of **D** and $(1, 2, 3, \cdots, 6m+1, 1)$ be a Hamiltonian dicycle in **D**. Then, for $1 \le i \le 6m+1$, $(i, i+3m+1)$ (since $-(3m+1-1) \equiv 3m+1 \bmod (6m+1)$) is an arc in $\overline{\mathbf{D}}$. If not, then $C = (1, 2, 3, \cdots, 3m+1, 1)$ is a $3m+1$-dicycle in **D**, a contradiction. Hence **D** is $3m+1$-good, a contradiction. Similarly, for $1 \le i \le 6m+1$, $(i, i+(3m+1)^2)$ is an arc in **D**. If not, then $C' = (1, 3m+1+1, 2, \cdots, 3m+1, 1)$ is a $3m+1$-dicycle in $\overline{\mathbf{D}}$. Thus $\overline{\mathbf{D'}}$ is $3m+1$-good. Hence **D** is $3m+1$-good. Proceeding similarly we notice that for $1 \le i \le 6m+1$ $(i, i+(3m+1)^x)$ is an arc in $\overline{D}$ whenever x is odd and is an arc in D whenever x is even. By Lemma 4(b) there exists some t such that $(i, i+4)$ is an arc in $\overline{D}$ whenever t is odd and is an arc in D whenever t is even. If t is odd, $t+2$ is also odd and by Lemma 4(a) $(3m+1)^{t+2} \equiv 1 \bmod (6m+1)$. Thus $(i, i+1)$ is an arc in $\overline{D}$, a contradiction, as $(i, i+1)$ is an arc in **D**
If t is even then as above $(i, i+4)$ is an arc in **D**. Hence $(1, 1+4, 1+2*4, \cdots, 1+m*4, 1+m*4+1, 1+m*4+2, \cdots, 1+m*4+k-m-1 = 6m+1, 1)$ is a k-dicycle in **D**. Hence $6m+1$ is $3m+1$-good.

Let $6m+2$ not be $3m+1$-good. Now we consider a Hamiltonian digraph D' on $6m+2$-vertices. Clearly for $1 \le i \le 6m+2$ $(i, i+2)$ is an arc in $\overline{D}$, since $6m+1$ is $3m+1$-good. As above $C = (1, 3, 5, \cdots, 6m+1, 1)$ is a $3m+1$-dicycle in $\overline{D}$. Hence **D** or $\overline{D}$ is $3m+1$-good. Hence $6m+2$ is $3m+1$-good, a contradiction.
Hence by Lemma 1, $f(3m+1) \le 6m+1$.
This completes the proof of Lemma 5. □

It follows from Lemma 3 and Lemma 5 that

Lemma 6. $f(6m+1) = 12m+1$, *for every* m, $m \ge 1$. □

The next two lemmas determine the value of $f(6m)$.

Lemma 7A. $f(6m) = 6m, m \ge 2$.

Proof. Obviously $6m$ is $6m$-good. Let $n = 6m+1$. Let **D** be a Hamiltonian digraph on $6m+1$ vertices which is not $6m$-good. Let $V = \{1, 2, \cdots, 6m+1\}$ be the vertex set of **D** and $C = (1, 2, 3, \cdots, 6m+1, 1)$ be a Hamiltonian dicycle in **D**. for every $i, 1 \le i \le 6m+1$, $(i, i+4)$ is an arc in **D** and $(i, i+2)$ is an arc in $\overline{\mathbf{D}}$. Wlg, $i = 6m-1$ (say), then $(i, i+3)$ is an arc in $\overline{\mathbf{D}}$, else $C' = (1, 2, 3, 7, 8, 9, \cdots, 6m+1, 4, 5, 6, 10, 11, 12, \cdots, 6m-2, 6m-1, 1)$ omits only 6m, hence is a $6m$-dicycle in **D** which is $6m$-good. Hence for every i, $1 \le i \le 6m+1$, $(i, i+3)$ is an arc in $\overline{\mathbf{D}}$. $C'' = (1, 3, 5, \cdots, 6m+1, 2, \cdots, 6m, 1)$ is a $6m+1$ dicycle in $\overline{\mathbf{D}}$. Hence $(1, 7)$ is an arc in **D**, hence $(i, i+6)$ is an arc in **D**. Similarly $(i, i+12)$ is an arc in $\overline{\mathbf{D}}$. Consider the cycle $C''' = (1, 2, 3, 4, 5, 9, 10, 11, 12, 13, 17, 18, 19, 23, 24, 25, \cdots, 6m-1, 6m, 6m+1, 6, 7, 8, 14, 15, 16, 20, 21, 22, 26, 27, 28, \cdots, 6m-4, 6m-3, 6m-2, 1)$ is a $6m+1$-cycle in **D**. Since $(i, i+6)$ is an arc in **D**, $(1, 10)$ is an arc in **D**. Similarly for every i, $1 \le i \le 6m+1$ $(i, i+9)$ is an arc in **D**. Again since $(i, i+9)$ is an arc in **D**, $(1, 13)$ is an arc in **D**, which is a contradiction because we have already proved that $(1, 13)$ is an arc in $\overline{\mathbf{D}}$. Hence $6m+1$ is $6m$-good. Hence by Lemma 1, it follows that $f(6m) = 6m$. □

Lemma 7B. $f(6) = 6$.

Proof. Clearly 6 is 6-good.
Let $n = 7$. Let **D** be a Hamiltonian digraph on 7 vertices which is not 6-good. Let $V = \{1, 2, 3, 4, 5, 6, 7\}$ be the vertex set of **D**. for every i, $1 \le i \le n$, $(i, i+2^t)$ is an arc in **D**, if t is even, and $(i, i+2^t)$ is an arc in $\overline{\mathbf{D}}$, if t is odd. We note that $(i, i+1)$

is an arc in $\mathbf{D}$. However $(i, i+2^3) = (i, i+1)$ is an arc in $\overline{\mathbf{D}}$, a contradiction. Hence 7 is 6-good.
By Lemma 1 it follows that $f(6) = 6$. □

It follows from Lemma 7A and 7B that

Lemma 7. *$f(6m) = 6m$, for every m.* □

Lemma 8. *$f(6m+2) = 6m+2$, $m \geq 1$.*

Proof. Obviously, $6m+2$ is $6m+2$-good. We prove that $6m+3$ is $6m+2$-good. Then the Lemma 8 follows from Lemma 1.
Let $\mathbf{D}$ be a Hamiltonian digraph on $6m+3$ vertices that is not $6m+2$-good. $V = \{1, 2, \ldots, 6m+3\}$ be the vertex set of D. Let $C = (1, 2, 3, \ldots, 6m+3, 1)$ is an Hamiltonian dicycle in $\mathbf{D}$. For every i, $1 \leq i \leq 6m+3$, $(i, i+4)$ is an arc in $\mathbf{D}$. Hence $C' = (i, i+1, i+2, i+6, i+7, i+8, i+12, i+13, i+14, \ldots .i+6m,$ $i+6m+1, i+6m+2, i+3, i+4, i+5, i+9, i+10, i+11, \ldots., i+6m-3, i+6m-2, i+6m-1, i)$ is a $6m+3$-dicycle in $\mathbf{D}$. We define three sets B, R and U as follows:
Let $B = \{b : b$ is of the form $3k+1, 0 \leq k \leq m\}$.
Let $R = \{r : r$ is of the form $3k+2, 0 \leq k \leq m\}$.
Let $U = \{u : u$ is of the form $3k, 0 \leq k \leq m\}$.
Hence $R \bigcup B \bigcup U = V$ and $R \bigcap B \bigcap U = \phi$.

Claim. For every $b \in B$ and for every i, $1 \leq i \leq 6m+3$, $(i, i+b)$ is an arc in $\mathbf{D}$.

Proof. We see that $4 \in B$, and that $i+7$ is that 4th vertex from i in C' and $i+12$ is the 4th vertex from $i+2$ in C'. Hence $(i, i+7)$ and $(i+2, i+12)$ are arcs of the subdigraph induced on the vertices of C'. Hence, $(i, i+7)$, $(i, i+10)$ are arcs in $\mathbf{D}$. Similarly, $i+13$ is the 7th vertex from i in C' and $i+18$ is the 7th vertex from $i+2$ in C'. Since $(i, i+7)$ is an arc in $\mathbf{D}$, $(i, i+13)$ and $(i+2, i+18)$ are arcs of the subdigraph induced on the vertices of C'. Hence, $(i, i+13)$ and $(i, i+16)$ are arcs in $\mathbf{D}$. Proceeding similarly, let the proposition hold for all integers $t \leq r$. Then if t is even then, $3(t/2)+4 \in B$ and if t is odd, $3(t+1)/2+1 \in B$. Hence $3(t+1)+1 \in B$. $(i+6(t+1)+1)$ and $(i+6(t+1)+5)$ are the $3(t+1)+1$ th vertex from i and $i+2$ in C' respectively. Hence $(i, i+6(t+1)+1)$, $(i, i+6(t+1)+4)$ are arcs on the subdigraph induced on the vertices of C'. Hence $(i, i+6(t+1)+1)$, $(i, i+6(t+1)+4)$ are arcs in $\mathbf{D}$.
This proves the claim.

Claim. For every $r \in R$ and for every i, $1 \leq i \leq 6m+3$, $(i, i+r)$ is an arc in $\overline{D}$.

Proof. We know that $(i, i+2)$ is an arc in $\overline{D}$. As $i+7$ is the second vertex from $i+2$, $(i+2, i+7)$ is an arc in $\overline{D}$. Let the proposition hold for all integers $r \leq t$. Then as in the above proof, we have $3(t+1)+2 \in R$. Hence $(i, i+6(t+1)+2)$, $(i, i+6(t+1)+5)$ are arcs on the subdigraph induced on the vertices of C' and are arcs in $\overline{D}$. This proves the claim.
From the claim above it follows that all arcs from B to R, R to U and U to B are all in $\mathbf{D}$, whereas all arc from R to B, U to R, and B to U are in $\overline{D}$. Notice that for every b, $b \in B$, r, $r \in R$, u, $u \in U$. (b, r), (r, u), (u, b) are arcs in $\mathbf{D}$. Therefore by Pigeon Hole Principle, there exists two sets X, and Y of R, U, B, such that, the subgraph induced on the vertices of X and that on Y have arcs (e, f) and (g, h) both belonging to either $\mathbf{D}$ or $\overline{D}$. Let (e, f) and (g, h) be arcs in $\mathbf{D}$. Without loss of generality assume that $e = 1$, $f = 4$, $g = 2$, $h = 5$. Then we can find a Hamiltonian dicycle $C = (1, 4, 2, 5, 6, 7, 8, 9, \cdots, 6r+1, 6r+2, 6r+3, \cdots, 6m+1, 6m+2, 6m+3, 1)$ of length $6m+2$ which is $6m+2$-good. Hence $6m+3$ is $6m+2$-good. □

The method of proof of the following lemma is similar to that of Lemma 8 and also uses Lemma 4(b).

Lemma 9. *$f(6m+5) = 12m+9$, $m \geq 1$.*

Proof. We prove that $12m+9$ is $6m+5$-good.

Let **D** be a Hamiltonian digraph on $12m+9$ vertices that is not $6m+5$ good. $V = \{1, 2, \cdots, 12m+9\}$ be the vertex set of D. Let $C = (1, 2, 3, \cdots, 12m+9, 1)$ is an Hamiltonian dicycle in **D**. If t is odd, $t+2$ is also odd and by Lemma 4(b) we have $k^{t+2} \equiv 1 \bmod (2k-1)$. Thus $(i, i+1)$ is an arc in $\overline{D}$, a contradiction, as $(i, i+1)$ is an arc in **D**.
If t is even then $(i, i+1)$ and $(i, i+4)$ are arcs in **D**. Hence $C' = (i, i+1, i+2, i+6, i+7, i+8, i+12, i+13, i+14, \cdots, i+12m+6, i+12m+7, i+12m+8, i+3, i+4, i+5, i+9, i+10, i+11, \cdots, i+12m+3, i+12m+4, i+12m+5, i)$ is a $12m+9$-dicycle in **D**. Proceeding as in the proof of Lemma 8, we have sets B, R and U, such that $R \bigcup B \bigcup U = V$ and $R \bigcap B \bigcap U = \phi$. We also find arcs (e, f) and (g, h) in **D**. Without loss of generality assume that $e = 1$, $f = 4$, $g = 2$, $h = 5$. Then we can find a Hamiltonian dicycle $C = (1, 4, 2, 5, 6, 7, 8, 9, \cdots, 6r+1, 6r+2, 6r+3, \cdots, 6m+4, 6m+5, 6m+6, 1)$ of length $6m+5$ which is $6m+5$-good. Hence $12m+9$ is $6m+5$-good.

Now proceeding as in the proof of Lemma 5, we have $12m+10$ is $6m+5$-good. Hence By Lemma 1 and Lemma 3, $f(6m+5) = 12m+9$. □

Lemma 10A. *$f(6m+4) = 6m+4$, $m \geq 2$.*

Proof. Obviously $6m+4$ is $6m+4$-good. Let $n = 6m+5$. Let **D** be a Hamiltonian digraph on $6m+5$ vertices which is not $6m+4$-good. Let $V = \{1, 2, \cdots, 6m+5\}$ be the vertex set of **D** and $C = (1, 2, 3, ..., 6m+5, 1)$ be a Hamiltonian dicycle in **D**. for every $i, 1 \leq i \leq 6m+4$, $(i, i+4)$ is an arc in **D** and $(i, i+2)$ is an arc in $\overline{\mathbf{D}}$. $(i, i+7)$ is an arc in $\overline{\mathbf{D}}$, else $C' = (1, 2, 3, 7, 8, 9, \cdots, 6m+1, 6m+2, 6m+3, 6m+4, 6m+5, 4, 5, 6, 10, 11, 12, \cdots, 6m-2, 6m-1, 1)$ is a $6m+4$-dicycle in **D** which is $6m+4$-good. $C'' = (1, 3, 5, 7, 9, ..., 6m+5, 2, ...6m, 1)$ is a $6m+5$ dicycle in $\overline{\mathbf{D}}$. Hence $(1, 15)$ is an arc in **D**, hence for every i, $1 \leq i \leq 6m+5$, $(i, i+14)$ is a arc in **D**. If $(6, 9)$ is an arc in **D**, then $C''' = (1, 2, 3, 7, 21, 22, 23, \cdots, 6m+3, 6m+4, 6m+5, 4, 5, 6, 9, 10, 11, \cdots, 20, \cdots, 6m, 6m+1, 6m+2, 1)$ omits only 8 and is a $6m+4$-dicycle in **D**, hence is $6m+4$-good. Thus for every i, $1 \leq i \leq 6m+5$, $(i, i+3)$ is an arc in $\overline{\mathbf{D}}$. Considering the cycle C'' we can say that for every $i, 1 \leq i \leq 6m+5$, $(i, i+6)$ is an arc in **D**, and $(i, i+12)$ is an arc in $\overline{\mathbf{D}}$. Consider the dicycle $C'''' = (1, 2, 3, 7, 8, 9, 13, 14, 15, \cdots, 6m+1, 6m+2, 6m+3, 6m+4, 6m+5, 4, 5, 6, 10, 11, 12, \cdots, 6m, 1)$ is a $6m+5$ dicycle in **D**. Since $(i, i+6)$ is an arc in **D**, $(1, 13)$ is an arc in **D**, which is a contradiction because we have already proved that $(1, 13)$ is an arc in $\overline{\mathbf{D}}$. Hence $6m+5$ is $6m+4$-good. Hence by Lemma 1, it follows that $f(6m+4) = 6m+4$. □

Lemma 10B. *$f(10) = 10$.*

Proof. 10 is obviously 10-good. Let **D** be a Hamiltonian digraph on 11 vertices that is not 10-good. $V = \{1, 2, \cdots, 11\}$ be the vertex set of D. Let $C = (1, 2, 3, \cdots, 11, 1)$ is an Hamiltonian dicycle in **D**. For every i, $1 \leq i \leq 11$, $(i, i+2^t)$ is an arc in **D**, whenever t is even, and $(i, i+2^t)$ is an arc in $\overline{D}$, whenever t is odd. Thus we see that $(i, i+1)$, $(i, i+3)$, $(i, i+4)$, $(i, i+5)$, $(i, i+9)$ are arc in **D** and $(i, i+2)$, $(i, i+6)$, $(i, i+7)$, $(i, i+8)$, $(i, i+10)$, are arcs in $\overline{D}$. Hence **D** is an Hamiltonian tournament, so it has a dicycle of length 10, by [14]. Thus 11 is 10-good. Now, from Lemma 1 it follows that $f(10) = 10$.

The following Lemma 10 follows from the Lemmas 10A and 10B. It follows from Lemma 10A and Lemma 10B that

Lemma 10. *$f(6m+4) = 6m+4$, for every m, $m \geq 1$.* □

Lemma 11. *$f(6m+3) = 12m+5$, for every m. $m \geq 2$.*

Proof. We show that $12m+5$ is $6m+3$-good.
Let **D** be a Hamiltonian digraph on $12m+5$ vertices that is not $6m+3$ good. $V = \{1, 2, \cdots, 12m+5\}$ be the vertex set of D. Let $C = (1, 2, 3, \cdots, 12m+5, 1)$ is an Hamiltonian dicycle in **D**. If t is odd, $t+2$ is also odd and by Lemma 4(b) $k^{t+2} \equiv 1 \bmod (2k-1)$. Thus $(i, i+1)$ is an arc in $\overline{D}$, a contradiction , as $(i, i+1)$ is an arc in **D**
If t is even then $(i, i+1)$ and $(i, i+4)$ are arcs in **D**. Let $(12m+3, 1)$ be an arc in **D**. Then $C = (1, 5, 9, \cdots, 1+8m, 1+8m+1, 1+8m+2, \cdots, 12m+3, 1)$ is an $6m+5$-dicycle in **D**. Hence $(12m+3, 1)$ is an arc in $\overline{D}$. Similarly, for every i, $(i, i+3)$ is an arc in $\overline{D}$. Hence, $C' = (1, 4, 7, \cdots, 12m+4, 2, \cdots, 12m+3, 1)$ is a $12m+5$ dicycle in $\overline{D}$. Hence, $(1, 6)$ is an arc in **D**.

Let $(12m, 1)$ be an arc in **D**. Then $C = (1, 5, 9, \cdots, 1+8m-4, 1+8m-3, 1+8m-2, \cdots, 12m, 1)$ is an $6m+3$-dicycle in **D**. Hence $(12m, 1)$ is an arc in $\overline{D}$. Similarly, for every i, $(i, i+6)$ is an arc in $\overline{D}$, which is a contradiction since $(1, 6)$ is an arc in **D**.

Now proceeding as in the proof of Lemma 5 $12m+6$ is $6m+3$-good. Hence By Lemma 1 and Lemma 3, $f(6m+3) = 12m+5$. □

Lemma 12. $f(4) = 6$.

Proof.
Part 1. To prove 6 is 4-good. Let **D** be a directed Hamiltonian graph on 6 vertices. Suppose **D** is not 4-good. Let $V = \{1, 2, 3, 4, 5, 6\}$ be the vertex set of **D** for every i, $1 \leq i \leq 6$, $(i, i+3)$ be an arc in $\overline{D}$, otherwise $C = (1, 4, 5, 6, 1)$ would form a 4-cycle in **D**. Hence, **D** is 4-good. Hence $(1, 4)$, $(2, 5)$, $(3, 6)$, $(4, 1)$, $(5, 2)$, $(6, 3)$ are arcs in $\overline{D}$.

Claim. $(1, 3)$ is an arc in $\overline{D}$. Suppose not. Then $(1, 3)$ be an arc in **D**. Hence 1, 3, 4, 5 is a path in **D**. Hence $(5, 1)$ is an arc in $\overline{D}$, otherwise $(1, 3, 4, 5, 1)$ would be a 4-cycle. 2, 5, 1, 4 is a path in $\overline{D}$. Hence $(4, 2)$ is an arc in **D**. 5, 6, 1, 3 is a path in **D**. Hence $(3, 5)$ is an arc in $\overline{D}$. 3, 5, 1, 4 is a path in $\overline{D}$.Hence $(4, 3)$ is an arc in **D**. 6, 1, 3, 4 is an arc in **D**. Hence $(4, 6)$ is an arc in $\overline{D}$. 4, 6, 3, 5 is a path in $\overline{D}$.Hence $(5, 4)$ is an arc in **D**. 6, 3, 5, 1 is a path in $\overline{D}$.Hence $(1, 6)$ is an arc in **D**. 5, 1, 4, 6 is a path in $\overline{D}$.Hence $(6, 5)$ is an arc in **D**. 6, 3, 5, 2 is a path in $\overline{D}$.Hence $(2, 6)$ is an arc in **D**. Hence $(2, 6, 5, 4, 2)$ is a 4-cycle in **D**. Hence **D** is 4-good, a contradiction. Hence $(1, 3)$ is an arc in $\overline{D}$.This proves the claim.
Similarly we see that $(2, 4)$, $(3, 5)$, $(4, 6)$, $(5, 1)$, $(6, 2)$ are arcs in $\overline{D}$. 1, 4, 6, 2 is a path in $\overline{D}$. Hence $(2, 1)$ is an arc in **D**. Similarly $(3, 2), (4, 3), (5, 4), (6, 5)$ are arcs in **D**. $(3, 1)$, $(4, 2)$, $(5, 3)$, $(6, 4)$, $(1, 5)$, $(2, 6)$ are arcs in $\overline{D}$. Hence $(3, 1, 5, 4, 3)$ is a 4-cycle in **D**. Hence **D** is 4-good a contradiction. Hence 6 is 4-good.

Part 2. To prove 7 is 4-good.
Let D' be a Hamiltonian digraph on 7 vertices. Suppose D' is not 4-good. Let $V' = \{1, 2, \cdots, 7\}$ be the vertex set of D'. As in Lemma 5, for $1 \leq i \leq 7$ $(i, i+2^x \bmod 7)$ is an arc in $\overline{D'}$ whenever x is odd and is an arc in D' whenever x is even. However $2^3 \equiv 1 \bmod 7$ a contradiction. Hence D' is a 4-good. Hence 7 is 4-good.

Hence by Lemma 1 $f(4) = 6$. □

Lemma 13. $f(5) = 9$.

Proof. By Lemma 9, $f(5) \geq 9$. We first prove that 7 is 5-good. Then use this result to prove that 9 and 10 are 5-good. Hence by Lemma 1 it we conclude that $f(5) = 9$.

Part 1. To prove 7 is 5-good.
Let **D** be a Hamiltonian digraph on 7 vertices. Suppose **D** is not 5-good. Let $V = \{1, 2, \cdots, 7\}$ be the vertex set of D. As in the proof of Lemma, $1 \leq i \leq 7$ $(i, i + (-2)^x \bmod 7)$ is an arc in $\overline{D}$ whenever x is odd and is an arc in D whenever x is even. We observe that for every i, $1 \leq i \leq 7$, $(i, i+1)$, $(i, i+4)$ and $(i, i+2)$ are arcs in **D**, whereas $(i, i+6)$, $(i, i+3)$, $(i, i+5)$ are arcs in $\overline{D}$. Hence **D** is a Hamiltonian tournament on 7 vertices. Hence **D** contains a 5-dicycle. Hence 7 is 5-good.

Part 2. To prove that 9 is 5-good.
We proved that 7 is 5-good. Then use this result to prove that 9 and 10 are 5-good. Hence by Lemma 1 it we conclude that $f(5) = 9$.
Let **D** be a Hamiltonian digraph on 9 vertices. Suppose **D** is not 5-good. Let $V = \{1, 2, \cdots, 9\}$ be the vertex set of D. As in the proof of Lemma 5, $1 \leq i \leq 9$ $(i, i + 5^x \bmod 9)$ is an arc in $\overline{D}$ whenever x is odd and is an arc in D whenever x is even. We observe that for every i, $1 \leq i \leq 7$, $(i, i+1)$, $(i, i+4)$ and $(i, i+7)$ are arcs in **D**, whereas $(i, i+2)$, $(i, i+5)$, $(i, i+8)$ are arcs in $\overline{D}$. Suppose $(i, i+3)$ is an arc in **D**. Then $C = (1, 4, 5, 6, 7, 8, 9, 1)$ is a 7-dicycle, which has been shown (by *Part 1*) to be 5-good. Hence **D** is 5-good, a contradiction. So, $(i, i+3)$ is an arc in $\overline{D}$. Since $(i, i+2)$ is an arc in $\overline{D}$, $C = (1, 3, 5, 7, 9, 2, 4, 6, 8, 1)$ is a 9-dicycle in $\overline{D}$. Since $(i, i+3)$ is an arc in $\overline{D}$, $(1, 7)$ is an arc in **D**. Hence for every i, $(i, i+6)$ is an arc in **D**. Hence we conclude that for every i, $1 \leq i \leq 9$, $(i, i+1)$, $(i, i+4)$ and $(i, i+7)$, $(i, i+6)$ are arcs in **D**, whereas $(i, i+2)$, $(i, i+3)$, $(1, i+5)$, $(i, i+8)$ are arcs in $\overline{D}$. Hence **D** is a Hamiltonian tournament on 9 vertices. Hence **D** contains a 5-dicycle. Hence 9 is 5-good.
Now proceeding as in the proof of Lemma 5, 10 is 5-good. Hence Lemma 13 follows from Lemma 1 and the above results. □

Proof of Theorem I. By lemma 2 it follows that $f(3) = 6$, by Lemma 12 it follows that $f(4) = 6$ and by Lemma 13 it follows that $f(5) = 9$. By Lemmas 5, 9 and 11 we get $f(k) = 2k - 1$, k odd and $k \geq 7$. By Lemmas 6, 7, 8 we get $f(k) = k$, k even and $k \geq 6$. It follows that $f(k)$ exists for every $k \leq 3$. □

Proof of Theorem II. The Theorem follows from Theorem I $k \geq 3$ and the fact that almost all digraphs are Hamiltonian [9] and [15].
For $k = 2$, it is easy to see that if **D** is a digraph, such that neither **D** nor $\overline{D}$ has C_2, then **D** is a tournament and the probability that a digraph is a tournament is zero. This completes the proof. □

4 An Application of Ramsey Theory for Directed Graphs

As mentioned before directed graphs find application in social and communication networks. One application of Ramsey theorem for cycles is in behavioral sciences. Harary raised the problem of defining the complement S^c of a given signed graph S (called here sigraph) while discussing "structural duality" in general, arguing the concept to be a "philosophical necessity" towards developing eventually conceptual resources towards a theoretical understanding of the structural dynamics of a social system. As had been expected, the concept is of fundamental importance in probing dynamic states of structural stability of a social system arising in epochs of its evaluation, the authors in [1] introduced the notion of the unary operation of obtaining the complement S^c of a sociogram S, that is a selfloop-free signed multigraph in which every pair of distinct vertices is joined by at most one arc of each sign in each direction or what has been referred to as an *ambisidigraph*: S^c has the vertex set $V(S)$ of S and two vertices u and v are joined by a positive(negative) arc $(u,v)^+$ (respectively, $(u,v)^-$) whenever it is not an arc in S. Continuing the study of this notion it has been established in [1] an extension of digraphs to the class of strongly connected ambisidigraphs. In particular, a very interesting connection between Heider's theory of *social balance* and the notion of *complementation* has been discussed by showing that for any strongly connected ambisidigraphs S of order at least six, S or its complement S^c must contain a positive 3-dicycle, the Ramsey theorems for ambisidigraphs. Further a complete characterization of an ambisidigraph S for which neither S nor S^c has a 3-dicycle has been given. All such ambisidigraphs of order $p \leq 5$ have been identified.

5 Conclusion and Future Work

In this paper in Theorem I and II, we have used Hamiltonian digraphs. Since determining the existence of a Hamiltonian dicycle in a graph is an NP-Complete problem, we can look for other properties like strong connectivity. We know that almost all digraphs are strongly connected and polynomial time algorithms exist for determining if a digraph is strongly connected. Hence we can attempt to find similar results for k-dicycles for strongly connected digraphs instead of Hamiltonian digraphs. See [1] for $k = 3$.

However, we note that if there is a dicycle of length at least $2k-1$ or k, then there exists a dicycle of length k in the digraph or its complement (depending on k is odd or even respectively). Thus, it is sufficient to find a dicycle of length $2k-1$ or k respectively instead of finding a Hamiltonian dicycle.

Another problem is to find the smallest number of vertices n_0, such that every Hamiltonian digraph (or strongly connected digraph) $\mathbf{D}$ on $n \geq n_0$ vertices contains a m_1 dicycle or the complement $\overline{D}$ contains a m_2 dicycle.

Probabilistic Ramsey theory has been investigated by Graham, Rödl and Ruciński in [4] and [5]. Some of these results can be easily adopted for digraphs using structural Ramsey theory results due to Nešetřil and Rödl to obtain a transition from undirected to directed graphs (as these results claim Ramsey results for undirected graphs with a given ordering). Particularly one sees that for every directed graph G almost all large digraphs H satisfy the property that either H or its complement contains G.

Acknowledgements

We thank the referee for valuable comments. We are very thankful to Prof. J. Nešetřil for his valuable suggestions.

References

[1] B. D. Acharya, S. Joshi and S. B. Rao, A Ramsey Type theorem for strongly connected ambisidigraphs, in *Proceedings of thr Second International Conference on Social Network Analysis: Theory, Methods and Applications (ISI, Kolkata, Dec. 2006)* World Scientific, USA (2008), In Press.

[2] B. Bollobas, *Random Graphs, Second Edition*, Academic Press, New York (2001).

[3] D.M. Burton, *Elementary Number Theory, Second Edition*, Universal Book Stall, (1980).

[4] R.L. Graham, V. Rödl and A. Ruciński, On bipartite graphs with linear Ramsey numbers, *Combinatorica*, **21(2)** (2001), pp. 199–209.

[5] R.L. Graham, V. Rödl and A. Ruciński, On graphs with linear Ramsey numbers, *Journal of Graph theory*, **35(3)** (2000), pp. 176–192.

[6] R.L. Graham, B.L. Rothschild and J.H. Spencer, *Ramsey Theory*, Wiley-Interscience Series in Discrete Mathematics and Optimization, (1980).

[7] A. Gyafras and J. Lehel, A Ramsey-type problem in directed and bipartite graphs, *Periodica Math. Hungar.*, **3** (1973), pp. 299–304.

[8] P. Erdos and G. Szekeres, A Combinatorial problem in Geometry, *Compositio Mathematica*,**2** (1935), pp. 464–470. Academic Press, New York (1973).

[9] F. Harary and E. M. Palmer, *Graphical Enumeration*, Academic Press, New York (1973).

[10] F. Harary and P. Hell, Generalized Ramsey theory V., *Bull. London Math. Soc.*, **6** (1974), pp. 175–182.

[11] I.N. Herstein, *Topics in Algebra, Second Edition*, Wiley Eastern Limited, (1975).

[12] G. Karolyi and V. Rosta, Generalized and geometric Ramsey numbers for cycles, *Theoretical Computer Science*, **263** (2000), pp. 87–98.

[13] Y. Manaussakis and Z. Tuza, Ramsey number for tournaments, *Theoretical Computer Science*, **263** (2001), pp. 75–85.

[14] J.W. Moon, *Topics on Tournaments*, Holt, Rinehart, Winston, New York (1968).

[15] J.W. Moon, Almost all graphs have a spanning cycle, *Canadian Math. Bull.*, **15** (1972), pp. 39–41.

[16] S. Radziszowski, Small Ramsey Numbers, Dynamic Surveys of, *Electronic Journal of Combinatorics*, (2004), **10**

[17] V. Rosta, Ramsey Theorem: Application. Dynamic Surveys of, *Electronic Journal of Combinatorics*, (2004), **13**

[18] D.B. West, *Introduction to Graph Theory*, Prentice Hall, India (1968).

Proc. of ICDM 2006, Pages 55–61

Some Recent Results on Domination in Graphs (An Extended Abstract)

Michael D. Plummer*
Department of Mathematics, Vanderbilt University, Nashville, Tennessee 37240, USA.
Email: michael.d.plummer@vanderbilt.edu

1 Introduction

A set of vertices $S \subseteq V(G)$ *dominates* the graph G if every vertex in $V(G)$ either belongs to S or is adjacent to a vertex in S. The size of any smallest dominating set for G is denoted by $\gamma(G)$.

2 3-Critical Graphs

A graph G is *k-critical* if $\gamma(G) = k$, but $\gamma(G + e) = k - 1$, for every edge $e \in \overline{G}$.

The first so-called "toughness result" for 3-critical graphs is due to Sumner and Blitch [22].

Theorem 2.1. *Let G be a connected 3-critical graph. Then if S is a vertex cutset in G, $G - S$ has at most $|S| + 1$ components.* □

The next theorem represents a sharpening the Sumner-Blitch result. Let $\omega(G - S)$ denote the number of components in the graph $G - S$.

Theorem 2.2. *Let G be a connected 3-critical graph and let S be a vertex cutset in G. Then*

(a) *if $|S| \geq 6$, $\omega(G - S) \leq |S| - 2$;*

(b) *if $|S| \geq 4$, $\omega(G - S) \leq |S| - 1$;*

(c) *if $|S| = 3$, then $\omega(G - S) \leq 3$, and if $G - S$ has exactly three components, each component is complete and at least one of them is a singleton;*

(d) *if $|S| = 2$, then $\omega(G - S) \leq 3$, and if $G - S$ has exactly three components, then G must the the graph shown in Figure 2.1 with $n = 2$; and*

(e) *if $|S| = 1$, then $\omega(G - S) = 2$, and exactly one of the components of $G - S$ is a singleton. Furthermore, G has at most three cutvertices. If it has two, then G is a graph of the type shown in Figure 2.1 with $n \geq 2$, while if it has three, it is the graph shown in Figure 2.1 with $n = 1$.* □

*The complete version of this paper will appear in the December, 2006, issue of *Discussiones Mathematicae Graph Theory*

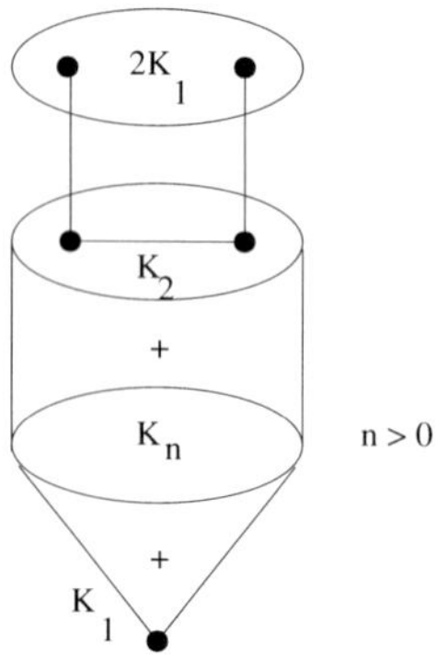

Figure 2.1:

Part (a) of the preceeding theorem is Theorem 2.1 of [2]. Part (b) follows immediately from Lemma 6 of [22] and Lemma 3 of [8]. Parts (c) and (d) are proved in Theorem 2.1 of [1]. As for the "missing case" in the Theorem 2.2 above, we have the following result [2, Theorem 2.2].

Theorem 2.3. *If G is a connected 3-critical graph, S is a vertex cutset with $4 \leq |S| \leq 5$, and if each component of $G - S$ has at least three vertices, then $\omega(G - S) \leq |S| - 2$.* □

Conjecture 2.4. Let G be a connected 3-critical graph, S, a vertex cutset in G and t, a positive integer. Then if $|S| \geq 2(t+1)$, $\omega(G - S) \leq |S| - t$.

The conjecture is true for $t = 1$ by part (b) of Theorem 2.2 and for $t = 2$ by part (a) of Theorem 2.2.

If u, v and w are vertices of graph G and u and v dominate $G - w$, we will write $[u, v] \longrightarrow w$.

Theorem 2.5. *Let G be a connected 3-critical graph and let S be an independent set of $n \geq 2$ vertices in $V(G)$.*

(i) *Then the vertices of S can be ordered as $a_1, a_2, \ldots, a_n$ in such a way that there exists a sequence of distinct vertices $x_1, x_2, \ldots, x_{n-1}$ so that $[a_i, x_i] \longrightarrow a_{i+1}$ for $i = 1, 2, \ldots, n-1$.*

(ii) *If, in addition, $n \geq 4$, then the x_i's can be chosen so that $x_1 x_2 \cdots x_{n-1}$ is a path and $S \cap \{x_1, \ldots, x_{n-1}\} = \emptyset$.* □

The case $n \geq 4$ was proved in [22]; the cases $n = 2$ and 3 were proved in [11].

Part (a) of the next theorem is historically the first result on matchings in 3-critical graphs and is due to Sumner and Blitch [22]. Part (b) is an easy consequence of the Gallai-Edmonds theorem on matchings.

Theorem 2.6. *Let G be a connected 3-critical graph. Then*

(a) *if $|V(G)|$ is even, G contains a perfect matching, while*

(b) *if $|V(G)|$ is odd, G contains a near-perfect matching.* □

A graph G of odd order is said to be *factor-critical* if $G - v$ contains a perfect matching, for all $v \in V(G)$ and a graph G of even order is called *bicritical* if $G - u - v$ contains a perfect matching, for all pairs of distinct vertices $u, v \in V(G)$. A 3-connected bicritical graph is called a *brick*.

Theorem 2.7 ([3, Theorem 2.1]). *If G is an even 3-connected 3-critical graph with* $\text{mindeg } G \geq 4$, *then G is a brick.* □

If the graph under consideration is claw-free, we can relax both the connectivity and minimum degree hypotheses somewhat as seen in the next theorem [3, Theorem 3.3].

Theorem 2.8. *Let G be a 3-critical 2-connected claw-free graph of even order. Then if* $\text{mindeg } G \geq 3$, *G is bicritical.* □

For graphs of odd order, we have the following result [Theorem 2.4][3].

Theorem 2.9. *Let G be a 2-connected 3-critical graph of odd order. Then G is factor-critical.* □

Let G be a graph on n vertices and suppose k is a positive integer with $k < n/2$. Then G is *k-factor-critical* if $G - S$ has a perfect matching for every set of vertices $S \subset V(G)$ with $|S| = k$. The next theorem is found in [4].

Theorem 2.10. *Let G be a 4-connected 3-critical graph of odd order and suppose* $\text{mindeg } G \geq 5$. *Then G is 3-factor-critical.* □

If the graph G is claw-free, we can relax both the connectivity and the minimum degree hypotheses of Theorem 2.10 slightly and still guarantee 3-factor-criticality.

Theorem 2.11 ([4, Theorem 3.4]). *Let G be a 3-connected claw-free 3-critical graph of odd order. Then if* $\text{mindeg } G \geq 4$, *G is 3-factor-critical.* □

3 3-Vertex-Critical Graphs

Graph G is *k-vertex-critical* if $\gamma(G) = k$, but $\gamma(G - v) = k - 1$, for every vertex $v \in V(G)$.

The next result is due independently to Sumner [21] and Las Vergnas [16].

Theorem 3.1. *Every connected claw-free graph of even order has a perfect matching.* □

Our next two results [5, 6] can be considered as a variations on the above classical result.

Theorem 3.2. *If G is a $K_{1,5}$-free 3-vertex-critical graph of even order, then G has a perfect matching.* □

Theorem 3.3. *If G is a $K_{1,5}$-free 3-vertex-critical graph of odd order at least 11 with* $\text{mindeg } G > 0$, *then G has a near-perfect matching.* □

In the odd order case, we also have the following result [6].

Theorem 3.4. *If G is a $K_{1,4}$-free 3-vertex-critical graph of odd order with minimum degree at least 3, then G is factor-critical.* □

We present a new family of 3-vertex-critical graphs. (See [5].) Let k be any integer with $k \geq 5$. We proceed to construct a graph which we will call $H_{k,\binom{k}{2}-k}$.

The vertex set will consist of two disjoint subsets of vertices called *central* and *peripheral* respectively. Let $\{v_1, \ldots, v_k\}$ denote the set of central vertices. The subgraph induced by these central vertices will be the complete graph K_k with the Hamilton cycle $v_1 \cdots v_k$ deleted. The peripheral vertices will be $\binom{k}{2}$ in number and will be denoted by the symbol $\sim \{i, j\}$, where the (unordered) pair $\{i, j\}$ $(i \neq j)$ ranges over all the $\binom{k}{2}$ subsets of size 2 of the set $1, \ldots, k$, *except* those having $j = i + 2$ where $i + 2$ is read modulo k. The neighbor set of peripheral vertex $\sim \{i, j\}$ will be precisely the set of all central vertices, except i and j. Figure 3.1 shows the graph $H_{6,9}$.

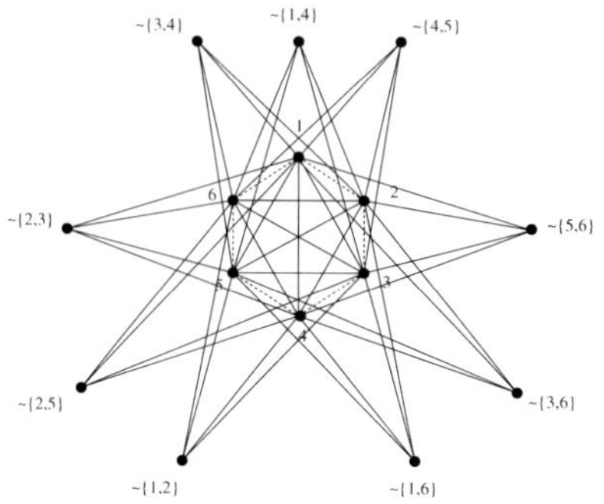

Figure 3.1:

We offer the following two conjectures. (See [5] and [6] respectively.)

Conjecture 3.5. If G is a 3-vertex-critical graph of even order and $K_{1,7}$-free, then G contains a perfect matching.

Conjecture 3.6. If G is a 2-connected 3-vertex-critical graph of odd order with minimum degree at least 3, then G is factor-critical.

If a 3-vertex-critical graph is also claw-free, more can be said. The following theorem, is due independently to Favaron, Flandrin and Ryjáček [10] and to Liu and Yu [17].

Theorem 3.7. *If G is a $(k+1)$-connected claw-free graph of order n, and if $n-k$ is even, then G is k-factor-critical.* □

In [7] the following result is found.

Theorem 3.8. *Let G be a connected claw-free 3-vertex-critical graph. Then*

(a) *G is 2-connected;*

(b) *if G is of even order or if* mindeg $G \geq 3$, *then G is 3-connected; and*

(c) *if* mindeg $G \geq 5$, *then G is 4-connected.* □

The next result is then immediate by Theorems 3.7 and 3.8.

Theorem 3.9. *Let G be a connected claw-free 3-vertex-critical graph. Then*

(a) *if G has odd order, G is factor-critical;*

(b) *if G has even order, G is bicritical; and*

(c) *if G has odd order and* mindeg $G \geq 5$, *G is 3-factor-critical.* □

4 Efficient Domination of Cubic Graphs

Some ten years ago, Reed [20] made the following interesting conjecture.

Conjecture 4.1. If G is a connected cubic graph on n vertices, then $\gamma(G) \leq \lceil n/3 \rceil$.

This conjecture was recently shown to be false by A.V. Kostochka and B.Y. Stodolsky [13] who produced an infinite family of counterexamples, the smallest of which has 60 vertices and $\gamma = 21$.

In the same paper containing Conjecture 4.1, Reed proved that if G has mindeg $G \geq 3$, then $\gamma(G) \leq (3/8)|V(G)|$. In a more recent paper, Kawarabayashi, Saito and the author [12] have obtained another result related to this conjecture for graphs of large girth.

Theorem 4.2. *Let G be a connected graph with a 2-factor F and let k be any positive integer. If F has at least two components and the order of each component is at least $3k$, then $\gamma(G) \leq \frac{3k+2}{9k+3}|V(G)|$.* □

Theorem 4.3. *Let k be any positive integer. Then every 2-edge-connected cubic graph G of girth at least $3k$ satisfies $\gamma(G) \leq \frac{3k+2}{9k+3}|V(G)|$.* □

If a cubic graph has a Hamilton cycle, then it trivially satisfies Reed's bound of $\lceil n/3 \rceil$.

Theorem 4.4 ([9]). *Let G be a cubic graph on $n \geq 4$ vertices which has a Hamilton cycle. Then it is true that $\gamma(G) \leq \lfloor n/3 \rfloor$, when $n \equiv 0 (mod\ 3)$ and when $n \equiv 1(mod\ 3)$, but not necessarily true when $n \equiv 2(mod\ 3)$.* □

Note added in proof: As this paper was going to press, the author learned of the following two even more recent results by Kostochka and Stodolsky [15].

Theorem 4.5. *If G is a connected cubic graph with $n > 8$ vertices, then $\gamma(G) \leq 4n/11$.* □

Theorem 4.6. *If G is a connected cubic graph with n vertices and girth at least g, then*

$$\gamma(G) \leq n\left(\frac{1}{3} + \frac{8}{3g^2}\right).$$

□

For graphs of large girth, the bound in Theorem 4.6 is better than that in Theorem 4.2. However Theorem 4.2 is more general in its application than Theorem 4.6 in that in Theorem 4.2 it is only the girth of a 2-factor which is required to be large. Short cycles, on the other hand, are permitted to exist.

5 Domination of Graphs Embedded in Surfaces

Theorem 5.1 ([19]). *Let G be a 3-connected graph and $\Phi : G \longrightarrow T$ be a 3-representative embedding of G on the torus T. Let f be any face of the embedding. Then $\Phi(G)$ contains a spanning cylinder Y which contains face f and is bounded by two disjoint homotopic cycles C_1 and C_2 such that*

(1) *$T - Y$ only contains edges each of which joins a vertex on C_1 to a vertex on C_2.*

(2) *Y contains a closed disc D_f such that D_f contains f and is bounded by a null homotopic cycle $P = C_1[x_2x_1] \cup P_{x_1y_1} \cup C_2[y_2y_1] \cup P_{x_2y_2}$, where x_1 and x_2 are two vertices on C_1, y_1 and y_2 are two vertices on C_2 and $P_{x_1y_1}$ and $P_{x_2y_2}$ are paths joining C_1 and C_2. All bridges of D_f in Y are either edges contained in $Y - D_f$ joining $P_{x_1y_1}$ to $P_{x_2y_2}$ or possibly in D_1 or D_2 where D_1 has $\{x_1, x_2\}$ as its vertices of attachment and D_2 has $\{y_1, y_2\}$ as its vertices of attachment. Moreover, both D_i's are discs and each is bounded by a null homotopic cycle of G.*

(3) *The underlying graph of Y is 2-connected.*

(4) *If, in addition, Φ is a triangulation, then $Y - D_f$ contains edges x_1x_2 and y_1y_2.* □

Theorem 5.2 ([18]). *Every triangulated disc on n vertices has a partition of its vertex set into three dominating sets (and hence can be dominated using no more than $\lfloor n/3 \rfloor$ vertices). Moreover, every two consecutive vertices on the boundary cycle belong to different sets of this partition.* □

An analog of Theorem 5.1 for the Klein bottle is also proved in [19]. One then combines these two theorems with Theorem 5.2 to prove the following.

Theorem 5.3. *If G is a triangulation of the torus or of the Klein bottle, then $\gamma(G) \leq \lceil |V(G)|/3 \rceil$.* □

Conjecture 5.4. [18] If $|V(G)|$ is sufficiently large and G is a planar triangulation, then $\gamma(G) \leq \lfloor |V(G)|/4 \rfloor$.

We end with an extension of their conjecture.

Conjecture 5.5. [19] If $|V(G)|$ is sufficiently large and graph G triangulates some surface (orientable or non-orientable), then $\gamma(G) \leq \lfloor |V(G)|/4 \rfloor$.

References

[1] N. Ananchuen and M.D. Plummer, Some results related to the toughness of 3-domination-critical graphs, *Discrete Math.* **272** (2003), pp. 5–15.

[2] N. Ananchuen and M.D. Plummer, Some results related to the toughness of 3-domination critical graphs, II, *Utilitas Math.*, **70** (2006), pp. 11–32.

[3] N. Ananchuen and M.D. Plummer, Matching properties in domination critical graphs, *Discrete Math.* **277** (2004), pp. 1–13.

[4] N. Ananchuen and M.D. Plummer, 3-factor-criticality in domination critical graphs, *Discrete Math.*, **307** (2007), pp. 3006–3015.

[5] N. Ananchuen and M.D. Plummer, Matchings in 3-vertex-critical graphs: the even case, *Networks* **45** (2005), pp. 210–213.

[6] N. Ananchuen and M.D. Plummer, Matchings in 3-vertex-critical graphs: the odd case, *Discrete Math.*, **307** (2007), pp. 1651–1658.

[7] N. Ananchuen and M.D. Plummer, On the connectivity and matchings in 3-vertex-critical claw-free graphs, *Utilitas Math.*, **69** (2006), pp. 85–96.

[8] Y. Chen, F. Tian and B. Wei, The 3-domination-critical graphs with toughness one, *Utilitas Math.* **61** (2002), pp. 239–253.

[9] M. Cropper, D. Greenwell, A.J.W. Hilton and A. Kostochka, The domination number of cubic Hamiltonian graphs, *AKCE Int. J. Graphs Comb.*, **2** (2005), pp. 137–144.

[10] O. Favaron, E. Flandrin and Z. Ryjáček, Factor-criticality and matching extension in DCT-graphs, *Discuss. Math. Graph Theory* **17** (1997), pp. 271–278.

[11] E. Flandrin, F. Tian, B. Wei and L. Zhang, Some properties of 3-domination-critical graphs, *Discrete Math.* **205** (1999), pp. 65–76.

[12] K. Kawarabayashi, A. Saito and M.D. Plummer, Domination in a graph with a 2-factor, *J. Graph Theory*, **52** (2006), pp. 1–6.

[13] A.V. Kostochka and B.Y. Stodolsky, On domination in connected cubic graphs, Communication: *Discrete Math.* **304** (2005), pp. 45–50.

[14] A.V. Kostochka and B.Y. Stodolsky, personal communication, November, (2005).

[15] A.V. Kostochka and B.Y. Stodolsky, An upper bound on the domination number of n-vertex connected cubic graphs, December, (2005) (submitted).

[16] M. Las Vergnas, A note on matchings in graphs, *Colloque sur la Théorie des Graphes (Paris, 1974)* Cahiers Centre Études Rech. Opér. **17** (1975), pp. 257–260.

[17] G. Liu and Q. Yu, On n-edge-deletable and n-critical graphs, *Bull. Inst. Combin. Appl.* **24** (1998), pp. 65–72.

[18] L. Matheson and R. Tarjan, Dominating sets in planar graphs, *European J. Combin.* **17** (1996), pp. 565–568.

[19] M.D. Plummer and X. Zha, On certain spanning subgraphs of embeddings with applications to domination, *Discrete Math.*, (2008), To appear.

[20] B. Reed, Paths, stars and the number three, *Combin. Probab. Comput.* **5** (1996), pp. 277–295.

[21] D.P. Sumner, 1-factors and anti-factor sets, *J. London Math. Soc.* **13** (1976), pp. 351–359.

[22] D.P. Sumner and P. Blitch, Domination critical graphs, *J. Combin. Theory Ser. B* **34** (1983), pp. 65–76.

Proc. of ICDM 2006, Pages 63–72

Resistance Matrix and q-Laplacian of a Unicyclic Graph

R.B. Bapat
Indian Statistical Institute, New Delhi, 110016, India
Email: rbb@isid.ac.in

Abstract. The resistance distance between two vertices of a graph can be defined as the effective resistance between the two vertices, when the graph is viewed as an electrical network with each edge carrying unit resistance. The concept has several different motivations. The resistance matrix of a graph is a matrix with its (i, j)-entry being the resistance distance between vertices i and j. We obtain an explicit formula for the determinant of the resistance matrix of a unicyclic graph. Some properties of a q-analogue of the Laplacian are also studied, with special attention to the limiting behaviour as q approaches 1. An expression for the inverse of the q-Laplacian of a unicyclic graph is derived.

1 Introduction and Preliminaries

The classical definition of distance between two vertices in a graph is the length of a shortest path between the two vertices. This definition is well-known and the concept is widely studied. However, it does not capture some features of "distance", such as the degree of communication between the vertices. For example, if there is a multitude of paths between two vertices, intuitively the two vertices should be thought of as having shorter distance.

The concept of *resistance distance*, introduced by Klein and Randić [8], arises naturally from several different considerations and is also mathematically more attractive than the classical distance. For more background information about resistance distance we refer to [2, 5, 8, 14].

It may be remarked that in the case of a tree, the concepts of classical distance and resistance distance coincide. This perhaps explains some very attractive properties of distances in trees which do not carry over to arbitrary graphs. However, if one uses the notion of resistance distance, then some of these attractive properties do have analogs for arbitrary graphs.

We consider only simple graphs, i.e., those with no loops or parallel edges. The results that we obtain can easily be extended to weighted graphs, though we consider only unweighted graphs for simplicity. Now we introduce some notation. Let $G = (V, E)$ be a graph with n vertices, labeled $\{1, 2, \ldots, n\}$. The Laplacian matrix L of G is defined as follows. For $i \neq j$, the (i, j)-entry of L is zero, if vertices i and j are not adjacent, while it is -1, if i and j are adjacent. The (i, i)-entry of L is defined to make the i-th row-sum equal to zero, $i = 1, 2, \ldots, n$. Thus L is a singular matrix. For basic properties of the Laplacian matrix, see [1, 10].

Let G be a connected graph with vertex set $\{1, 2, \ldots, n\}$. The resistance distance between two vertices can be defined in a number of different, equivalent ways.

We give two definitions here. Let L be the Laplacian of G. Let i, j be vertices of G. The resistance distance r_{ij} between i and j is zero if $i = j$, and if $i \neq j$, then

$$r_{ij} = \frac{\det L(i,j;i,j)}{\det L(i,i)},$$

where $L(i,i)$ is the submatrix obtained by deleting row i, column i, of L; while $L(i,j;i,j)$ is the submatrix obtained by deleting rows i, j and columns i, j of L. As usual, det denotes determinant. Let $\chi(G)$ denote the complexity, that is, the number of spanning trees of G. We remark that by the well-known Matrix-Tree theorem [13], $\det L(i,i)$ equals $\chi(G)$ for $i = 1, 2, \ldots, n$.

The second definition is in terms of electrical networks. Think of G as an electrical network in which a unit resistance is placed along each edge. Current is allowed to enter the network only at vertex i and leave the network only at vertex j. Then the resistance distance between i and j is the "effective resistance" between i and j.

If there is a unique (ij)-path in G, then it is clear from the second definition that the resistance distance between i and j equals the length (that is, the number of edges) of the path. This explains why the resistance distance and the classical distance coincide when the graph is a tree.

We now state some known results which will be required. The proofs are omitted and can be found in [2]. Denote the vector of all ones by $\mathbf{1}$, the size of which will be clear from the context. The $n \times n$ matrix of all ones will be denoted J_n. Let G be a connected graph with vertex set $\{1, 2, \ldots, n\}$, and let L be the Laplacian of G. Then L is singular, has rank $n-1$, and any vector in the null space of L is a scalar multiple of $\mathbf{1}$. The matrix $L + \frac{1}{n}J_n$ is nonsingular. We set

$$X = \left(L + \frac{1}{n}J_n\right)^{-1}. \tag{1}$$

The resistance matrix R of G is the $n \times n$ matrix with its (i,j)-entry equal to 0 if $i = j$, and r_{ij}, the resistance distance between i and j, otherwise. Let τ be a vector of order $n \times 1$ with its components defined by

$$\tau_i = 2 - \sum_{j:j\sim i} r_{ij}, i = 1, 2, \ldots, n, \tag{2}$$

where $j \sim i$ denotes that j is adjacent to i.

Theorem 1.1. *Let G be a connected graph with vertex set $\{1, 2, \ldots, n\}$, let L be the Laplacian of G, let R be the resistance matrix of G, and let $X = (x_{ij})$ and τ be defined as in (1),(2). Let $\tilde{X}$ be the $n \times n$ diagonal matrix with $x_{11}, x_{22}, \ldots, x_{nn}$ along the diagonal. Then the following assertions are true:*

(i) $r_{ij} = x_{ii} + x_{jj} - 2x_{ij}, i, j = 1, 2, \ldots, n$

(ii) $R = \tilde{X}J + J\tilde{X} - 2X$

(iii) $\tau = L\tilde{X}\mathbf{1} + \frac{2}{n}\mathbf{1}$

(iv) $\mathbf{1}'\tau = 2$

(v) R is nonsingular and $R^{-1} = -\frac{1}{2}L + \frac{1}{\tau' R \tau}\tau\tau'$

(vi) $\det R = (-1)^{n-1}2^{n-3}\frac{\tau' R\tau}{\chi(G)}$

For a graph G with n vertices, D will denote the (classical) distance matrix of G; thus the (i,j)-entry of D is 0, if $i = j$ and it is the length of the shortest path between i and j, if $i \neq j$. When the graph is a tree, one gets the following consequence of Theorem 1.1. Assertions (iii) and (iv) are well-known results due to Graham and Pollak [7] and Graham and Lovász [6] respectively.

Theorem 1.2. *Let G be a tree with vertex set $\{1, 2, \ldots, n\}$, let L be the Laplacian of G, let D be the distance matrix of G, and let τ be defined as in (2). Then the following assertions are true:*

(i) $\tau_i = 2 - \delta_i$, *where* δ_i *is the degree of vertex* $i, i = 1, 2, \ldots, n$

(ii) $\tau' D\tau = 2(n-1)$

(iii) D *is nonsingular and* $D^{-1} = -\frac{1}{2}L + \frac{1}{2(n-1)}\tau\tau'$

(iv) $\det D = (-1)^{n-1}(n-1)2^{n-2}$

2 Determinant of the Resistance Matrix of a Unicyclic Graph

We now consider unicyclic graphs. Recall that a graph is unicyclic if it is connected and has a unique cycle. The resistance distance between two vertices in a unicyclic graph G is particularly easy to determine. Clearly, if there is a unique path between vertices i and j of G, then the resistance distance r_{ij} between i and j is the length of the path between i and j.

Suppose a path between vertices i and j meets the cycle in G in at least two vertices. Let u and v be vertices on the cycle such that there is a unique path between i and u, and a unique path between j and v (see Figure). Let the length of the iu-path be a, that of the jv-path be b, and suppose the two paths between u and v have lengths c and d, where $c + d = k$, the length of the cycle. By the interpretation of resistance distance as effective resistance in an electrical network, it follows that r_{ij} is the sum of a, b and the "parallel" sum of c and d. Thus

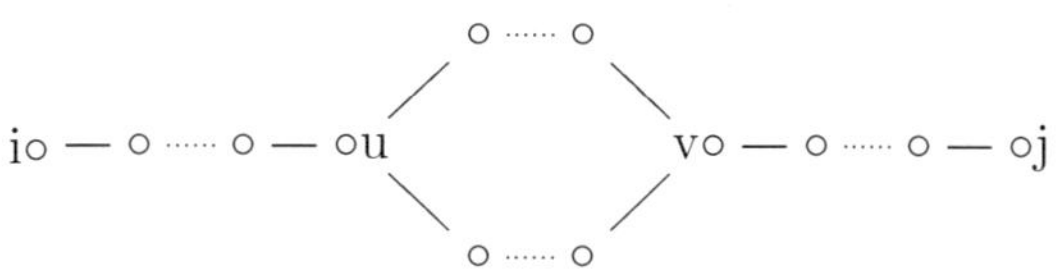

$$r_{ij} = a + b + \frac{1}{\frac{1}{c} + \frac{1}{d}} = a + b + \frac{cd}{c+d}. \tag{3}$$

Since the resistance matrix R of a unicyclic graph has simple structure, it is natural to seek a precise formula for the determinant of R, which we now proceed to obtain.

We first consider the case of a cycle. Let $G = C_n$, the cycle on the n vertices $\{1, 2, \ldots, n\}, n \geq 3$, (where, of course, i and $i+1$ are adjacent, $i = 1, 2, \ldots, n-1$ and 1 is adjacent to n). If $i, j \in \{1, 2, \ldots, n\}$, then there are two paths between vertices i and j, of lengths $|i - j|$ and $n - |i - j|$. If R denotes the resistance matrix of G, then in view of the preceding discussion, $r_{ii} = 0$, $i = 1, 2, \ldots, n$; $r_{ij} = r_{ji}$ for all i and j, and

$$r_{ij} = \frac{(j-i)(n-j+i)}{n} \quad \text{if } i < j. \tag{4}$$

Let τ be defined as in (2). Then

$$\tau_i = 2 - r_{i-1,i} - r_{i,i+1}, i = 1, 2, \ldots, n;$$

where the subscripts are interpreted modulo n. It follows from (4) that

$$r_{i-1,i} = r_{i,i+1} = \frac{n-1}{n}, i = 1, 2, \ldots, n,$$

and hence $\tau_i = \frac{2}{n}, i = 1, 2, \ldots, n$. This last observation also follows from the fact that $\mathbf{1}'\tau = 2$ (see (iv), Theorem 1.1) and by symmetry.

Theorem 2.1. *Let C_n be the cycle on the vertices $\{1, 2, \ldots, n\}, n \geq 3$, and let R be the resistance matrix of C_n. Then*

(i) $\tau' R \tau = \frac{2(n^2-1)}{3n}$

(ii) $detR = (-1)^{n-1} 2^{n-2} \frac{(n^2-1)}{3n^2}$.

Proof. (i). We first consider the case when n is even, say $n = 2m$. It is clear by symmetry that each row-sum of R is the same. Using the expression (4) we see that the first row of R has sum

$$\begin{aligned}
r_{11} + r_{12} + \cdots + r_{1n} &= (r_{12} + r_{13} + \cdots + r_{1m}) \\
&\quad + r_{1,m+1} + (r_{1,m+2} + r_{1,m+3} + \cdots + r_{1,2m}) \\
&= 2\Big(\frac{1(n-1)}{n} + \frac{2(n-2)}{n} + \cdots + \frac{(m-1)(n-m+1)}{n}\Big) \\
&\quad + \frac{m(n-m)}{n} \\
&= \frac{2}{n} \cdot n\big(1 + 2 + \cdots + (m-1)\big) - \frac{2}{n}\Big(1^2 + 2^2 + \cdots + (m-1)^2\Big) \\
&\quad + \frac{m(n-m)}{n} \\
&= \frac{2}{n}\Big(\frac{nm(m-1)}{2} - \frac{(m-1)m(2m-1)}{6}\Big) + \frac{m(n-m)}{n} \\
&= \frac{m-1}{2}\Big(2m - \frac{2m-1}{3}\Big) + \frac{m^2}{2m} \\
&= \frac{n^2-1}{6}.
\end{aligned}$$

Thus each row-sum of R is $\frac{n^2-1}{6}$. Since $\tau_i = \frac{2}{n}, i = 1, 2, \ldots, n$, it follows that

$$\tau' R \tau = \sum_{i=1}^{n} \sum_{j=1}^{n} r_{ij} \tau_i \tau_j = \frac{4}{n^2} \cdot n \cdot \frac{n^2-1}{6} = \frac{2(n^2-1)}{3n}.$$

The proof when n is odd is similar and we omit the details.

(ii). By (vi), Theorem 1.1, $detR = (-1)^{n-1}2^{n-3}\frac{\tau' R\tau}{\chi(C_n)}$. Since $\tau' R\tau = \frac{2(n^2-1)}{3n}$ by the first part and since $\chi(C_n) = n$, the result follows. □

We give an example. The resistance matrix of C_5 is

$$\frac{1}{5}\begin{bmatrix} 0 & 1\cdot 4 & 2\cdot 3 & 3\cdot 2 & 4\cdot 1 \\ 1\cdot 4 & 0 & 1\cdot 4 & 2\cdot 3 & 3\cdot 2 \\ 2\cdot 3 & 1\cdot 4 & 0 & 1\cdot 4 & 2\cdot 3 \\ 3\cdot 2 & 2\cdot 3 & 1\cdot 4 & 0 & 1\cdot 4 \\ 4\cdot 1 & 3\cdot 2 & 2\cdot 3 & 1\cdot 4 & 0 \end{bmatrix}.$$

According to Theorem 2.1, (ii), the determinant of the matrix is $\frac{2^3(5^2-1)}{3.5^2} = \frac{64}{25}$. This matrix is a symmetric Toeplitz matrix and a formula for its determinant may be of independent interest.

The following is the main result of this section.

Theorem 2.2. *Let G be a unicyclic graph with vertices $\{1, 2, \dots, n\}$, suppose the unique cycle C_k of G has length k, and that it is formed by the vertices $\{1, 2, \dots, k\}$; $n \geq k \geq 3$. Let R be the resistance matrix of G. Then*

$$\det R = (-1)^{n-1}2^{n-2}\frac{3kn - 2k^2 - 1}{3k^2}. \tag{5}$$

Proof. If $k = n \geq 3$, then G is C_n and the result follows by Theorem 2.1, (ii). So let $n > k \geq 3$. We prove the result by induction on n. So suppose that the result is true for a graph with $n-1$ vertices. Since $n > k$, then G has at least one pendant vertex. We assume, without loss of generality, that vertex n is pendant and that it is adjacent to vertex $n-1$.

Partition R as

$$\begin{bmatrix} R_{n-1} & z \\ z' & 0 \end{bmatrix},$$

where R_{n-1} is of order $n-1$ and z is $(n-1)\times 1$. If $1 \leq i, j \leq n-1$, then an (ij)-path does not pass through n and therefore R_{n-1} is in fact the resistance matrix of $G \setminus \{n\}$.

Perform the following row and column operations on R. From row n, subtract row $n-1$ and from column n, subtract column $n-1$. For any $1 \leq i \leq n-1$, the resistance distance between i and n is 1 plus the resistance distance between i and $n-1$. Therefore after the row and column operations, the resulting matrix, which has the same determinant as R, is

$$\begin{bmatrix} R_{n-1} & \mathbf{1} \\ \mathbf{1}' & -2 \end{bmatrix}.$$

By Theorem 1.1, (v), R_{n-1} is nonsingular. Using the well-known Schur formula for the determinant,

$$\det R = (\det R_{n-1})(-2 - \mathbf{1}' R_{n-1}^{-1}\mathbf{1}). \tag{6}$$

Let $\tilde{L}$ be the Laplacian of $G \setminus \{n\}$ and let $\tilde{\tau}$ be defined for $G \setminus \{n\}$ as in (2). By Theorem 1.1, (v),

$$R_{n-1}^{-1} = -\frac{1}{2}\tilde{L} + \frac{1}{\tilde{\tau}' R_{n-1}\tilde{\tau}}\tilde{\tau}\tilde{\tau}'. \tag{7}$$

Since $\tilde{L}$ has row-sums zero, (7) and Theorem 1.1, (iv) give

$$\mathbf{1}'R_{n-1}^{-1}\mathbf{1} = \frac{4}{\tilde{\tau}'R_{n-1}\tilde{\tau}}. \tag{8}$$

It follows from (6) and (8) that

$$\det R = \Big(\det R_{n-1})(-2 - \frac{4}{\tau'R_{n-1}\tau}\Big) = -2\det R_{n-1} - 4\frac{\det R_{n-1}}{\tau'R_{n-1}\tau}. \tag{9}$$

Again, (9) and Theorem 1.1, (vi) lead to

$$\det R = -2\det R_{n-1} - 4\frac{(-1)^{n-2}2^{n-4}}{k}. \tag{10}$$

By induction assumption,

$$\det R_{n-1} = (-1)^{n-2}2^{n-3}\frac{3k(n-1) - 2k^2 - 1}{3k^2}. \tag{11}$$

Substituting (11) in (10) we get

$$\begin{aligned}\det R &= (-1)^{n-1}2^{n-2}\Big(\frac{3k(n-1) - 2k^2 - 1}{3k^2} + \frac{1}{k}\Big)\\ &= (-1)^{n-1}2^{n-2}\frac{3kn - 2k^2 - 1}{3k^2}\end{aligned}$$

and (5) is proved. □

We conclude this section with the remark that a formula for the determinant of the (classical) distance matrix of a unicyclic graph has been obtained in [3].

3 The q-Laplacian

Let G be a graph with n vertices, labeled $\{1, 2, \ldots, n\}$. Let A be the adjacency matrix of G. Thus A is an $n \times n$ matrix with (ij)-entry equal to 1 if i and j are adjacent, and zero if i and j are not adjacent. Also, $a_{ii} = 0, i = 1, 2, \ldots, n$. Let Δ be the $n \times n$ diagonal matrix with its i-th diagonal entry equal to the degree of vertex $i, i = 1, 2, \ldots, n$. Then $L = \Delta - A$ is the Laplacian of G. For a parameter q, the q-Laplacian L_q of G is defined as

$$L_q = I - qA + q^2(\Delta - I) = qL + (1 - q^2)I + q(q-1)\Delta. \tag{12}$$

The matrix L_q has been called the generalized Laplacian of G in [9]. The matrix was also introduced in [4] for the case of a tree, in the context of a formula for the inverse of a q-analogue of the distance matrix.

The following result has been obtained by Northshield [11].

Theorem 3.1. *Let G be a graph with n vertices and m edges, and let L_q be the q-Laplacian of G. If $f(q) = \det L_q$, then $f'(q)|_{q=1} = 2(m-n)\chi(G)$.*

Lemma 3.2. *Let G be a graph with n vertices and m edges, and let L_q be the q-Laplacian of G. Suppose G is connected and not unicyclic. Then*

$$\lim_{q\to 1}(1-q)L_q^{-1}L = 0. \tag{13}$$

Proof. Let adj L_q be the adjoint of L_q, so that $L_q^{-1} = \frac{\text{adj } L_q}{\det L_q}$.

Now

$$\lim_{q\to 1}(1-q)L_q^{-1}L = \lim_{q\to 1}(1-q)\frac{(\text{adj } L_q)L}{\det L_q}. \tag{14}$$

Since $\lim_{q\to 1}(1-q)(\text{adj } L_q)L = 0$ and $\lim_{q\to 1}\det L_q = 0$, we may apply L'Hospital's rule and then the limit in (14) equals

$$\lim_{q\to 1}\frac{(1-q)\frac{d}{dq}(\text{adj } L_q)L - (\text{adj } L_q)L}{\frac{d}{dq}(\det L_q)}. \tag{15}$$

Note that

$$\lim_{q\to 1}(1-q)\frac{d}{dq}(\text{adj } L_q)L = 0, \tag{16}$$

and

$$\lim_{q\to 1}(\text{adj } L_q)L = (\text{adj } L)L = 0. \tag{17}$$

Furthermore, by Theorem 3.1,

$$\lim_{q\to 1}\frac{d}{dq}(\det L_q) = 2(m-n)\chi(G), \tag{18}$$

which is nonzero, since G is connected and not unicyclic.

Substituting (16), (17) and (18) in (15) we get (13). □

We now obtain a limiting property of the inverse of the q-Laplacian, which is motivated by the expression for resistance distance contained in Theorem 1.1,(i).

Theorem 3.3. *Let G be a graph with n vertices and m edges, let L_q be the q-Laplacian of G and let $K_q = L_q^{-1}$. Suppose G is connected and not unicyclic. Then for $i, j = 1, 2, \ldots, n$,*

$$\lim_{q\to 1}((K_q)_{ii} + (K_q)_{jj} - 2(K_q)_{ij}) = r_{ij}, \tag{19}$$

the resistance distance between i and j.

Proof. Since $L_q L_q^{-1} L = L$, therefore, using the definition of L_q,

$$(I - qA + q^2\Delta - q^2 I)L_q^{-1}L = L. \tag{20}$$

After a routine simplification, (20) leads to

$$(1-q^2)L_q^{-1}L + (q^2-q)\Delta L_q^{-1}L + qLL_q^{-1}L = L. \tag{21}$$

By Lemma 3.2 we see that the first two terms on the left hand side in (21) approach 0 as q approaches 1. Therefore

$$\lim_{q\to 1} LL_q^{-1}L = \lim_{q\to 1} LK_qL = L. \tag{22}$$

We now introduce some notation. Fix vertices $i \neq j$ of G. Let e_{ij} be the $n \times 1$ vector with i-th coordinate 1, j-th coordinate -1 and zeros elsewhere. Since $\mathbf{1}'e_{ij} = 0$, e_{ij} is in the column space of L and hence there exists a vector w_{ij} such that $Lw_{ij} = e_{ij}$.

Let X be defined as in (1). Then $L\mathbf{1} = 0$ implies $(L + \frac{1}{n}J_n)\mathbf{1} = \mathbf{1}$ and hence $X\mathbf{1} = \mathbf{1}$. Therefore it easily follows that $LXL = L$. Hence

$$e'_{ij}Xe_{ij} = w'_{ij}LXLw_{ij} = w'_{ij}Lw_{ij}. \tag{23}$$

Now

$$\begin{aligned}\lim_{q\to 1}((K_q)_{ii} + (K_q)_{jj} - 2(K_q)_{ij}) &= \lim_{q\to 1} e'_{ij}K_q e_{ij}\\ &= \lim_{q\to 1} w'_{ij}LK_qLw_{ij}\\ &= w'_{ij}Lw_{ij} \text{ by (22)}\\ &= e'_{ij}Xe_{ij}\\ &= x_{ii} + x_{jj} - 2x_{ij}\\ &= r_{ij}, \text{ by Theorem 1.1, (i)},\end{aligned}$$

and the proof is complete. □

4 Inverse of the q-Laplacian of a Unicyclic Graph

A walk without backtracking is a walk in which any two consecutive edges are distinct. Let G be a graph with the vertices $\{1, 2, \ldots, n\}$. For $m \geq 1$, let A_m be the $n \times n$ matrix whose (i, j)-entry is the number of walks in G of length m with no backtracking from i to j, $i, j = 1, \ldots, n$. We also set $A_0 = I$. Note that $A_1 = A$, the adjacency matrix of G. The following identity has been proved in [12, p.139].

Lemma 4.1. *Let G be a graph with the vertices $\{1, 2, \ldots, n\}$, and let L_q be the q-Laplacian of G. Then*

$$(\sum_{m=0}^{\infty} A_m q^m)L_q = (1 - q^2)I. \tag{24}$$

For a tree or a unicyclic graph, matrices A_m can be described explicitly and then Lemma 4.1 may be employed to get an expression for L_q^{-1}.

First consider the case of a tree. As usual, we denote the distance between vertices i and j as r_{ij}. Clearly, $(A_m)_{ii} = 0$ for $m \geq 1$ and $(A_0)_{ii} = 1, i = 1, 2, \ldots, n$. If i, j are distinct vertices of the tree with $d = r_{ij}$, then $(A_d)_{ij} = 1$, whereas $(A_m)_{ij} = 0$ for $m \neq d$. It follows from (24) that

$$(L_q^{-1})_{ii} = \frac{1}{1 - q^2} \text{ and } (L_q^{-1})_{ij} = \frac{q^d}{1 - q^2}, i, j = 1, 2, \ldots, n. \tag{25}$$

The matrix $(q^{r_{ij}})$ has been termed the "exponential distance matrix" of a tree in [4], where its relation with L_q^{-1} is proved in a different way.

We return to unicyclic graphs. For any vertex i of a unicylic graph G, denote by α_i the distance from α_i to the cycle in G. Thus α_i is the least distance from i to a vertex in the cycle.

Theorem 4.2. *Let G be a unicyclic graph with vertices $\{1, 2, \ldots, n\}$ and suppose the unique cycle C_k of G has length $k, n \geq k \geq 3$. Let α_i denote the distance of i*

from the cycle, $i = 1, 2, \ldots, n$. Let L_q be the q-Laplacian of G. Let $d(i,j)$ denote the classical distance between i and j. Then for $i, j = 1, 2, \ldots, n$;

$$(1-q^2)(L_q^{-1})_{ij} = q^{d(i,j)} + \frac{2q^{\alpha_i+\alpha_j+k}}{1-q^k}, \tag{26}$$

if an ij-path does not meet the cycle and

$$(1-q^2)(L_q^{-1})_{ij} = \frac{q^{d(i,j)} + q^{2(\alpha_i+\alpha_j)+k-d(i,j)}}{1-q^k}, \tag{27}$$

if an ij-path meets the cycle.

Proof. Let i, j be vertices of G and suppose an ij-path does not meet the cycle. There is an ij-path, which is also an ij-walk with no backtracking, of length $d(i,j)$. There are two ij-walks with no backtracking of length $\alpha_i + \alpha_j + k$, two ij-walks with no backtracking of length $\alpha_i + \alpha_j + 2k$, and, in general, two ij-walks with no backtracking of length $\alpha_i + \alpha_j + sk, s \geq 1$. Therefore, by (24), it follows that

$$(1-q^2)(L_q^{-1})_{ij} = q^{d(i,j)} + 2\sum_{s=1}^{\infty} q^{\alpha_i+\alpha_j+sk} = q^{d(i,j)} + \frac{2q^{\alpha_i+\alpha_j+k}}{1-q^k}, \tag{28}$$

and (26) is proved. It may be remarked that when $i = j$, (26) is applicable (whether or not i is on the cycle) and we have

$$(1-q^2)(L_q^{-1})_{ii} = 1 + \frac{2q^{2\alpha_i+k}}{1-q^k}. \tag{29}$$

The proof of (27) is similar. □

We are now in a position to complete the case of unicyclic graphs which was left out in Theorem 3.3. Interestingly, the conclusion is different in case of unicyclic graphs.

Theorem 4.3. *Let G be a unicyclic graph with vertices $\{1, 2, \ldots, n\}$ and suppose the unique cycle C_k of G has length $k, n \geq k \geq 3$. Let α_i denote the distance of i from the cycle, $i = 1, 2, \ldots, n$. Let L_q be the q-Laplacian of G and let $K_q = L_q^{-1}$. Then for $i, j = 1, 2, \ldots, n$,*

$$\lim_{q\to 1}((K_q)_{ii} + (K_q)_{jj} - 2(K_q)_{ij}) = r_{ij} + \frac{(\alpha_i - \alpha_j)^2}{k}, \tag{30}$$

where r_{ij} is the resistance distance between i and j.

Proof. Let i, j be vertices of G and suppose an ij-path does not meet the cycle. By Theorem 4.2,

$$(1-q^2)((K_q)_{ii} + (K_q)_{jj} - 2(K_q)_{ij}) = 2 + \frac{2q^{2\alpha_i+k}}{1-q^k} + \frac{2q^{2\alpha_j+k}}{1-q^k} - 2(q^{d(i,j)} + \frac{2q^{\alpha_i+\alpha_j+k}}{1-q^k}).$$

The result follows by computing the limit of $(K_q)_{ii} + (K_q)_{jj} - 2(K_q)_{ij}$ as $q \to 1$ by L'Hospital's rule. We omit the details. The proof follows similarly when an ij-path meets the cycle. □

References

[1] R.B. Bapat, The Laplacian matrix of a graph, *The Mathematics Student*, **65** (1996), 214–223.

[2] R.B. Bapat, Resistance distance in graphs, *The Mathematics Student*, **68** (1999), 87–98.

[3] R.B. Bapat, S. Kirkland and M. Neumann, On distance matrices and Laplacians, *Linear Algebra Appl.*, **401** (2005), 193–209.

[4] R.B. Bapat, A.K. Lal and Sukanta Pati, A q-analogue of the distance matrix of a tree, *Linear Algebra Appl.*, **416** (2006), 799–814.

[5] P.G. Doyle and J.L. Snell, *Random Walks and Electrical Networks*,, Math. Assoc. Am. Washington, (1984).

[6] R.L. Graham and L. Lovász, Distance matrix polynomials of trees, *Adv. in Math.*, **29(1)** (1978), 60–88.

[7] R.L. Graham and H.O. Pollak, On the addressing problem for loop switching, *Bell System Tech. J.*, **50** (1971), 2495–2519.

[8] D.J. Klein and M. Randić, Resistance distance, *Journal of Mathematical Chemistry*,, **12** (1993), 81–95.

[9] Jin Ho Kwak and Iwao Sato, Zeta functions of line, middle, total graphs of a graph and their coverings, *Linear Algebra Appl.*, **418** (2006), 234–256.

[10] R. Merris, Laplacian matrices of graphs: a survey, *Linear Algebra Appl.*, **197–198** (1994), 143–176.

[11] Sam Northshield, A note on the zeta function of a graph, *Journal of Combinatorial Theory, Series B*, **74** (1998), 408–410.

[12] H.M. Stark and A.A. Terras, Zeta functions of finite graphs and coverings, *Advances in Math.*, **121** (1996), 124–165.

[13] D.B. West, *Introduction to Graph Theory*, 2nd Edition, Prentice-Hall Inc., (2001).

[14] Wenjun Xiao and Ivan Gutman, On resistance matrices, *MATCH Commun. Math. Comput. Chem.*, **49** (2003), 67–81.

Proc. of ICDM 2006, Pages 73–85

A Discussion of Constrained Binary Embeddings with Applications to Cryptanalysis of Irregularly Clocked Stream Ciphers

DIANE M. DONOVAN AND JAMES LEFEVRE
Centre for Discrete Mathematics and Computing, University of Queensland, Brisbane, 4072 Australia
LEONIE SIMPSON
School of Software Engineering and Data Communications, Queensland University of Technology, Brisbane, 4001 Australia

Abstract. A given binary string of length n can be obtained from a random binary string of length $(d+1)n$ by constrained deletion - deleting at most d bits of the longer string before obtaining each successive bit of the given string, and deleting an appropriate number of bits at the end of the string. Bounds are discussed for the probability that a given binary string of length n can be obtained from a random binary string of length $(d+1)n$. When the given string is either the constant string or the alternating string these bounds are shown to be exact. Further, the constant string is shown to achieve the lower bound, and for $d = 1$ it is shown that the alternating string uniquely achieves the upper bound. The upper bound on the probability is particularly important in the cryptanalysis of binary stream ciphers in which a shift register is clocked at least once and at most $d+1$ times per output symbol. If the probability decreases sufficiently fast as the given string length increases then reconstruction of the shift register initial state is possible given a sufficiently large value of n.

1 Introduction

With the increasing use of computer and communications systems, large amounts of data are transmitted and stored by electronic means. For sensitive data, such as financial information or personal medical details, maintaining confidentiality is important, and may be a legal requirement. Cryptographic controls are often employed for this. Cryptography transforms a plaintext message into ciphertext using some secret knowledge called a key so that an unauthorized observer of the ciphertext, who is not in possession of the key, is unable to deduce the original plaintext. The legitimate receiver of the message, using the key, can perform the inverse transformation and recover the plaintext.

A cryptographic algorithm is called a cipher. A stream cipher is an encryption algorithm which encrypts a plaintext message one character at a time, with each character encrypted under a time-varying function of the key. This means that different occurrences of the same plaintext character may encrypt to different ciphertext characters. The most common type of stream cipher is the binary additive stream cipher, where a binary keystream is generated independently of the plaintext and ciphertext, and combined with the plaintext or ciphertext using bitwise addition modulo 2 to form the ciphertext or plaintext, repectively. Let $p(t), c(t)$ and $z(t)$ represent, respectively, the plaintext, ciphertext and keystream characters at time t, for $t \geq 0$. Then $c(t) = p(t) \oplus z(t)$, where $\oplus$ denotes addition modulo 2. Similarly, $p(t) = c(t) \oplus z(t)$. Under this system, the same device can be used for both encryption and decryption, as shown in Figure 1. Note that if some segment of the plaintext is known, then by combining it with the corresponding ciphertext, a segment of the keystream can be recovered, since $z(t) = p(t) \oplus c(t)$. The cipher cannot be considered secure if it is possible to predict the entire keystream from a known segment.

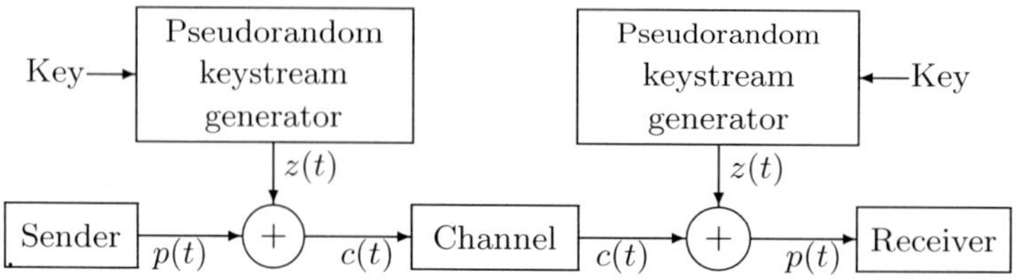

Figure 1.1: Encryption and decryption in binary additive stream ciphers.

Many keystream generators for stream cipher applications are based on feedback shift registers, particularly linear feedback shift registers (LFSRs) as these are fast, easily implemented in hardware and produce binary sequences with long periods and good statistical properties. For stream ciphers based on shift registers, we assume the structure of the keystream generator is known, and the key is the register initial states only.

A shift register may be clocked regularly, producing an output bit each time the register is stepped, or irregularly, with the register stepped a variable number of times before an output bit is produced. For keystream generators which employ irregular clocking, the construction may impose a constraint on the clock control, so that the register is stepped up to some preset maximum number of times, say $d+1$, before a keystream bit is produced. If no such constraint exists, the clocking is unconstrained. For example, the stop and go generator is clocked at most once per output symbol, and for the shrinking generator [1], the constraint is the length of one of the component shift registers.

For regularly clocked LFSRs, the linearity makes them susceptible to an attack [3] in which the entire keystream can be reproduced from a known keystream segment with length only twice that of the underlying LFSR. To make use of the properties of LFSR's but avoid this attack, LFSRs are often used as components in a keystream generator. If the LFSR's are regularly clocked, the linearity of the LFSR sequence can be disguised by using a nonlinear function to combine the outputs of several LFSR's or the outputs of several stages of one LFSR. However, these keystream generators have been shown to possess weaknesses which can be exploited to recover the key from a known segment of the keystream [5, 6]. Alternatively, nonlinearity can be introduced implicitly by using irregularly clocked LFSRs.

Some proposed keystream generators such as the shrinking generator [1] rely only on irregular clocking of an underlying LFSR to produce an unpredictable keystream. If the LFSR is clocked at least once per output symbol, the keystream may be viewed as an irregular decimation of the binary sequence which the LFSR would produce if regularly clocked. That is, if we take the binary sequence produced by the regularly clocked LFSR and delete some of the bits, we obtain successive bits of the keystream. Then the keystream segment is said to be embedded in the regularly clocked LFSR sequence. If a known segment of keystream can be used to identify the underlying regularly clocked binary sequence, then the initial state of the LFSR can be recovered.

In this paper we investigate the situation relating to constrained clocking where an LFSR is clocked at least once and at most $d+1$ times before an output bit is produced. For the cryptanalytic problem of reconstructing the initial state of the clock controlled LFSR from a known output segment, we are interested in the proportion of all binary strings of length $(d+1)n$ that a given string of length n can be obtained from by deleting at most d bits of the longer string before obtaining each successive bit of the keystream. For such strings, we say that the keystream can be $d-$embedded into the longer strings. If this proportion, which is the probability that the given binary string can be obtained from a random binary string of length $(d+1)n$, decreases sufficiently fast as the string length increases, then reconstruction of the initial state is possible, given enough keystream.

For the case where $d = 1$ an upper bound on the embedding probability is obtained by direct counting in [7] and by analytical methods in [2]. Both bounds are exponentially small for large string lengths, showing that the length of keystream required for successful reconstruction is linear in the shift register length. However the results apply only to the reconstruction of clock controlled LFSRs which are clocked once or twice per output symbol. That is, where $d = 1$. An extension of this work from binary strings to strings of symbols over a finite alphabet is presented in [4], also for the case $d = 1$. We present an approach which provides lower bounds for binary strings for all d, and verifies that this bound is uniquely achieved by the constant string. We calculate the embedding probability for the alternating string for all d and show that this embedding probability is an upper bound on the probability for clock controlled shift registers for the constrained case where $d = 1$. We conjecture that for all d the embedding probability of the alternating string is a unique upper bound. Empirical results for $d = 2$ are given in the final section of this paper.

2 Notation and Terminology

Let n and d be positive integers. Consider the binary strings $\widetilde{j}$, of length n, and $\widetilde{i}$, of length $n(d+1)$. The binary string $\widetilde{j} = [j_0 j_1 \dots j_{n-2} j_{n-1}]$, of length n can be represented by an integer j, $0 \leq j \leq 2^n - 1$; that is, $j = 2^{n-1}j_0 + 2^{n-2}j_1 + \cdots + 2^1 j_{n-2} + j_{n-1}$. For example, the binary string $\widetilde{j} = [0111]$, of length $n = 4$ has $j_0 = 0, j_1 = 1, j_2 = 1$ and $j_3 = 1$ and $j = 7$. Similarly the binary string $\widetilde{i} = [i_0 i_1 \dots i_{m-2} i_{m-1}]$, where $m = n(d+1)$, can be represented by the integer $i = 2^{m-1}i_0 + 2^{m-2}i_1 + \cdots + 2^1 i_{m-2} + i_{m-1}$.

At times it will be necessary to differentiate between the start and finish of a string. So for the binary string $[r_0 r_1 \dots r_{s-2} r_{s-1}]$ we define the start to be the right hand end of the string corresponding to the bit r_{s-1} and the finish to be the left hand end corresponding to the bit r_0. Specific strings of importance throughout this paper are:

$\widetilde{1}_k$	a string of k 1s
$\widetilde{0}_k$	a string of k 0s
C_n	the constant string $\widetilde{c}_n$ where $c \in \{0,1\}$
$\widetilde{r}_k$	an unconstrained string of k bits
A_n^0	the alternating string $[\ldots 101010]$ of length n
A_n^1	the alternating string $[\ldots 010101]$ of length n

It is useful to consider a binary string as the concatenation of several shorter binary strings. For example, $\widetilde{j} = [0111]$, of length $n = 4$, could be considered as $\widetilde{j} = \widetilde{a}\widetilde{b}$, the concatenation of the binary string $\widetilde{a} = [01]$ with the binary string $\widetilde{b} = [11]$. It is also useful to think of string $\widetilde{j}$ as the concatenation of constant substrings. For instance, $\widetilde{j} = [0111] = [0_1][1_3]$. Here the substrings will be termed *blocks*.

Given the binary strings $\widetilde{i}$, of length $n(d+1)$, and $\widetilde{j}$, of length n, we say that $\widetilde{j}$ can be *d–embedded* in $\widetilde{i}$ if $\widetilde{j}$ can be obtained by deleting bits of $\widetilde{i}$ according to the following rules:

- At most d consecutive deletions are allowed at the start of $\widetilde{i}$ before obtaining the bit j_{n-1}.
- At most d consecutive deletions from the string $\widetilde{i}$ are allowed before successive bits of $\widetilde{j}$ are obtained.
- After obtaining the bit j_0 of $\widetilde{j}$, the remaining bits at the end of $\widetilde{i}$ are deleted (up to dn bits will be deleted at this stage).

The notation $\widetilde{j} \Rightarrow \widetilde{i}$ will be used to indicate that the binary string $\widetilde{j}$ can be d–embedded into the binary string $\widetilde{i}$ and $j_p \longrightarrow i_q$ will indicate that the bit j_p is embedded onto the bit i_q. For ease of explanation, at times, it will be useful to highlight those bits of $\widetilde{i}$ which have bits of $\widetilde{j}$ embedded onto them. In this case we underline each such bit of $\widetilde{i}$. For example, let $d = 2$ and $n = 5$. The binary string $\widetilde{j} = [01001]$ can be 2–embedded in the binary string $\widetilde{i} = [100010000000011]$, and we write $[01001] \Rightarrow [100010000000011]$. To observe the embedding delete bits i_{14}, i_{12}, i_{11}, i_9, i_8, i_6 and i_5 and note that $j_4 \longrightarrow i_{13}$, $j_3 \longrightarrow i_{10}$, $j_2 \longrightarrow i_7$, $j_1 \longrightarrow i_4$ and $j_0 \longrightarrow i_3$. This leaves $3 \leq 10$ deletions at the end of $\widetilde{i}$. We can represent $\widetilde{i}$ with this embedding by underlining the bits of $\widetilde{i}$ on which bits of $\widetilde{j}$ are embedded: $\widetilde{i} = [100\underline{0}\underline{1}00\underline{0}00\underline{0}0\underline{1}1]$.

We say a d–embedding of the string $\widetilde{j}$, of length n, into the string $\widetilde{i}$, of length $m = n(d+1)$, is minimal if the first possible position, subject to the need to embed the remaining bits of the string, is always used. That is, assume $\widetilde{j} \Rightarrow \widetilde{i}$, and so for each $p \in \{0, \ldots, n-1\}$ there exists a $q \in \{0, \ldots, m-1\}$ such that $j_p \longrightarrow i_q$. The d–embedding is said to satisfy the *minimality condition* if for all such p there does not exist a d–embedding of $\widetilde{j}$ into $\widetilde{i}$ such that $j_p \longrightarrow i_s$, where $s > q$. The associated d–embedding is said to be *minimal*. If there are multiple possible embeddings, then there will be a unique minimal embedding. Note that the above example is a minimal embedding.

For a given string $\widetilde{j}$, of length n, we let $\mathcal{R}_n[\widetilde{j}]$ represent the set of all strings of length $m = n(d+1)$ which $\widetilde{j}$ can d-embed in, and let $R_n[\widetilde{j}] = |\mathcal{R}_n[\widetilde{j}]|$.

Given any binary string $\widetilde{i}$, define $\widetilde{i}^C$ to be the string obtained by replacing each 0 in $\widetilde{i}$ with a 1, and each 1 with a 0. We refer to the binary string $\widetilde{i}^C$ as the binary complement of the binary string $\widetilde{i}$. Similarly, if $\mathcal{X}$ is a set of binary strings, define $\mathcal{X}^C = \{\widetilde{i}^C \mid \widetilde{i} \in \mathcal{X}\}$ (this is the set of strings which are the binary complements of the strings in $\mathcal{X}$, not the set complement). Note that $(\widetilde{i}^C)^C = \widetilde{i}$, $|\mathcal{X}| = |\mathcal{X}^C|$ and $\mathcal{R}_n[\widetilde{j}^C] = (\mathcal{R}_n[\widetilde{j}])^C$.

Finally, in this paper we let $*$ indicate an unknown or unspecified bit of the string $\widetilde{i}$.

3 Embeddings of the Constant String

Consider the constant strings $\widetilde{1}_n$ and $\widetilde{0}_n$. Note that $\widetilde{1}_n^C = \widetilde{0}_n$, and so by symmetry, $R_n[\widetilde{1}_n] = R_n[\widetilde{0}_n]$. In this section we calculate the value of $R_n[\widetilde{0}_n]$ for every n and d and also show that this is a strict minimum: for every d, the constant strings of length n embed in less strings than do any other strings of length n.

Let $\widetilde{j}$ and $\widetilde{i}$ be strings of length n and $n(d+1)$ respectively. For the purposes of this section, we define a *greedy* d-embedding of $\widetilde{j}$ onto $\widetilde{i}$ to be one in which each bit of $\widetilde{j}$ in turn is embedded onto the first possible bit of $\widetilde{i}$, regardless of whether the remaining bits of $\widetilde{j}$ may be embedded onto the remaining bits of $\widetilde{i}$. Recall the 2-embedding example given above, which we denote $[100\underline{01}00\underline{0}00\underline{0}0\underline{1}1]$. Although this is minimal, it is not greedy. A greedy embedding would require $j_4 \rightarrow i_{14}$, $j_3 \rightarrow i_{12}$, and $j_2 \rightarrow i_{11}$, giving the partial embedding $[1000100000\underline{00}1\underline{1}]$. At this point the embedding cannot proceed, so there is no greedy embedding in this case.

Given a string $\widetilde{j}$ of length n, define $\mathcal{G}_n[\widetilde{j}]$ to be the set of strings of length $n(d+1)$ for which there exists a greedy d-embedding of $\widetilde{j}$. A string may admit an embedding but no greedy embedding (as in the example above); but if a greedy embedding exists, then it is the minimal embedding. Thus $\mathcal{G}_n[\widetilde{j}] \subseteq \mathcal{R}_n[\widetilde{j}]$, and for each $\widetilde{i} \in \mathcal{G}_n[\widetilde{j}]$, the minimal embedding (which is permanently associated with the string) is greedy. Our strategy is to calculate $|\mathcal{G}_n[\widetilde{j}]|$, then show that $\mathcal{G}_n[\widetilde{j}] = \mathcal{R}_n[\widetilde{j}]$ if and only if $\widetilde{j} = \widetilde{1}_n$ or $\widetilde{j} = \widetilde{0}_n$.

Lemma 3.1. *For every binary string $\widetilde{j}$ of length n, $|\mathcal{G}_n[\widetilde{j}]| = (2^{d+1} - 1)^n$.*

Proof. Let $\widetilde{i} \in \mathcal{G}_n[\widetilde{j}]$. Assume without loss of generality that $\widetilde{j} = \widetilde{j}'\,[0]$. Suppose that the first bit of $\widetilde{j}$, which we have just assumed is a 0, is embedded on the kth bit of $\widetilde{i}$, $1 \leq k \leq d+1$; that is, $0 = j_{n-1} \rightarrow i_{m-k}$. Then the first $k-1$ bits of $\widetilde{i}$ must be equal to 1 by the greedy property, and by the d-embedding property no bits of $\widetilde{j}$ are embedded on the last $d-k+1$ bits of $\widetilde{i}$ (so these bits are unconstrained). Thus $\widetilde{i}$ has the form

$$\widetilde{i} = \widetilde{r}_{(d-k+1)}\,\widetilde{i}'\,[\underline{0}]\,\widetilde{1}_{(k-1)},$$

where $\widetilde{i}' \in \mathcal{G}_{n-1}[\widetilde{j}']$; conversely, any string of this form is in $\mathcal{G}_n[\widetilde{j}]$.

There are 2^{d-k+1} possible values of $\widetilde{r}_{(d-k+1)}$, so summing across k we have

$$|\mathcal{G}_n[\widetilde{j}]| = \sum_{k=1}^{d+1} 2^{d-k+1} |\mathcal{G}_{n-1}[\widetilde{j}']| = (2^{d+1} - 1)|\mathcal{G}_{n-1}[\widetilde{j}']|.$$

Since $\widetilde{j}'$ is a string of length $n-1$, the result follows by induction. □

Lemma 3.2. *Let $\widetilde{j}$ be any string of length n. Then $\mathcal{G}_n[\widetilde{j}] = \mathcal{R}_n[\widetilde{j}]$ if and only if $\widetilde{j} \in \{\widetilde{1}_n, \widetilde{0}_n\}$.*

Proof. If $\widetilde{j} = \widetilde{1}_n$ then $\widetilde{j} \Rightarrow \widetilde{i}$ if and only if there is no block of $d+1$ or more consecutive 0s before the nth 1 in $\widetilde{i}$. If this condition holds, then there also exists a greedy embedding of $\widetilde{j}$ onto $\widetilde{i}$; thus $\mathcal{G}_n[\widetilde{1}_n] = \mathcal{R}_n[\widetilde{1}_n]$. Similarly $\mathcal{G}_n[\widetilde{0}_n] = \mathcal{R}_n[\widetilde{0}_n]$. It remains to be shown that if $\widetilde{j} \notin \{\widetilde{1}_n, \widetilde{0}_n\}$, then $\mathcal{R}_n[\widetilde{j}] \setminus \mathcal{G}_n[\widetilde{j}] > 0$; that is, that there exists at least one string of length $n(d+1)$ which admits an embedding of $\widetilde{j}$ but not a greedy embedding.

Assuming that $\widetilde{j} \notin \{\widetilde{1}_n, \widetilde{0}_n\}$, then $n \geq 2$ and we can assume without loss of generality that $\widetilde{j} = \widetilde{a}\,[0\,1]\,\widetilde{b}$. The required string is

$$\widetilde{i} = \widetilde{r}_{nd-d-1}\,\widetilde{a}\,[0]\,\widetilde{1}_{d+2}\,\widetilde{b}.$$

□

From the above lemmas and the property that $\mathcal{G}_n[\widetilde{j}] \subseteq \mathcal{R}_n[\widetilde{j}]$, we have the following result:

Theorem 3.3. *Let $\widetilde{j}$ be a string of length n. Then $R_n[\widetilde{j}] \geq (2^{d+1}-1)^n$, with equality if and only if $\widetilde{j} \in \{\widetilde{1}_n, \widetilde{0}_n\}$.*

4 Embeddings of Alternating Strings

Consider the alternating strings A_n^1 and A_n^0. By symmetry, $R_n[A_n^1] = R_n[A_n^0]$ for every n and d. In this section, we calculate this value for each $d \geq 0$ and each $n \geq 0$. We use a similar (but slightly more complicated) approach to that used in Section 3.

Let $\widetilde{i} \in \mathcal{R}_n[A_n^1]$. For every string $\widetilde{i} \in \mathcal{R}_n[A_n^1]$, e assume that A_n^1 is minimally embedded on $\widetilde{i}$ (this makes the embedding unique). We consider $\widetilde{i}$ as a sequence of blocks of either 1s or 0s alternating until the end. We define $\mathcal{X}_n$ to be the subset of $\mathcal{R}_n[A_n^1]$ in which the first block is of 1s; that is, $\mathcal{X}_n = \{\widetilde{i} \mid \widetilde{i} \in \mathcal{R}_n[A_n^1] \text{ and } i_{m-1} = 1\}$. Given the regular nature of A_n^1, we have the following result:

Lemma 4.1. *Let $\widetilde{i} \in \mathcal{X}_n$. Then in the minimal embedding of A_n^1 into $\widetilde{i}$, the first bit ($a_{n-1} = 1$) of A_n^1 must be embedded onto the first block of 1s in $\widetilde{i}$, the next bit ($a_{n-2} = 0$) must be embedded onto the first block of 0s in $\widetilde{i}$, and in general the kth bit of A_n^1 must be embedded onto the kth block of $\widetilde{i}$.*

We partition $\mathcal{X}_n$ according to the position of the first bit in the minimal embedding. Let $1 \leq p \leq d+1$ and $A_n^1 = [j_0\,j_1\,\ldots\,j_{n-1}]$, and define $\mathcal{X}_{n,p} = \{\widetilde{i} \mid \widetilde{i} \in \mathcal{X}_n \text{ and } j_{n-1} \rightarrow i_{m-p}\}$. Thus $\mathcal{X}_{n,p}$ is the subset of $\mathcal{X}_n$ in which the first bit of A_n^1 is embedded in position p of $\widetilde{i}$. Define $X_n = |\mathcal{X}_n|$ and $X_{n,p} = |\mathcal{X}_{n,p}|$. In calculating the values of $R_n[A_n^1]$ we are particularly interested in the cases $p = 1, 2$.

Lemma 4.2. $R_n[A_n^1] = X_{n+1,1}/2^d$, $n \geq 1$.

Proof. Let $\widetilde{x}_{n+1,1} \in \mathcal{X}_{n+1,1}$; this string has $(n+1)(d+1)$ bits. By definition, $\widetilde{x}_{n+1,1}$ contains a minimal embedding of A_{n+1}^1 in which the first bit ($a_n = 1$) of A_{n+1}^1 is embedded on the first bit of $\widetilde{x}_{n+1,1}$. By the d-embedding condition and a simple counting argument, no bits of A_{n+1}^1 may be embedded in the last d bits of $\widetilde{x}_{n+1,1}$. Thus

$$\widetilde{x}_{n+1,1} = \widetilde{r}_d\,\widetilde{s}\,[\underline{1}],$$

where $\widetilde{r}_d$ is an unconstrained string of length d and $\widetilde{s}$ is some string of length $n(d+1)$ on which the remaining n bits of A_{n+1}^1 must be embedded. Note that A_{n+1}^1 is the concatenation of the binary string A_n^0 and the bit 1. That is, $\widetilde{s} \in \mathcal{R}_n[A_n^0]$. Now $\widetilde{s}$ may be any string in $\mathcal{R}_n[A_n^0]$ and the d bits of $\widetilde{r}_d$ are unconstrained; thus

$$|\mathcal{X}_{n+1,1}| = |\mathcal{R}_n[A_n^0]| \cdot 2^d = R_n[A_n^1] \cdot 2^d$$

□

Lemma 4.3. *Let $2 \leq p \leq d+1$. Then $X_{n,p} = X_{n,2}/2^{p-2}$.*

Proof. Let $\widetilde{x}_{n,p} \in \mathcal{X}_{n,p}$. By definition, $\widetilde{x}_{n,p}$ contains a minimal embedding of A_n^1 in which the first bit of A_n^1 ($a_{n-1} = 1$) is embedded on bit p of $\widetilde{x}_{n,p}$. That is, $a_{n-1} \rightarrow x_{m-p-1}$. By Lemma 4.1 the first p bits of $\widetilde{x}_{n,p}$ are equal to 1. Let $\widetilde{s} = [x_0 \dots x_{m-p+1}]$; that is $\widetilde{x}_{n,p}$ with the first $p-2$ bits removed. Then

$$\begin{aligned} \widetilde{x}_{n,p} &= [* \dots * \underline{1}\,1\,1 \dots 1] \\ &= [* \dots * \underline{1}\,1]\,[1 \dots 1] \\ &= \widetilde{s}\,\widetilde{1}_{p-2} \end{aligned}$$

No bit of A_n^1 was embedded on the first $p-2$ 1s, thus the embedding of the alternating string A_n^1 is entirely within $\widetilde{s}$, with the first bit of A_n^1 being embedded on the second bit of $\widetilde{s}$. Hence there exist binary strings $\widetilde{r}_{p-2}\widetilde{s} \in \mathcal{X}_{n,2}$ (that is, if we remove the first $p-2$ bits of $\widetilde{x}_{n,p}$ and add an unconstrained string $\widetilde{r}_{p-2}$ of length $p-2$ bits to the other end, the resultant string will be in $\mathcal{X}_{n,2}$). Finally, we show that it is not possible to embed the first bit of A_n^1 on the first bit of $\widetilde{r}_{p-2}\,\widetilde{s}$. Suppose such an embedding is possible; then it follows that we could also embed the first bit of A_n^1 on bit $p-1$ of $\widetilde{x}_{n,p}$, contradicting minimality. This embedding cannot use any of the bits in $\widetilde{r}_{p-2}$ (by the minimality condition and a simple counting argument); the embedding is entirely in $\widetilde{s}$.

Since there are 2^{p-2} possibilities for $\widetilde{r}_{p-2}$, we have

$$2^{p-2} \cdot |\mathcal{X}_{n,p}| \leq |\mathcal{X}_{n,2}|$$

The reverse argument holds similarly; given any string in $\mathcal{X}_{n,2}$, we can obtain a string in $\mathcal{X}_{n,p}$ by removing the last $p-2$ bits and adding $p-2$ 1s to the start (this maps 2^{p-2} strings in $\mathcal{X}_{n,2}$ to one string in $\mathcal{X}_{n,p}$). Thus equality holds above, and we are done. □

For all p, we can now express $X_{n,p}$ (as well as $R_n[A_n^1]$) in terms of $X_{n,1}$ and $X_{n,2}$. We now derive a recursive formula giving $X_{n+1,1}$ and $X_{n+1,2}$ in terms of $X_{n,1}$ and $X_{n,2}$.

Lemma 4.4. *For every $n \geq 1$, $X_{n+1,1} = X_{n,1}(2^{d+1} - 1) + X_{n,2}(2^{d+2} - 2d - 4)$.*

Proof. Let $\widetilde{x}_{n+1,1} \in \mathcal{X}_{n+1,1}$, and put $\widetilde{x}_{n+1,1} = [x_0 \dots x_{m-1}]$, where $m = (n+1)(d+1)$ is the length of $\widetilde{x}_{n+1,1}$. Suppose that the second bit of A_{n+1}^1 ($a_{n-1} = 0$) is embedded on bit q of $\widetilde{x}_{n+1,1}$ (x_{m-q}). By definition the first bit of A_{n+1}^1 ($a_n = 1$) is embedded on the first bit of $\widetilde{x}_{n+1,1}$, and so $2 \leq q \leq d+2$. By Lemma 4.1 the second bit of A_{n+1}^1 is embedded on the first block of 0s in $\widetilde{x}_{n+1,1}$.

$$\widetilde{x}_{n+1,1} = [* \dots * \underline{0}\,0 \dots 0\,1 \dots \underline{1}]$$

Let k be the size of the first block of 1s, where $1 \leq k \leq q-1 \leq d+1$; then (for a given q) the first q bits of $\widetilde{x}_{n+1,1}$ are defined by k.

Define $\widetilde{s} = [x_{d-k+1} \ldots x_{m-k-1}]$; that is, $\widetilde{s}$ is the string consisting of bits $k+1$ through to $k + n(d+1)$ of $\widetilde{x}_{n+1,1}$. The string $\widetilde{s}$ can be obtained from $\widetilde{x}_{n+1,1}$ by removing the initial block of k 1s together with a further $d+1-k$ bits from the left hand end of $\widetilde{x}_{n+1,1}$. A simple counting argument shows that no bits can be embedded in the last $d+1-k$ bits of $\widetilde{x}_{n+1,1}$. Therefore these bits are unconstrained, and A^1_{n+1} is embedded entirely in $\widetilde{s}$, except for the first bit.

$$\begin{aligned}\widetilde{x}_{n+1,1} &= [* \ldots * \, \underline{0}\, 0 \ldots 0\, 1 \ldots \underline{1}] \\ &= [* \ldots *]\, [* \ldots * \, \underline{0}\, 0 \ldots 0]\, [1 \ldots \underline{1}] \\ &= \widetilde{r}_{d+1-k}\, \widetilde{s}\, \widetilde{1}_k\end{aligned}$$

It follows that $\widetilde{s}$ contains a minimal embedding of A^0_n in which the first bit is embedded in position $q-k$; that is, $\widetilde{s} \in \mathcal{X}^C_{n,q-k}$. Conversely, $\widetilde{s}$ may be any string in $\mathcal{X}^C_{n,q-k}$. The first k bits of $\widetilde{x}_{n+1,1}$ are fixed, but the last $d+1-k$ bits are unconstrained. Thus for a given q and k the number of strings in $\mathcal{X}_{n+1,1}$ is $2^{d+1-k} X_{n,q-k}$. Summing over q and k gives

$$\begin{aligned}X_{n+1,1} &= |\mathcal{X}_{n+1,1}| \\ &= \sum_{q=2}^{d+2} \sum_{k=1}^{q-1} 2^{d+1-k} X_{n,q-k} \\ &= \sum_{q=2}^{d+2} \sum_{k=1}^{q-2} 2^{d+1-k} X_{n,q-k} + \sum_{q=2}^{d+2} 2^{d+1-(q-1)} X_{n,q-(q-1)} \\ &= \sum_{q=2}^{d+2} \sum_{k=1}^{q-2} 2^{d+1-k} (X_{n,2}/2^{q-k-2}) + \sum_{q=2}^{d+2} 2^{d-q+2} X_{n,1} \\ &\qquad \text{(by Lemma 4.3)} \\ &= X_{n,2} \sum_{q=2}^{d+2} 2^{d-q+3}(q-2) + X_{n,1}(2^{d+1}-1) \\ &= X_{n,2} \sum_{i=1}^{d} \sum_{j=1}^{i} 2^j + X_{n,1}(2^{d+1}-1) \\ &= X_{n,1}(2^{d+1}-1) + X_{n,2}(2^{d+2} - 2d - 4)\end{aligned}$$

□

Lemma 4.5. *For every $n \geq 1$, $X_{n+1,2} = X_{n,1}/2 + dX_{n,2}$.*

Proof. Let $\widetilde{x}_{n+1,2} \in \mathcal{X}_{n+1,2}$, and put $\widetilde{x}_{n+1,2} = [x_0 \ldots x_{m-1}]$, where $m = (n+1)(d+1)$ is the length of $\widetilde{x}_{n+1,2}$. The proof is similar to above, except that the first bit of A^1_{n+1} ($a_n = 1$) is embedded on the second bit of $\widetilde{x}_{n+1,2}$, and the second bit of A^1_{n+1} ($a_{n-1} = 0$) is embedded in position $d+3$ (x_{m-d-3}). The embedding begins as follows.

$$\widetilde{x}_{n+1,2} = [* \ldots * \, \underline{0}\, 0 \ldots 0\, 1 \ldots \underline{1}\, 1]$$

To see this, note that if the position of this second bit was further left, the d-embedding condition would be violated, while if the position of this second bit was further right, the first bit of A^1_{n+1} could have been embedded on the first bit of $\widetilde{x}_{n+1,2}$ (which equals 1 by definition), violating the minimality condition. As in the previous lemma, we let k be the size of the initial block of 1s in $\widetilde{x}_{n+1,2}$; this parameter fixes the first $d+3$ bits of $\widetilde{x}_{n+1,2}$.

We have $2 \le k \le d+2$, but we will assume that $k \le d+1$ for now, since $k = d+2$ requires some special treatment. Define $\widetilde{s} = [x_0 \, \ldots \, x_{m-d-2}]$ to be the string consisting of bits $d+2$ through to m of $\widetilde{x}_{n+1,2}$; that is, $\widetilde{s}$ is the string obtained by removing the first $d+1$ bits from $\widetilde{x}_{n+1,2}$.

$$\begin{aligned}\widetilde{x}_{n+1,2} &= [* \ldots * \underline{0}\, 0 \ldots 0\, 1 \ldots \underline{1}\, 1] \\ &= [* \ldots * \underline{0}\, 0]\, [0 \ldots 0\, 1 \ldots \underline{1}\, 1] \\ &= \widetilde{s}\, \widetilde{0}_{d+1-k}\, \widetilde{1}_k\end{aligned}$$

It follows that $\widetilde{s}$ contains a minimal embedding of A_n^0 in which the first bit of A_n^0 is embedded in the second position of $\widetilde{s}$; that is, $\widetilde{s} \in \mathcal{X}_{n,2}^C$. Conversely, $\widetilde{s}$ may be any string in $\mathcal{X}_{n,2}^C$. The first $d+1$ bits of $\widetilde{x}_{n+1,2}$ are fixed, thus for a given k $(2 \le k \le d+1)$ the number of strings in $\mathcal{X}_{n+1,2}$ is $X_{n,2}$.

In the case $k = d+2$, we let $\widetilde{s} = [x_0 \, \ldots \, x_{m-d-2}]$ be the string consisting of bits $d+3$ through to m of $\widetilde{x}_{n+1,2}$.

$$\begin{aligned}\widetilde{x}_{n+1,2} &= [* \ldots * \underline{0}]\, [1 \ldots \underline{1}\, 1] \\ &= \widetilde{s}\, \widetilde{1}_{d+2}\end{aligned}$$

Now $\widetilde{s}$ contains a minimal embedding of A_n^0, with the first bit of A_n^0 embedded onto the first bit of $\widetilde{s}$; however, $\widetilde{s} \notin \mathcal{X}_{n,1}^C$ since $\widetilde{s}$ is of length $(n+1)(d+1) - (d+2) = n(d+1) - 1$. If we add either a 0 or 1 to the end of $\widetilde{s}$ we obtain a string in $\mathcal{X}_{n,1}^C$. Conversely, $\widetilde{s}$ can be equal to any string in $\mathcal{X}_{n,1}^C$ with the last bit removed (for $\widetilde{i} \in \mathcal{X}_{n,1}^C$, a simple counting argument shows that there cannot be a bit of A_n^0 embedded on the last bit of $\widetilde{i}$, so this last bit is unconstrained). Therefore if $k = d+2$ the number of strings in $\mathcal{X}_{n+1,2}$ is equal to $X_{n,1}/2$. Summing over k gives $X_{n+1,2} = X_{n,1}/2 + dX_{n,2}$. □

It remains for us to solve the recurrence relations, noting that $X_{1,1} = 2^d$ and $X_{1,2} = 0$.

Lemma 4.6. *For all* $n \ge 1$,

$$\begin{aligned}X_{n,1} &= \left(\frac{2^{d-1}(\alpha + 2^{d+1} - d - 1)}{\alpha}\right)\left(\frac{2^{d+1} + d - 1 + \alpha}{2}\right)^{n-1} \\ &\quad + \left(\frac{2^{d-1}(\alpha - 2^{d+1} + d + 1)}{\alpha}\right)\left(\frac{2^{d+1} + d - 1 - \alpha}{2}\right)^{n-1}, \\ X_{n,2} &= \left(\frac{2^{d-1}}{\alpha}\right)\left(\frac{2^{d+1} + d - 1 + \alpha}{2}\right)^{n-1} - \left(\frac{2^{d-1}}{\alpha}\right)\left(\frac{2^{d+1} + d - 1 - \alpha}{2}\right)^{n-1},\end{aligned}$$

and thus

$$\begin{aligned}R_n[A_n^1] &= \left(\frac{\alpha + 2^{d+1} - d - 1}{2\alpha}\right)\left(\frac{2^{d+1} + d - 1 + \alpha}{2}\right)^{n} \\ &\quad + \left(\frac{\alpha - 2^{d+1} + d + 1}{2\alpha}\right)\left(\frac{2^{d+1} + d - 1 - \alpha}{2}\right)^{n},\end{aligned}$$

where

$$\alpha = \sqrt{(2^{d+1} - d + 1)^2 - 8}.$$

5 The Upper Bound in the Case $d = 1$

In this section we prove that the result in Section 4 represents an upper bound when $d = 1$ (the smallest non-trivial case). That is, A_n^1 and A_n^0 1-embed in more strings of length $2n$ than do any other strings of length n. We use a similar approach to Section 4, but we fix $d = 1$ and consider an arbitrary string.

Given a string $\widetilde{j}$ of length n (and letting $m = 2n$), define $\mathcal{X}_n[\widetilde{j}] = \{\widetilde{i} \mid \widetilde{i} \in \mathcal{R}_n[\widetilde{j}]$ and $i_{m-1} = j_{n-1}\}$. For $p \in \{1, 2\}$, let $\mathcal{X}_{n,p}[\widetilde{j}] = \{\widetilde{i} \mid \widetilde{i} \in \mathcal{X}_n[\widetilde{j}]$ and $j_{n-1} \to i_{m-p}\}$; that is, the subset of $\mathcal{X}_n[\widetilde{j}]$ in which the first bit of $\widetilde{j}$ is embedded in the pth position of $\widetilde{i}$. As usual, $X_{n,p}[\widetilde{j}] = |\mathcal{X}_{n,p}[\widetilde{j}]|$ and $X_n[\widetilde{j}] = |\mathcal{X}_n[\widetilde{j}]|$. We now show that the recurrence relations derived in Section 4 hold, except that here one of them is an inequality. Hence we prove inductively that $X_{n,1}[\widetilde{j}] \leq X_{n,1}[A_n^1]$, $X_{n,2}[\widetilde{j}] \leq X_{n,2}[A_n^1]$, and thus $R_n[\widetilde{j}] \leq R_n[A_n^1]$, with equality if and only if $\widetilde{j} \in \{A_n^1, A_n^0\}$. We begin by proving the equivalent of Lemma 4.2.

Lemma 5.1. *Let $\widetilde{j}$ be an arbitrary string of length n and let $x \in \{0, 1\}$. Then $R_n[\widetilde{j}] = X_{n+1,1}[\widetilde{j}\,[x]]/2$.*

Proof. Consider an arbitrary string $\widetilde{i} \in \mathcal{X}_{n+1,1}[\widetilde{j}\,[x]]$, and let $m = 2(n+1)$ be the length of $\widetilde{i}$. By definition, $i_{m-1} = x$ and the initial x in $\widetilde{j}\,[x]$ is embedded on i_{m-1}. By standard counting arguments, the bit i_0 of $\widetilde{i}$ cannot have a bit of $\widetilde{j}$ embedded on it, consequently it is unconstrained. The bits of $\widetilde{j}$ must be embedded on the remaining $2n$ bits of $\widetilde{i}$. Thus

$$\widetilde{i} = [r]\,\widetilde{i'}\,[\underline{x}],$$

where $r \in \{0, 1\}$ and $\widetilde{i'} \in \mathcal{R}_n[\widetilde{j}]$. Conversely, any string of this form is in $\mathcal{X}_{n+1,1}[\widetilde{j}\,[x]]$. Since there are two choices of r, the result follows. □

We do not have to prove an analogue of Lemma 4.3, since this is trivial when $d = 1$. The following lemma is equivalent to Lemma 4.4.

Lemma 5.2. *Let $\widetilde{j}$ be an arbitrary string of length n and let $x \in \{0, 1\}$. Then $X_{n+1,1}[\widetilde{j}\,[x]] = 3X_{n,1}[\widetilde{j}] + 2X_{n,2}[\widetilde{j}]$.*

Proof. Let $j_{n-1} = y$, where $y \in \{0, 1\}$, and $\widetilde{i} \in X_{n+1,1}[\widetilde{j}\,[x]]$. Then $\widetilde{i}$ (with the minimal embedding of $\widetilde{j}[x]$) must have the form

$$\begin{aligned} \widetilde{i} &= [* \ldots * \ \underline{y}\,\underline{x}], \quad \text{or} \\ \widetilde{i} &= [* \ldots * \ \underline{y}\,(1-y)\,\underline{x}], \quad \text{or} \\ \widetilde{i} &= [* \ldots * \ \underline{y}\,y\,\underline{x}]. \end{aligned}$$

In the first case $\widetilde{i} = [r]\,\widetilde{i'}\,[\underline{x}]$, where $r \in \{0, 1\}$ and $\widetilde{i'} \in \mathcal{X}_{n,1}[\widetilde{j}]$, so there are $2X_{n,1}[\widetilde{j}]$ possibilities. In the second case $\widetilde{i} = \widetilde{i'}\,[(1-y)\,\underline{x}]$, where $\widetilde{i'} \in \mathcal{X}_{n,1}[\widetilde{j}]$, and there are $X_{n,1}[\widetilde{j}]$ possibilities. In the third case $\widetilde{i} = [r]\,\widetilde{i'}\,[\underline{x}]$, where $r \in \{0, 1\}$ and $\widetilde{i'} \in \mathcal{X}_{n,2}[\widetilde{j}]$, and there are $2X_{n,2}[\widetilde{j}]$ possibilities.

Conversely, every string which has one of these 3 forms is in $\mathcal{X}_{n+1,1}[\widetilde{j}\,[x]]$, and no string can have more than one of these forms. The result follows. □

The following lemma is equivalent to Lemma 4.5.

Lemma 5.3. *Let $\widetilde{j}$ be an arbitrary string of length n and let $x \in \{0, 1\}$. Then $X_{n+1,2}[\widetilde{j}\,[x]] \leq X_{n,1}[\widetilde{j}]/2 + X_{n,2}[\widetilde{j}]$.*

Proof. Let $j_{n-1} = y$, where $y \in \{0,1\}$, and let $\widetilde{i} \in \mathcal{X}_{n+1,2}[\widetilde{j}\,[x]]$. Then $\widetilde{i}$ (with the minimal embedding of $\widetilde{j}$) must have the form

$$\widetilde{i} = [* \ldots * \; \underline{y}\, y\, \underline{x}\, x], \qquad \text{or}$$
$$\widetilde{i} = [* \ldots * \; \underline{y}\, (1-y)\, \underline{x}\, x].$$

Note that y cannot be embedded in the third position of $\widetilde{i}$, because then the bit x adjoined to $\widetilde{j}$ could be embedded on i_{m-1}, contradicting minimality.

In the first case $\widetilde{i} = \widetilde{i'}\,[\underline{x}\, x]$, where $\widetilde{i'} \in \mathcal{X}_{n,2}[\widetilde{j}]$; there are $X_{n,2}[\widetilde{j}]$ possibilities. In the second case $\widetilde{i} = \widetilde{i'}\,[(1-y)\,\underline{x}\, x]$, where $\widetilde{i'}$ can be regarded as a string in $\mathcal{X}_{n,1}[\widetilde{j}]$ with the bit i_0 removed (the last bit of a string in $\mathcal{X}_{n,1}[\widetilde{j}]$ is unused, and therefore arbitrary); thus there are $X_{n,1}[\widetilde{j}]/2$ possibilities.

If $x \neq y$, then every $\widetilde{i}$ which has one of these 2 forms is in $\mathcal{X}_{n+1,2}[\widetilde{j}\,[x]]$. But if $x = y$, it is possible that the first two bits of $\widetilde{j}\,[x]$ (both equal to x) could be embedded on the first two bits of $\widetilde{i}$, and hence by definition $\widetilde{i} \notin \mathcal{X}_{n+1,2}[\widetilde{j}\,[x]]$. Thus if $x = y$ we have an inequality. □

Since any string of length 1 is equal to A_1^1 or A_1^0 (so the basis case holds), it follows inductively that $X_{n,1}[\widetilde{j}] \leq X_{n,1}[A_n^1]$ and $X_{n,2}[\widetilde{j}] \leq X_{n,2}[A_n^1]$. Thus from Lemma 5.1, $R_n[\widetilde{j}] \leq R_n[A_n^1]$. All that is left to prove is that equality holds if and only if $\widetilde{j}$ is the alternating string. To do this, all we need to do is show that if $\widetilde{j} \notin \{A_n^1, A_n^0\}$, then the inequality in Lemma 5.3 is strict at some point in the induction. We consider the first point at which two consecutive bits are the same.

Lemma 5.4. *Given* $x \in \{0,1\}$, *then* $X_{n+1,2}[A_n^x\,[x]] < X_{n,1}[A_n^x]/2 + X_{n,2}[A_n^x]$.

Proof. Consider the proof of Lemma 5.3. We have $\widetilde{j} = A_n^x$, and so this proof shows that every $\widetilde{i} \in \mathcal{X}_{n+1,2}[A_n^x\,[x]]$ has the form $\widetilde{i} = \widetilde{i'}\,[\underline{x}\, x]$, where $\widetilde{i'} \in \mathcal{X}_{n,2}[A_n^x]$, or the form $\widetilde{i} = \widetilde{i'}\,[(1-x)\,\underline{x}\, x]$, where $\widetilde{i'}$ can be regarded as a string in $\mathcal{X}_{n,1}[A_n^x]$ with bit i_0 removed. We need to show that the converse is not true; that there is at least one string which has one of these two forms but is not contained in $\mathcal{X}_{n+1,2}[A_n^x\,[x]]$. In fact, no string of the form $\widetilde{i} = \widetilde{i'}\,[(1-x)\,\underline{x}\, x]$ is in $\mathcal{X}_{n+1,2}[A_n^x\,[x]]$. We can embed the first three bits of $A_n^x\,[x]$ on the first three bits of $\widetilde{i}$, so $\widetilde{i} = \widetilde{i'}\,[\underline{(1-x)\,\underline{x}\,\underline{x}}]$. The remaining $n-2$ bits of $A_n^x\,[x]$ form the string A_{n-2}^x, which can be embedded on $\widetilde{i'}$. Therefore $\widetilde{i} \in \mathcal{X}_{n+1,1}[A_n^x\,[x]]$, and we are done. □

Using the value for $R_n[A_n^1]$ from the previous section (calculated with $d = 1$), we have the following result:

Theorem 5.5. *Let* $\widetilde{j}$ *be a string of length* n, *and let* $d = 1$. *Then*

$$R_n[\widetilde{j}] \leq (2+\sqrt{2})^{n+1}/4 + (2-\sqrt{2})^{n+1}/4,$$

with equality if and only if $\widetilde{j} \in \{A_n^1, A_n^0\}$.

Table 1: Given a binary string $\tilde{j}$ of length n: the number of possible 1-embedding of $\tilde{j}$ into strings length $2n$ and the number of possible 2-embedding of $\tilde{j}$ into strings of length $3n$.

$d = 1$					
	n				
$\tilde{j}$	1	2	3	4	5
0	**3**	9	27	81	243
1	**3**	**10**	30	90	270
2		**10**	**34**	102	306
3		9	31	96	288
4			31	106	327
5			**34**	**116**	348
6			30	104	324
7			27	97	306
8				97	334
9				104	356
10				**116**	**396**
11				106	362
12				96	340
13				102	354
14				90	326
15				81	298

$d = 2$					
	n				
$\tilde{j}$	1	2	3	4	5
0	**7**	49	343	2401	16807
1	**7**	**53**	378	2646	18522
2		**53**	**407**	2947	20825
3		49	378	2779	19698
4			378	2912	21511
5			**407**	**3133**	22932
6			378	2949	21966
7			343	2691	20209
8				2691	20789
9				2949	22761
10				**3133**	**24127**
11				2912	22428
12				2779	21821
13				2947	23071
14				2646	20884
15				2401	18986

6 Conclusion

In conclusion, lower bounds have been established for the embedding probability for clock controlled shift registers where the linear feedback shift register is clocked at least one and at most $d + 1$ times before an output bit is produced. It has been verified that this bound is uniquely achieved by the constant string. The embedding probability has been calculated for the alternating string for all d and when $d = 1$ it has been shown that this embedding probability is an upper bound.

The analysis conducted in this paper leads us to conjecture that for all d the embedding probability of the alternating string is a unique upper bound.

In support of this conjecture we present the values of $R_n[\tilde{j}]$ for the cases $d = 1$ and $d = 2$ in Table 1. Note that for $n = 5$, entries cover only $0 \leq j \leq 15$; values for $16 \leq j \leq 31$ may be obtained using the symmetry of the table. The values of $R_n[\tilde{j}]$ for alternating strings are shown in bold type in Table 1 and we see that the alternating string uniquely achieves the upper bound. This empirical evidence supports the above conjecture. However it is still an open problem to establish embedding probabilities for $d \geq 2$ and relatively small values of n, say $n \leq 10$, as is the problem of establishing the exact form of the string which achieves the upper bound.

References

[1] D. Coppersmith and H. Krawczyk and Y. Mansour, The shrinking generator, Advances in Cryptology – CRYPTO '93, Ed. D. R. Stinson, 773 Lecture Notes in Computer Science, 1993, Springer-Verlag, pp. 22–39.

[2] J. Dj. Golić, Constrained embedding probability for two binary strings, *Society for Industrial and Applied Mathematics Journal of Discrete Mathematics*, **9(3)** (1996), pp. 360–364.

[3] J.L. Massey, Shift register synthesis and BCH decoding, *IEEE Trans. Inform. Theory*, **IT-15** (1969), pp. 122–127.

[4] V.G. Mikhailov and N. M. Mezhennaya, Bounds for the probability of a constrained embedding of one discrete sequence into another, *Discrete Math. Appl.*, **15(4)** (2005), pp. 377–386.

[5] T. Siegenthaler, Decrypting a class of stream ciphers using ciphertext only, *IEEE Trans. Comput.*, **C-34** (1985), pp. 81–85.

[6] T. Siegenthaler, Cryptanalyst's representation of nonlinearly filtered ML-sequences, *Advances in Cryptology – EUROCRYPT '85, Lecture Notes in Computer Science*, **219** Ed. F. Pichler, Springer-Verlag, (1986), pp. 103–110.

[7] M. Živković, An Algorithm for the Initial State Reconstruction of the Clock-Controlled Shift Register, *IEEE Transactions on Information Theory*, **IT-37** (1991), pp. 1488–1490.

Proc. of ICDM 2006, Pages 87–100

Disjoint Dominating Sets in Graphs

SANDRA M. HEDETNIEMI AND STEPHEN T. HEDETNIEMI
Department of Computer Science, Clemson University, Clemson, SC 29634
RENU C. LASKAR
Department of Mathematical Sciences, Clemson University, Clemson, SC 29634
LISA MARKUS
Department of Mathematics, De Anza College, Cupertino, CA 95014
PETER J. SLATER
Department of Mathematical Sciences, University of Alabama in Huntsville, Huntsville, AL 35899

Abstract. The *disjoint domination number* of a graph G is the minimum cardinality of the union of two disjoint dominating sets in G. In this paper, we introduce the disjoint domination number and study its graph theoretical and algorithmic properties.

---

1 Introduction

Let $G = (V, E)$ be a graph with a vertex set V of order n and an edge set E. Let v be a vertex in V. The *open neighborhood* $N(v)$ of v is the set of all vertices adjacent to v. The *closed neighborhood* $N[v]$ of v is the set $N(v) \cup \{v\}$. For a set S of vertices, the *open neighborhood* $N(S)$ *of* S is defined by $N(S) = \cup_{v \in S} N(v)$ and the *closed neighborhood* $N[S]$ *of* S is the set $N(S) \cup S$. A set $S \subseteq V$ is a *dominating set* if every vertex in $V - S$ is adjacent to a vertex in S, that is, $N[S] = V$. The *domination number* $\gamma(G)$ is the minimum cardinality of a dominating set S of G; a dominating set of minimum cardinality is called a γ*-set* of G.

In 1991, Kulli and Sigarkanti [13] introduced the concept of the inverse domination number of a graph, as follows. Let S be a γ-set of a graph $G = (V, E)$. If G has no isolated vertices, then it is well known (cf. Ore [14],p.207) that the complement $V - S$ of every minimal dominating set S contains a minimal dominating set. Such a set, say $S' \subseteq V - S$ is called an *inverse dominating set* with respect to S. The *inverse domination number*, denoted $\gamma^{-1}(G)$, is the minimum cardinality of an inverse dominating set with respect to a γ-set of G.

A set $S \subseteq V$ is said to be *independent* if no two vertices in S are adjacent. The maximum cardinality of an independent set S in a graph G is called the *independence number* and is denoted $\beta_0(G)$, while the minimum cardinality of a maximal independent set is called the *independent domination number* and is denoted $i(G)$. Note that every maximal independent set is also a minimal dominating set. Finally, the *upper domination number* $\Gamma(G)$ equals the maximum cardinality of a minimal dominating set of G.

In [13], Kulli and Sigarkanti attempt to prove that for any graph G, $\gamma^{-1}(G) \leq \beta_0(G)$. Unfortunately, their proof of this inequality contains an error. As far as we know, no proof of this inequality exists, nor does a counterexample. We will henceforth refer to this as the *Kulli-Sigarkanti Conjecture*, that is,

Conjecture 1.1. For every graph G without isolated vertices,

$$\gamma^{-1}(G) \leq \beta_0(G).$$

We note that this conjecture is true for several classes of graphs, for the following reason.

Proposition 1.2. *Let G be any graph without isolated vertices. If $\beta_0(G) = \Gamma(G)$, then $\gamma^{-1}(G) \leq \beta_0(G)$.*

Proof. Let S be any γ-set of G and let S' be a minimal dominating set in $V - S$.
By definition, $\gamma^{-1}(G) \leq |S'| \leq \Gamma(G)$. But $\beta_0(G) = \Gamma(G)$, and hence, $\gamma^{-1}(G) \leq \beta_0(G)$. □

In the book *Fundamentals of Domination in Graphs* by Haynes, Hedetniemi and Slater [10] one can find a discussion of classes of graphs for which $\beta_0(G) = \Gamma(G)$. We include many of these graphs in the following corollary.

Corollary 1.3. *The Kulli-Sigarkanti Conjecture is true for each of the following classes of graphs: trees, bipartite graphs, chordal graphs, circular arc graphs, claw-free graphs, C_4-free graphs, net-free graphs, upper bound graphs, trestled graphs, cographs, permutation graphs, comparability graphs, co-chordal graphs, peripheral graphs, parity graphs, Gallai graphs, perfectly orderable graphs and Meyniel graphs.*

Since the publication of [13], a second paper on the inverse domination number was published by Domke, Dunbar and Markus [4], in which the class of graphs for which $\gamma(G) + \gamma^{-1}(G) = n$ is characterized; a lower bound for $\gamma^{-1}(T)$ of a tree is given, and a constructive characterization is given of the class of trees that achieve this lower bound.

It is the Kulli-Sigarkanti Conjecture and the concept of the inverse domination number which inspires the following concept.

The *disjoint domination number* $\gamma\gamma(G)$ is defined as follows: $\gamma\gamma(G) = \min\{|S_1| + |S_2| : S_1, S_2$ are disjoint dominating sets of $G\}$. We say that two disjoint dominating sets, whose union has cardinality $\gamma\gamma(G)$, is a *$\gamma\gamma$-pair* of G.

To date many papers have been published which study the existence of disjoint dominating sets in graphs. Perhaps most notable among these are the approximately 72 papers (too numerous to mention here) which study various types of domatic numbers. In this case one seeks to partition the vertex set of a graph into as many disjoint dominating sets of a given type as possible. For a review of this literature the reader is referred to the survey paper on domatic numbers by Zelinka [16].

More closely related to the concept of the disjoint domination number are the following two papers. In [1], Bange, Barkauskas and Slater provide a characterization of the class of trees having two disjoint minimum dominating sets, and in a related paper Grinstead and Slater [7] study the existence of two minimum dominating sets having a minimum intersection.

Several papers have been written on the subject of disjoint independent dominating sets in graphs and when they exist, cf. Cockayne and Hedetniemi [3], Grinstead and Slater [7, 8], and Dunbar et al [5]. Also in [11], Haynes and Henning provide a constructive characterization of the class of trees having two disjoint minimum independent dominating sets.

Proposition 1.4. *For every graph G without isolated vertices,*

$$2\gamma(G) \leq \gamma\gamma(G) \leq \gamma(G) + \gamma^{-1}(G) \leq n.$$

We define the *disjoint independent domination number* $ii(G)$ to equal the minimum cardinality of the union of two disjoint independent dominating sets in a graph G. Notice that not all graphs have a disjoint independent domination number. For example, the graph obtained from the complete graph K_3 on three vertices, by making each vertex adjacent to a vertex of degree one, does not have two disjoint independent dominating sets. The following result establishes when a corona of the form $H \circ K_1$ has two disjoint independent dominating sets. The *corona* $H \circ K_1$ is the graph constructed from a copy of H by adding a new vertex v' and a pendant edge vv' to each vertex $v \in V(H)$.

Theorem 1.5. *If $G = H \circ K_1$, then $ii(G)$ is defined if and only if H is bipartite.*

Proof. Assume that H is a bipartite graph. Let the vertices in the bipartition of H be the two independent sets X and Y. Let X' and Y' be the sets of leaves of G adjacent to the vertices in X and Y, respectively. Then it follows that $X \cup Y'$ and $X' \cup Y$ are two disjoint independent dominating sets of G. Hence, $ii(G)$ is defined.

Conversely, assume that $ii(G)$ is defined for the corona $G = H \circ K_1$, and let D_1 and D_2 be two disjoint, independent dominating sets of G. Because every leaf and every vertex adjacent to a leaf must be in D_1 or D_2, it follows that $V(G) = D_1 \cup D_2$. Thus, G, and hence H, must be bipartite. □

We will also consider an invariant intermediate to both $\gamma\gamma(G)$ and $ii(G)$, namely, the minimum cardinality of the disjoint union of a dominating set S and an independent dominating set S_i, which we will denote $\gamma i(G)$. We will call such a pair of dominating sets (S, S_i) a *γi-pair.* We note that every graph G without isolated vertices has a γi-pair, which can be found by letting S_i be any maximal independent set, and then noting that the complement $V - S_i$ is a dominating set, and therefore contains a minimal dominating set, say S.

When the disjoint independent domination number exists, the following inequalities hold.

Proposition 1.6. *For any graph G having two disjoint independent dominating sets,*

$$\gamma\gamma(G) \leq \gamma i(G) \leq ii(G).$$

The following three results indicate the disjoint domination numbers and disjoint independent domination numbers of paths, cycles and complete graphs; since their proofs are immediate, they are omitted.

Proposition 1.7. *For the path P_n of order $n \geq 2$,*

(i) $\gamma\gamma(P_{3k}) = ii(P_{3k}) = 2\gamma(P_{3k}) + 1 = 2k + 1.$

(ii) $\gamma\gamma(P_{3k+1}) = ii(P_{3k+1}) = 2\gamma(P_{3k+1}) = 2k + 2.$

(iii) $\gamma\gamma(P_{3k+2}) = ii(P_{3k+2}) = 2\gamma(P_{3k+2}) = 2k + 2.$

Proposition 1.8. *For the cycle C_n, $n \geq 3$,*

$$\gamma\gamma(C_n) = ii(C_n) = 2\gamma(C_n) = 2\lceil\frac{n}{3}\rceil.$$

Proposition 1.9. *For the complete graph K_n, $n \geq 2$,*

$$\gamma\gamma(K_n) = ii(K_n) = 2.$$

We say that a graph G is *$\gamma\gamma$-minimum* if it has two disjoint γ-sets, that is, $\gamma\gamma(G) = 2\gamma(G)$. Similarly, a graph G is called *$\gamma\gamma$-maximum* if $\gamma\gamma(G) = n$. It follows from the previous three results that all paths except those of the form P_{3k} are $\gamma\gamma$-minimum, and so are all cycles C_n and all complete graphs K_n. Grinstead and Slater [7, 8] define $M_\gamma(G)$ to be the minimum cardinality of the intersection of two γ-sets. Note that G is $\gamma\gamma$-minimum if and only if $M_\gamma(G) = 0$.

One can also see that the paths P_2, P_3 and P_4 are all $\gamma\gamma$-maximum. In fact, we can completely characterize the $\gamma\gamma$-maximum graphs. We say that a vertex which is adjacent to a leaf is a *stem*.

Theorem 1.10. *A connected graph G is $\gamma\gamma$-maximum if and only if either $G = C_4$ or every vertex is a leaf or a stem.*

Proof. If every vertex of a graph G is either a leaf or a stem, then clearly $\gamma\gamma(G) = n$. Also, $\gamma\gamma(C_4) = 4 = n$.

We will show that if G has a vertex that is neither a leaf nor a stem, then either $\gamma\gamma(G) < n$ or $G = C_4$. Suppose that there exists a graph G with n vertices having a vertex x that is neither a leaf nor a stem, and $\gamma\gamma(G) = n$. We will consider two cases:

Case 1. There exist two adjacent vertices $y, z \in N(x)$, that is, x, y, z form a triangle in G. Extend $\{y\}$ to a maximal independent set in G. Call this set A. Note that $x, z \notin A$. It follows that the set $V - A - \{x\}$ is a dominating set of G, since if $u \in N(x) \cap A$, then because x is not a stem, u must have a neighbor other than x which is not in A.

Case 2. The set $N(x)$ is independent. Let $N(x) = \{x_1, x_2, ..., x_k\}$, where $k \geq 2$.

Case 2.1. For some i, $1 \leq i \leq k$, there is a $z \neq x \in N(x_i)$ with $|N(x) - N(z)| > 0$. Extend the set $\{z\} \cup (N(x) - N(z))$ to a maximal independent set; call this set A. The set A is an independent dominating set of G, and the set $V - A - \{x\}$ is a dominating set of G. Therefore, $\gamma\gamma(G) < n$.

Case 2.2. For all i, if $z \in N(x_i)$, then $N(x) \subseteq N(z)$. Then $|N(x_i)| = |N(x_j)|$ for all $1 \leq i, j \leq k$. We consider two cases.

Case 2.2.1. $|N(x_1)| \geq 3$. Let $z \in N(x_1) - \{x\}$. The set $B = \{x_1, z\} \cup (V - (N(x_1) \cup N(z)))$ is a dominating set of G. Let $A \subseteq B$ be a minimal dominating set of G. Note that $x \notin A$ and $x_2, x_3, ..., x_k \notin A$, and therefore $x_1 \in A$. The set $V - A - \{x\}$ is also a dominating set of G. Thus, $\gamma\gamma(G) < n$.

Case 2.2.2. $|N(x_1)| = 2$. We consider two subcases:

Case 2.2.2.1. $|N(x)| \geq 3$. Let $N(x_1) = \{x, z\}$. The set $B = \{x_1\} \cup (V - N[x])$ is a dominating set of G. Let $A \subseteq B$ be a minimal dominating set of G. Note that $x_1, z \in A$. Since the set $V - A - \{x_2\}$ is also a dominating set of G, it follows that $\gamma\gamma(G) < n$.

Case 2.2.2.2. $|N(x)| = 2$. Let $y \in N(x_1) - \{x\}$. If $V = \{x_1, x_2, x, y\}$, then $G = C_4$ and $\gamma\gamma(G) = 4 = n$. If $n \geq 5$, let $z \in N(y) - \{x_1, x_2\}$. The set $B = V - \{x, x_1, z\}$ is a dominating set of G. Let $A \subseteq B$ be a minimal dominating set of G. Note that $x_2, y \in A$ and $x, x_1, z \in V - A$. Then the set $V - A - \{x_1\}$ is a dominating set for G, so that $\gamma\gamma(G) < n$.

This completes the proof of the theorem. □

Since we have characterized the class of all $\gamma\gamma$-maximum graphs, we turn our attention to the problem of characterizing the class of $\gamma\gamma$-minimum graphs, that is graphs having two disjoint γ-sets.

The following classes of graphs can be seen to be $\gamma\gamma$-minimum:

(i) paths P_{3n+1} and P_{3n+2}, $n \geq 1$ and P_2, but not paths of the form P_{3n}.

(ii) all cycles, C_n, $n \geq 1$.

(iii) graphs of the form $G = K_2 + H$, where H is an arbitrary graph, every vertex of which is adjacent in G to both vertices of the K_2. For this class of graphs, $\gamma\gamma(G) = 2$. This class includes all non-trivial complete graphs.

(iv) the complete bipartite graphs $K_{m,n}$, for $m, n \geq 2$, and the complete multipartite graphs $K_{m_1,m_2,\ldots,m_t}$, where $2 \leq m_1 \leq m_2 \ldots \leq m_t$. For these graphs, $\gamma(G) = 2$ and $\gamma\gamma(G) = 4$.

(v) all $2 \times n$ grid graphs are $\gamma\gamma$-minimum, where the $2 \times n$ grid graph is the Cartesian product graph $P_2 \Box P_n$.

(vi) all Rooks graphs, $R_n = K_n \Box K_n$ are $\gamma\gamma$-minimum. For these graphs, $\gamma(R_n) = i(R_n) = n$, and $\gamma\gamma(R_n) = 2n$.

(vii) all coronas of the form $G \circ K_1$. If G has order n, then $G \circ K_1$ has order $2n$, and $\gamma(G \circ K_1) = n$. In this case the leaves of $G \circ K_1$ form a γ-set, as do the vertices of G. Thus, $\gamma\gamma(G \circ K_1) = 2n$. We note that from this observation one can conclude that the only graphs that are both $\gamma\gamma$-minimum and $\gamma\gamma$-maximum are the coronas $G \circ K_1$ and C_4.

In the Open Problems section at the end of this paper we conjecture that several other well-known classes of graphs are $\gamma\gamma$-minimum.

We next consider the problem of deriving a mathematical characterization of the class of $\gamma\gamma$-minimum trees, which we denote $\mathcal{D}_2$. The following results establish some of the properties of this class of trees.

The class $\mathcal{D}_2$ contains:

(i) all paths P_{3k+1} and P_{3k+2};

(ii) all subdivided stars, which we denote $S(K_{1,s})$, for $s \geq 2$. For this class of trees T, $\gamma(T) = s$ and T has two disjoint γ-sets. In fact, $\gamma(T) = i(T)$.

(iii) all trees of the form $T_1 = T \circ K_1$. For this class of trees, $\gamma(T_1) = n/2$ and T_1 has two disjoint γ-sets. In fact, $\gamma(T_1) = i(T_1)$.

The class $\mathcal{D}_2$ of $\gamma\gamma$-minimum trees has been algorithmically characterized by Bange, Barkauskas and Slater, who in [1] present an algorithm to determine if any tree T is a member of $\mathcal{D}_2$. Their algorithm consists of a sequence of reductions which applied to T produce subtrees T' of T, which are in $\mathcal{D}_2$ if and only if T is in $\mathcal{D}_2$. The algorithm stops when T has been reduced to a path or another subtree which is known to be in or not in $\mathcal{D}_2$.

A vertex v in a graph $G = (V, E)$ is called *γ-required* if it is an element of every γ-set of G.

Lemma 1.11. *If G is $\gamma\gamma$-minimum, then G does not contain any γ-required vertices.*

Corollary 1.12. *No $\gamma\gamma$-minimum tree contains a stem with two or more leaves.*

In fact we make the following conjecture:

Conjecture 1.13. A tree T of order $n \geq 2$ is $\gamma\gamma$-minimum if and only if T has no γ-required vertices.

We note that, in general, a graph can have no γ-required vertices, and yet not be $\gamma\gamma$-minimum. The following is such a graph. Let the vertices of a K_3 be labelled u_1, u_2, u_3. Add two vertices v_1, v_2, each adjacent to u_1 and u_2. Add two vertices w_1, w_3, each adjacent to u_1 and u_3. Finally, add two vertices x_2, x_3, each adjacent to u_2 and u_3. It can be seen that this graph G has no γ-required vertices, $\gamma(G) = 2$, yet $\gamma\gamma(G) = 5$.

2 Inequalities Involving the Disjoint Domination Number

In this section we establish several inequalities involving the disjoint domination number of a graph.

Theorem 2.1. *Let G be a graph of order $n \geq 2$ with no isolated vertices. Then*

$$\gamma\gamma(G) \leq \gamma i(G) \leq i(G) + \beta_0(G).$$

Proof. Let I be a maximal independent set with $|I| = i(G)$, and let S be a maximal independent set in $\langle V - I \rangle$, the subgraph induced by $V - I$. Every vertex in $V - I$ is dominated by S. Let D be the vertices of I that are not dominated by S. Note that $S \cup D$ is a maximal independent set. Since I is an independent dominating set, every vertex in D has at least one neighbor in $(V - I) - S$. For each vertex in D, choose one vertex in $(V - I) - S$. Call this set T. Then the set $T \cup S$ is a dominating set for G, with $|T \cup S| = |T| + |S| \leq |D| + |S| \leq \beta_0(G)$, and $(T \cup S, I)$ is a γi pair for G. Thus, $\gamma\gamma(G) \leq \gamma i(G) \leq |I| + |T \cup S| \leq i(G) + \beta_0(G)$. □

Corollary 2.2. *If G is a graph of order $n \geq 2$ with no isolated vertices, and if $\gamma(G) = i(G)$, then $\gamma^{-1}(G) \leq \beta_0(G)$, that is, the Kulli-Sigarkanti Conjecture is true when $\gamma(G) = i(G)$.*

Corollary 2.3. *The Kulli-Sigarkanti Conjecture is true when G has no isolated vertices and no induced subgraph isomorphic to $K_{1,3}$. In particular, the Kulli-Sigarkanti Conjecture is true for line graphs.*

Theorem 2.4. *If G is a connected graph of order n and size m, then $m \geq 2n - 3\gamma\gamma(G)/2$.*

Proof. Let D_1, D_2 be two disjoint dominating sets with $|D_1| + |D_2| = \gamma\gamma(G)$. Then (i) one of D_1, D_2 has at least $\gamma\gamma(G)/2$ vertices, (ii) every vertex in $V - (D_1 \cup D_2)$ has a neighbor in D_1 and a neighbor in D_2, (iii) every vertex in D_1 has a neighbor in D_2, and (iv) every vertex in D_2 has a neighbor in D_1.

Thus, the number of edges is at least $2|V - (D_1 \cup D_2)| + \max\{|D_1|, |D_2|\} \geq 2(n - \gamma\gamma(G)) + \gamma\gamma(G)/2 = 2n - 3\gamma\gamma(G)/2$. □

Using the fact that every tree has $n - 1$ edges, we get the following corollary:

Corollary 2.5. *Let T be a tree. Then $\gamma\gamma(T) \geq 2(n + 1)/3$.*

Although both $\gamma\gamma(G)$ and $\gamma i(G)$ are bounded above by the sum $i(G) + \beta_0(G)$, the value $ii(G)$ is not bounded above by this sum. The following construction gives an infinite family of graphs G for which $i(G)+\beta_0(G) < ii(G)$. Let G be the complete k-partite graph $K_{t,t,\ldots,t}$ with the edges of one complete graph K_k removed, where $t > k$. It follows that: $n = tk$, $i(G) = k < \beta_0(G) = t$, and $ii(G) = 2t$.

However, for trees, we can prove that the disjoint domination number equals the disjoint independent domination number.

Theorem 2.6. *For any tree T of order $n \geq 2$, $\gamma\gamma(T) = ii(T)$.*

Proof. Let U be the union of disjoint dominating sets D_1 and D_2, which form a $\gamma\gamma$-pair, in a tree T. Assume that wv is in $E(T)$ with $\{w, v\}$ a subset of U. Assume also that w and v are in the same dominating set, say D_1, which implies that $\deg(w) \geq 2$ and $deg(v) \geq 2$. Let T_w be the subtree of $T - wv$ containing w, and let T_v be the subtree of the other component of $T - wv$. Let $D_1^* = (D_1 \cap T_v) \cup (D_2 \cap T_w)$, and let $D_2^* = (D_2 \cap T_v) \cup (D_1 \cap T_w)$. Then (D_1^*, D_2^*) is also a $\gamma\gamma$-pair.

In T_v, every edge that was (D_1, D_1) is still (D_1, D_1). Every edge that was (D_1, D_2) is still (D_1, D_2), while in T_w, every (D_1, D_1) edge becomes (D_2, D_2), and every (D_1, D_2) edge becomes (D_2, D_1). Thus, by doing this switch, the total number of (D_i, D_i) edges is reduced by one. Continuing in this manner, we can find two disjoint dominating sets which are also independent, the sum of whose cardinalities is $\gamma\gamma(T)$. □

Since $\gamma\gamma(T) = ii(T)$, it follows that $ii(T) \leq i(T) + \beta_0(T)$.

Corollary 2.7. *For any tree T of order $n \geq 2$,*

$$\gamma\gamma(T) = \gamma i(T) = ii(T) \leq i(T) + \beta_0(T).$$

As in the previous theorem, another equality exists for the disjoint domination number of a tree. A *double dominating set* is a set S such that for every vertex $v \in V$, $|N[v] \cap S| \geq 2$. The *double domination number* $dd(G)$ of a graph G, as introduced by Harary and Haynes [9], is the minimum cardinality of a double dominating set in G. Clearly, the union of two disjoint dominating sets is a double dominating set, so that $dd(G) \leq \gamma\gamma(G)$ for any graph G. Note that every leaf and every stem must be an element of every double dominating set.

Theorem 2.8. *For any tree T with $n \geq 2$, $dd(T) = \gamma\gamma(T)$.*

Proof. Let U be a $dd(T)$-set. Root T at an arbitrary vertex $v \in U$, and color v red. On each path from v to a leaf, alternately color the vertices that are in U red and blue, starting with v colored red. Note that every vertex in U will then be adjacent to a vertex of the other color. Assume that $u \notin U$ and that $U \cap N(u)$ is monochromatic, that is, assume that every colored vertex adjacent to u is red. Since U is a $dd(T)$-set, if the parent of u is colored, then one child of u will have the other color. If the parent of u is not colored, then at least two children of u are colored. Let $x \in U$ be a child of u, and reverse the colors of the vertices in $U \cap T_x$.

Now iterate this process for each vertex with a monochromatic neighborhood. The result is a red/blue coloring of U for which every closed neighborhood $N[w]$ contains two colors. Let D_1 and D_2, respectively, be the sets of red and blue vertices produced by this coloring process. Then (D_1, D_2) is a $\gamma\gamma$-pair, and hence, $\gamma\gamma(T) \leq dd(T)$, that is, $\gamma\gamma(T) = dd(T)$. □

Theorems 2.4 and 2.6 show that for any tree T we have $dd(T) = \gamma\gamma(T) = ii(T)$, but these equalities do not hold for all bipartite graphs [12]. For example, the complete bipartite graph $K_{2,r}$ with $r \geq 3$ has $dd(K_{2,r}) = 3$, $\gamma\gamma(K_{2,r}) = 4$ and $ii(K_{2,r}) = r + 2$.

Theorem 2.9. *Let G be a graph with domatic number $d(G)$. Then $\gamma\gamma(G) \leq 2n/d(G)$.*

Proof. Let $V_1, V_2, V_3, ..., V_d$ be a partition of G into $d(G)$ dominating sets. Without loss of generality, assume that $|V_1| \leq |V_2| \leq |V_3| \leq ... \leq |V_d|$. Then $|V_1| + |V_2| \leq 2n/d(G)$. □

Theorem 2.10. *Let G and $\overline{G}$ have no isolates. If the diameter of G is at least four, then $\gamma\gamma(G) + \gamma\gamma(\overline{G}) \leq n + 4$.*

Proof. Let G be as in the hypothesis of the theorem. Let x, y be such that $d(x, y)$ = diam(G). Let P be a shortest path from x to y, let a be the neighbor of x on P, and let b be the neighbor of y on P. Note that since diam(G) ≥ 4, we have $ab \notin E(G)$.

Certainly $\gamma\gamma(G) \leq n$. The two disjoint sets $\{x, b\}$, $\{y, a\}$ are each dominating sets of $\overline{G}$. Thus, $\gamma\gamma(\overline{G}) \leq 4$. □

3 Complexity Results

In this section we consider the algorithmic complexity of various decision problems related to disjoint domination. Consider, for example, the following decision problems:

DOMINATING SET
INSTANCE: Graph $G = (V, E)$, positive integer $k \leq |V|$.
QUESTION: Does G have a dominating set of cardinality at most k?

DISJOINT MINIMUM DOMINATING SETS
INSTANCE: Graph $G = (V, E)$
QUESTION: Does G have two disjoint minimum dominating sets, that is, is $M_\gamma(G) = 0$?

DISJOINT DOMINATING SETS
INSTANCE: Graph $G = (V, E)$, positive integer $k \leq |V|$.
QUESTION: Does G have two disjoint dominating sets, whose union has cardinality at most k?

DISJOINT INDEPENDENT DOMINATING SETS
INSTANCE: Graph $G = (V, E)$
QUESTION: Does G have two disjoint independent dominating sets?

Theorem 3.1 ([6]). *DOMINATING SET is NP-complete.*

Theorem 3.2 ([7]). *DISJOINT MINIMUM DOMINATING SETS is NP-hard, even when restricted to bipartite graphs.*

Theorem 3.3. *DISJOINT DOMINATING SETS is NP-complete, even when restricted to chordal graphs.*

Proof. We will use a transformation from DOMINATING SET. Let G be any graph. Form the graph $G^* = G + K_1$, by adding a new vertex x and making it adjacent to every vertex in G. Note that this construction can be done in polynomial time.

Claim. G has a dominating set of cardinality at most k if and only if G^* has disjoint domination number at most $j = k + 1$.

Proof of Claim. If G has a dominating set S with $|S| \leq k$, S is a dominating set of G^*. The set $\{x\}$ is also a dominating set for G^*, so that $\gamma\gamma(G^*) \leq k + 1$. Now suppose that G^* has disjoint domination number at most $j = k + 1$. Let S_1 and S_2 be disjoint minimal dominating sets for G^* with $|S_1| + |S_2| \leq k + 1$. If $x \in S_1$, by minimality, $S_1 = \{x\}$. Then the set S_2 gives a dominating set for G of cardinality at most k. So suppose $x \notin S_1 \cup S_2$. Each of S_1 and S_2 dominates G^*, and $S_i \cap G$ dominates G for $i = 1, 2$. Now at least one of $|S_1|$ and $|S_2|$ is at most k, which gives us a dominating set of G of cardinality at most k. This completes the proof of the Claim.

Since the join operation preserves chordality, if G is chordal, then so is $G + K_1$. Therefore, DISJOINT DOMINATING SETS is NP-complete even when restricted to chordal graphs, since DOMINATING SET is NP-complete when restricted to chordal graphs (cf. Booth and Johnson [2]). □

If a graph G has no isolated vertices, then the complement of every minimal dominating set contains a dominating set. Thus, it is easy to find two disjoint dominating sets in every graph without isolated vertices. It is also easy to find two disjoint dominating sets in such a graph, where one of the two dominating sets is independent. However, recall that not every graph contains two disjoint independent dominating sets. In fact, the problem of deciding if a graph G has two disjoint independent dominating sets is NP-complete.

Theorem 3.4. *DISJOINT INDEPENDENT DOMINATING SETS is NP-complete.*

Proof (outline). The proof is by a transformation from 3-SAT. Given an instance, $U = \{u_1, u_2, ..., u_n\}$ and $C = \{c_1, c_2, ..., c_m\}$ of 3-SAT, construct the following graph G. For each literal u_i construct a K_3 on vertices $u_i, \overline{u_i}$ and u_i^* (in the graph G vertex u_i^* will have degree two). For each clause c_j add a clause vertex c_j of degree three, where vertex c_j is adjacent to the literal vertices that correspond to the three literals it contains. For example, if $c_i = (u_4 \vee \overline{u_7} \vee \overline{u_9})$, then connect vertex c_i to vertices u_4, $\overline{u_7}$ and $\overline{u_9}$. Then add two vertices x and y. Vertex x has degree one and is adjacent to y. Vertex y has degree $2n + 1$ with $N(y) = \{x, u_1, u_2, \ldots, u_n, \overline{u_1}, \overline{u_2}, \ldots, \overline{u_n}\}$.

If G has disjoint independent dominating sets A and B, then because $deg(x) = 1$, we can assume that $x \in A$ and $y \in B$. It follows that we must have $B = \{y, c_1, c_2, \ldots, c_m, u_1^*, u_2^*, \ldots, u_n^*\}$, and the set $A \cap \{u_1, u_2, \ldots, u_n, \overline{u_1}, \overline{u_2}, \ldots, \overline{u_n}\}$ defines a satisfying truth assignment for C.

Conversely, if C has a satisfying truth assignment, then taking the vertices of G defined by the truth assignment, together with vertex x gives an independent dominating set for G, and the set $\{y, c_1, c_2, \ldots, c_m, u_1^*, u_2^*, \ldots, u_n^*\}$ is also an independent dominating set for G. □

In the remainder of this section we present a linear time algorithm for computing the value of $\gamma\gamma(T)$ for any tree T. We use the dynamic programming style first developed by Wimer [15].

We make use of the fact that the class of (rooted) trees can be constructed recursively from copies of the single vertex K_1, using the rule of composition which combines two rooted trees (T_1, r_1) and (T_2, r_2) by adding an edge between r_1 and r_2, and calling r_1 the root of the resulting larger tree. We denote this as follows: $(T, r_1) = (T_1, r_1) \circ (T_2, r_2)$.

	[1]	[2]	[3]	[4]	[5]	[6]	[7]	[8]
	$A(B)$	$A()$	$B(A)$	$B()$	$C(A)$	$C(B)$	$C(AB)$	$C()$
[1] $A(B)$	$A(B)$	$A()$	$B(A)$	$B(A)$	$C(A)$	$C(AB)$	$C(AB)$	$C(A)$
[2] $A()$	X	X	$B(A)$	$B(A)$	X	X	X	X
[3] $B(A)$	$A(B)$	$A(B)$	$B(A)$	$B()$	$C(AB)$	$C(B)$	$C(AB)$	$C(B)$
[4] $B()$	$A(B)$	$A(B)$	X	X	X	X	X	X
[5] $C(A)$	X	X	$B(A)$	$B()$	X	X	X	X
[6] $C(B)$	$A(B)$	$A()$	X	X	X	X	X	X
[7] $C(AB)$	$A(B)$	$A()$	$B(A)$	$B()$	$C(A)$	$C(B)$	$C(AB)$	$C()$
[8] $C()$	X	X	X	X	X	X	X	X

Figure 3.1: Possible tree compositions

In particular, if (A, B) is a $\gamma\gamma$-pair of T, then (A, B) splits into two pairs (A_1, B_1) and (A_2, B_2) according to this decomposition. We express this as follows: $(T, (A, B)) = (T_1, (A_1, B_1)) \circ (T_2, (A_2, B_2))$.

In order to construct an algorithm to compute $\gamma\gamma(T)$ for any tree T, we need to determine the classes of tree-subset pairs $(T, (A, B))$ which can occur for $\gamma\gamma$-sets of T. For this problem there are eight classes, where we assume that sets A and B are the disjoint dominating sets which form a $\gamma\gamma$-pair, and $C = V - (A \cup B)$ is the set of vertices which are not in either dominating set.

[1] :$A(B) = \{(T, (A, B)) | r \in A, (A, B)$ is a $\gamma\gamma$-pair of $T\}$.

[2] :$A() = \{(T, (A, B)) | r \in A$ but is not dominated by a vertex in $B\}$; (A, B) is a $\gamma\gamma$-pair if r is dominated by a vertex in $B\}$.

[3] :$B(A) = \{(T, (A, B)) | r \in B, (A, B)$ is a $\gamma\gamma$-pair of $T\}$.

[4] :$B() = \{(T, (A, B)) | r \in B$, but is not dominated by a vertex in $A\}$; (A, B) is a $\gamma\gamma$-pair if r is dominated by a vertex in $A\}$.

[5] :$C(A) = \{(T, (A, B)) | r \in C$ is dominated by a vertex in A but not by a vertex in $B\}$, (A, B) is a $\gamma\gamma$-pair of $T - r\}$.

[6] :$C(B) = \{(T, (A, B)) | r \in C$, is dominated by a vertex in B, but not by a vertex in A; (A, B) is a $\gamma\gamma$-pair of $T - r\}$.

[7] :$C(AB) = \{(T, (A, B)) | r \in C$, (A, B) is a $\gamma\gamma$-pair of $T\}$.

[8] :$C() = \{(T, (A, B)) | r \in C$, is not dominated by a vertex in A or in B; (A, B) is a $\gamma\gamma$-pair of $T - r\}$.

Next we must consider the effect of composing a tree T_1 having a pair (A_1, B_1) which is of class $[i]$ (column headings of Figure 1) with a tree T_2 having a pair (A_2, B_2) which is of class $[j]$ (row headings of Figure 1), for every possible combination of classes. That is, we must describe the appropriate class of the combined pairs $(A_1 \cup A_2, B_1 \cup B_2)$ in the composed tree $T = T_1 \circ T_2$. This is described in Figure 3.1 . An 'X' in the table signifies that this composition cannot happen, that is, no $\gamma\gamma$-pair (A, B) of T can ever decompose into two pairs (A_1, B_1) and (A_2, B_2) of the types indicated.

From Figure 3.1, we can now write out a system of recurrence relations, as follows.

$[1] = [1] \circ [1] \cup [3] \circ [1] \cup [4] \circ [1] \cup [6] \circ [1] \cup [7] \circ [1] \cup [3] \circ [2] \cup [4] \circ [2]$

$[2] = [1] \circ [2] \cup [6] \circ [2] \cup [7] \circ [2]$
$[3] = [1] \circ [3] \cup [2] \circ [3] \cup [3] \circ [3] \cup [5] \circ [3] \cup [7] \circ [3] \cup [1] \circ [4] \cup [2] \circ [4]$
$[4] = [3] \circ [4] \cup [5] \circ [4] \cup [7] \circ [4]$
$[5] = [1] \circ [5] \cup [7] \circ [5] \cup [1] \circ [8]$
$[6] = [3] \circ [6] \cup [7] \circ [6] \cup [3] \circ [8]$
$[7] = [3] \circ [5] \cup [1] \circ [6] \cup [1] \circ [7] \cup [3] \circ [7] \cup [7] \circ [7]$
$[8] = [7] \circ [8]$

To illustrate this, a tree-subset pair $(T, (A, B))$ which is of class $C(B)$ can be obtained only by composing a pair $(T_1, (A_1, B_1))$ of class $C(B)$ (column [6]) with a pair $(T_2, (A_2, B_2))$ of class $B(A)$ (row [3]) or by composing a class $C(B)$ (column [6]) with a class $C(AB)$ (row [7]), or a class $C()$ (column [8]) with a class $B(A)$ (row [3]).

To prove the correctness of this dynamic programming algorithm for computing $\gamma\gamma(T)$ for any tree T, we would have to prove a theorem asserting that each of these recurrences is correct. Space limitations prevent us from doing this here, but it is easy to do. It is even easier to verify the correctness of Figure 3.1, which can be done by inspection. The final step in specifying a $\gamma\gamma(T)$-algorithm is to define the initial vector. In this case, for trees, the only basis graph is the tree with single vertex K_1. We need to know the minimum cardinality of a pair (A, B) in a class of each of the eight types in the graph K_1, if any exists. It is easy to see that the initial vector is $[X, 1, X, 1, X, X, X, X]$, where X means *undefined*, for all leaves of T, and $[X, 1, X, 1, X, X, X, 0]$ for all non-leaves of T.

We now have all the ingredients for a $\gamma\gamma$-algorithm, where the input is the parent array $parent[1 \ldots n]$ for the input tree, and where the output is the 8-tuple corresponding to the root (that is, vertex 1) of T, which is computed by repeatedly applying the recurrence system to each vertex in the parent array, with the indicated initial vector being associated with every vertex in the parent array as the computation begins.

The basic structure for the algorithm is a simple iteration.

```
    procedure γγ;

for i:=1 to n do
    initialize vector [i,1], ..., [i,8];

for j:=n downto 2 do
begin
k:=parent[j];
vector[k,1]:= min {vector[j,1]+vector[k,1], vector[j,3]+vector[k,1],
                  vector[j,3]+vector[k,2], vector[j,4]+vector[k,1]
                  vector[j,4]+vector[k,2], vector[j,6]+vector[k,1]
                  vector[j,7]+vector[k,1]};
vector[k,2]:= min {vector[j,1]+vector[k,2], vector[j,6]+vectorjk,1],
                  vector[j,7]+vector[k,1]};
vector[k,3]:= min {vector[j,1]+vector[k,3], vector[j,2]+vector[k,3],
                  vector[j,3]+vector[k,3], vector[j,5]+vector[k,3]
                  vector[j,7]+vector[k,3], vector[j,1]+vector[k,4]
                  vector[j,2]+vector[k,4]};
```

```
vector[k,4]:= min {vector[j,3]+vector[k,4], vector[j,5]+vector[k,4],
                   vector[j,7]+vector[k,4]};
vector[k,5]:= min {vector[j,1]+vector[k,5], vector[j,7]+vector[k,5],
                   vector[j,1]+vector[k,8]};
vector[k,6]:= min {vector[j,3]+vector[k,6], vector[j,7]+vector[k,6],
                   vector[j,3]+vector[k,8]};
vector[k,7]:= min {vector[j,1]+vector[k,6], vector[j,1]+vector[k,7],
                   vector[j,3]+vector[k,5], vector[j,3]+vector[k,7],
                   vector[j,7]+vector[k,7]};
vector[k,8]:= vector[j,7]+vector[k,8]
end;
γγ(T) := min {vector[1,1], vector[1,3], vector[1,7]};
end; {γγ}
```

It is clear that procedure $\gamma\gamma$ has a linear execution time.

Note that the value recorded for a given class is the minimum possible sum obtained by adding the entries in two positions. For example, equation [6] can be re-written as:

$[6] = min\{[3] + [6], [7] + [6], [3] + [8]\}$,

where the left summand corresponds to the value of the current child vertex and the right summand corresponds to the current value of the parent vertex. □

4 Open Problems

Being new concepts, disjoint domination and disjoint independent domination are both invariants whose properties are relatively unknown. Many questions are suggested by this research, among them are the following:

1. Characterize the class of $\gamma\gamma$-minimum graphs.
2. Characterize the class of $\gamma\gamma$-minimum trees.
3. Conjecture. All Queens graphs Q_n are $\gamma\gamma$-minimum, for $n \geq 4$. Notice that the 3×3 Queens graph is not $\gamma\gamma$-minimum.
4. Determine which Kings graphs, which Knights graphs and which Bishops graphs are $\gamma\gamma$-minimum.
5. Conjecture. All n-cubes are $\gamma\gamma$-minimum, for $n \geq 2$.
6. Probabilistically, are almost no graphs (or trees) $\gamma\gamma$-minimum?
7. Conjecture. A tree T is $\gamma\gamma$-minimum if and only if T has no γ-required vertex.
8. For which class of trees T is $\gamma\gamma(T) = 2(n+1)/3$?
9. Prove or disprove the Kulli-Sigarkanti Conjecture.
10. Under what conditions does $ii(G)$ exist?
11. When is $\gamma\gamma(G) = \gamma i(G)$?
12. When is $\gamma i(G) = ii(G)$?

13. What is the complexity of the decision problem corresponding to $\gamma i(G)$?

14. When is any of $\{\gamma\gamma(G), \gamma i(G), ii(G)\} = i(G) + \beta_0(G)$?

15. When is $\gamma\gamma(G) + \gamma\gamma(\overline{G}) = n + 4$?

16. Is there a better bound for $\gamma\gamma(G) + \gamma\gamma(\overline{G})$ than $n + 4$? For example, if G is self-complementary, then $\gamma\gamma(G) + \gamma\gamma(\overline{G}) = 2\gamma\gamma(G)$.

17. Is DISJOINT DOMINATION NP-complete for bipartite graphs?

18. $\gamma\gamma(G) + \gamma\gamma(\overline{G}) \leq n + 4$ if $diam(G) \geq 4$. What if $diam(G) = 2$? or 3?

19. If T is $\gamma\gamma$-minimum, is $\gamma(T - v) = \gamma(T)$?

20. Which grid graphs are $\gamma\gamma$-minimum?

21. Does every tree of order $n \geq 2$ have a γ-set D, whose complement $V - D$ contains an independent dominating set of T? This is true for $n \leq 6$.

22. Among all trees of order n, for which ones is the ratio $\gamma\gamma(T)/n$ a minimum?

Finally, we observe that the study of the three new invariants $\gamma\gamma(G), \gamma i(G)$, and $ii(G)$ is just the beginning of the study of two disjoint sets of vertices in a graph having a given property, for example, of being (maximal) irredundant, (minimal) dominating, or (maximal) independent. Similar definitions can be given for the following invariants: $ir \cdot ir(G)$, $ir \cdot \gamma(G)$, $\beta \cdot \beta(G)$, $\beta \cdot \Gamma(G)$, $\Gamma \cdot \Gamma(G)$, $\Gamma \cdot IR(G)$, and $IR \cdot IR(G)$.

References

[1] D.W. Bange, A.E. Barkauskas and P.J. Slater, A constructive characterization of trees with two disjoint minimum dominating sets, *Congr. Numer.* **21** (1978), pp. 101–112.

[2] K.S. Booth and J.H. Johnson, Dominating sets in chordal graphs, *SIAM J. Comput.* **11** (1982), pp. 191–199.

[3] E.J. Cockayne and S.T. Hedetniemi, Disjoint independent dominating sets in graphs, *Discrete Math.* **15** (1976), pp. 213–222.

[4] G.S. Domke, J.E. Dunbar and L.R. Markus, The inverse domination number of a graph, *Ars Combin.* **72** (2004), pp. 149–160.

[5] J.E. Dunbar, S.M. Hedetniemi, S.T. Hedetniemi, D.P. Jacobs, J. Knisely, R.C. Laskar, and D.F. Rall, Fall colorings of graphs, *J. Combin. Math. Combin. Comput.* **33** (2000), pp. 257–273.

[6] M.R. Garey and D.S. Johnson, *Computers and Intractability: A Guide to the Theory of NP-Completeness*, Freeman, New York, (1979).

[7] D.L. Grinstead and P.J. Slater, On mimimum dominating sets with minimum intersection, *Discrete Math.* **86** (1990), pp. 239–254.

[8] D.L. Grinstead and P.J. Slater, On the minimum intersection of minimum dominating sets in series-parallel graphs. In Y. Alavi, G. Chartrand, O.R. Oellermann, and A.J. Schwenk, Eds., *Graph Theory, Combinatorics and Applications VI, Proc. Sixth Internat. Conf. on the Theory and Applications of Graphs*, Vol. 1, pp. 563–584. (Kalamazoo, MI, 1988), 1991, Wiley.

[9] F. Harary and T.W. Haynes, Double domination in graphs, *Ars Combin.* **55** (2000), pp. 201–213.

[10] T.W. Haynes, S.T. Hedetniemi, and P.J. Slater, *Fundamentals of Domination in Graphs*, Marcel Dekker, New York, (1998).

[11] T.W. Haynes and M.A. Henning, Trees with two minimum independent dominating sets, *Discrete Math.* **304** (2005), pp. 69–78.

[12] J. Knisely, personal communication, (2000).

[13] V.R. Kulli and S.C. Sigarkanti, Inverse domination in graphs. *Nat. Acad. Sci. Lett.* **14** (1991), pp. 473–475.

[14] O. Ore, *Theory of Graphs.* Amer. Math. Soc. Colloq. Publ. **38**, (1962), (Amer. Math. Soc., Providence, RI).

[15] T.V. Wimer, *Linear Algorithms on k-Terminal Graphs.* Ph.D. thesis, Clemson University, (1987).

[16] B. Zelinka, Domatic numbers of graphs and their variants: a survey. In *Domination in Graphs, Advanced Topics*, T.W. Haynes, P.J. Slater and S.T. Hedetniemi, Editors, Marcel Dekker, (1998).

Proc. of ICDM 2006, Pages 101–108

On the Complexity of the Geodetic and Convexity Numbers of a Graph

MITRE C. DOURADO
Instituto de Ciências Exatas (Universidade Federal Rural do Rio de Janeiro) and Núcleo de Computação Eletrônica (Universidade Federal do Rio de Janeiro), Brazil.
Email: `mitre@nce.ufrj.br`
FÁBIO PROTTI*
Instituto de Matemática and Núcleo de Computação Eletrônica (Universidade Federal do Rio de Janeiro), Brazil.
Email: `fabiop@nce.ufrj.br`
JAYME L. SZWARCFITER†
Instituto de Matemática, Núcleo de Computação Eletrônica, and COPPE-Sistemas (Universidade Federal do Rio de Janeiro), Brazil.
Email: `jayme@nce.ufrj.br`
Caixa Postal 2324, CEP 20010-974, Rio de Janeiro, RJ, Brazil

Abstract. Let G be a graph. If $u, v \in V(G)$, a $u - v$ *geodesic* of G is a shortest path linking u and v. The *closed interval* $I[u, v]$ consists of all vertices lying in some $u - v$ geodesic of G. For $S \subseteq V(G)$, the set $I[S]$ is the union of all sets $I[u, v]$ for $u, v \in S$. We say that a nonempty subset S is a *convex set* if $I[S] = S$. The maximum cardinality of a convex set, properly contained in $V(G)$, is the *convexity number* of G, $c(G)$. A set S is *geodetic* if $I[S] = V(G)$. The minimum cardinality of a geodetic set of G is the *geodetic number* of G, denoted by $g(G)$. In this work, we prove that the problems of determining whether $g(G) \leq k$ and whether $c(G) \geq k$ are NP-complete, for bipartite graphs. In addition, deciding whether $g(G) \leq k$ is also NP-complete, for chordal graphs. We show how to determine the geodetic and convexity numbers of a cograph in linear time.

1 Introduction

Let G be a graph with vertex set $V(G)$ and edge set $E(G)$. By $N(v)$ denote the set of neighbors of $v \in V(G)$, and $N[v] = N(v) \cup \{v\}$. We say that u and v are *twins* if $N(u) = N(v)$. A subset of vertices is *complete* (*independent*) if every pair of its vertices are adjacent (nonadjacent). If $u, v \in V(G)$, a $u - v$ *geodesic* of G is a minimum length path linking u and v. The *closed interval* $I[u, v]$ consists of all vertices lying in some $u - v$ geodesic of G. For $S \subseteq V(G)$, the set $I[S]$ is the union of all sets $I[u, v]$ for $u, v \in S$. We say that S is a *convex set* if $I[S] = S$. The *convex hull* of S, denoted $I_h[S]$, is the smallest convex set containing S. Clearly, $I[S] \subseteq I_h[S]$. A set S is a *geodetic set* if $I[S] = V(G)$. The minimum cardinality of a geodetic set of G is the *geodetic number* of G, denoted by $g(G)$; and the maximum cardinality of a convex set of G, properly contained in $V(G)$, is the *convexity number*

*Partially supported by the Conselho Nacional de Desenvolvimento Científico e Tecnológico - CNPq, and Fundação de Amparo à Pesquisa do Estado do Rio de Janeiro - FAPERJ, Brazil.
†Corresponding author

of G, denoted by $c(G)$. It is clear that $g(G) \leq c(G) + 1$, for any graph G. We say that a vertex v is a *simplicial* vertex of G if $N(v)$ is a complete set. It follows that every geodetic set of a graph G contains all simplicial vertices of G.

First, we consider the geodetic number. We describe a simple bound for $g(G)$, relating it to the cardinality of maximal matchings. Then we prove that deciding whether the geodetic number of a graph is at most k is NP-complete, both for bipartite graphs and chordal graphs (graphs containing no induced cycle with four or more vertices). We mention that the NP-completeness of the geodetic number problem has been established in [1], for general graphs. Next, we consider the convexity number. First, we describe a simple method for deciding whether $c(G) = k$, which terminates in polynomial time, whenever k is fixed. Next, we prove that deciding whether $c(G)$ is at least k is NP-complete for bipartite graphs. The NP-completeness of this problem for general graphs has been proved in [14]. Finally, we consider the class of cographs (graphs containing no induced paths of four vertices) and describe polynomial time algorithms to compute the geodetic and convexity numbers for this class of graphs.

For some related works on the convexity number, see [4, 5, 8, 15]. The papers [3, 6, 7] deal with geodetic number. Other related papers are [11–13]

2 Geodetic Number

In this section, we consider the determination of the geodetic number of a graph G. First, we mention a simple bound for it and then prove NP-completeness results, for the classes of bipartite graphs and chordal graphs.

Theorem 2.1. *Let G be a triangle-free graph with no vertices of degree one. Let M be an arbitrary maximal matching of G. Then $g(G) \leq 2|M|$.*

Proof. Let M be a maximal matching of G. If M saturates all vertices of G, there is nothing to prove. Otherwise, let $v \in V(G) \setminus M$ be a non-saturated vertex. Since G does not contain vertices of degree one, v has at least two neighbors w, z. We know that w, z are saturated vertices, otherwise M would not be maximal. We also know that the size of a $w - z$ geodesic is at least two, because G is triangle-free. Then v belongs to a $w - z$ geodesic and therefore the subset of vertices incident to the edges of M is a geodetic set. That is, $g(G) \leq 2|M|$. □

Below, we prove that the geodetic number problem is NP-complete for bipartite graphs. We employ the NP-completeness of the dominating set problem for bipartite graphs [10, 18].

Theorem 2.2. *Given a bipartite graph G and an integer k it is NP-complete to decide whether the geodetic number of G is less than or equal to k.*

Proof. This problem is in NP because the closed interval of a set can be easily done in polynomial time [11]. The hardness proof is based on a reduction from the dominating set problem for bipartite graphs. Given a bipartite graph G with bipartition (V_1, V_2) and an integer k, construct G' by adding the vertices a_1, a_2, b_1, b_2, where a_1 is adjacent to every vertex of V_2 and a_2 is adjacent to every vertex of V_1. Finally add the edges $\{b_1, a_2\}$, $\{a_1, b_2\}$, and $\{a_1, a_2\}$. Note that G' is a bipartite graph, see Figure 2.1.

Define $k' = k + 2$. We show that there is a dominating set with at most k vertices in G if and only if the geodetic number of G' is less than or equal to k'.

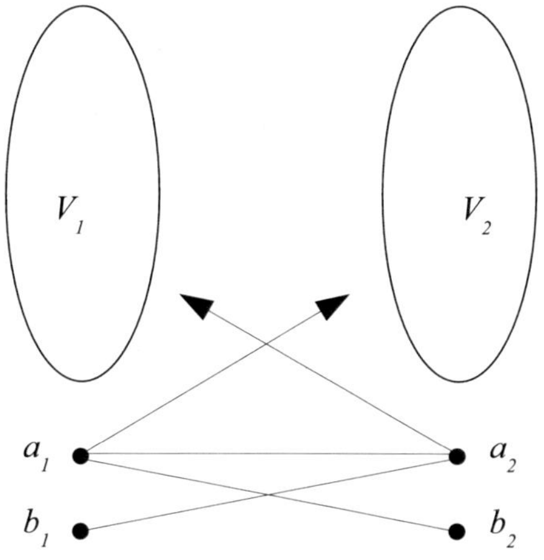

Figure 2.1: Graph G'.

Let D be a dominating set of G with k or less vertices. Consider the set $D' = D \cup \{b_1, b_2\}$. We need to show that $I[D'] = V(G')$. In order to do this, we need to show that every vertex of $V(G') \setminus D'$ is in a shortest path between two vertices of D'.

Consider a vertex $v_i \in D$. The vertices a_1 and a_2 are in $I[v_i, b_2]$ and $I[v_i, b_1]$, respectively. For every $v_j \in V(G) \setminus D$, there is a vertex $v_i \in D$ adjacent to v_j such that v_j is in $I[v_i, b_2]$ or $I[v_i, b_1]$. Then D' is a geodetic set of G'.

Conversely, let D' be a minimal geodetic set of G' with $k + 2$ or less vertices. Clearly, $b_1, b_2 \in D'$. Since $I[a_2, v] \subsetneq I[b_1, v]$ and $I[a_1, v] \subsetneq I[b_2, v]$, for any v, we conclude that $a_1, a_2 \notin D'$. Define $D = D' \cap V(G)$. We prove that D is a dominating set of G.

Let $v \in V(G) \setminus D$. Since D' is a geodetic set of G', v is in the closed interval of two vertices of D', say u and w. Observe that v is adjacent to u or w or both, because the diameter of G' is three. Since v is neither adjacent to b_1 nor b_2, v is adjacent to at least one vertex of D. Hence D is a dominating set of G. □

Next, we prove the NP-completeness of the geodetic number problem for chordal graphs. Again, we employ the dominating set problem, which is NP-complete for chordal graphs [2].

Theorem 2.3. *Given a chordal graph G and an integer k it is NP-complete to decide whether the geodetic number of G is less than or equal to k.*

Proof. This problem is in NP because the closed interval of a set can be easily done in polynomial time [11]. The hardness proof is based on a reduction from dominating set problem for chordal graphs. Given a chordal graph G and an integer k, construct G' by adding for each vertex $v_i \in V(G)$ the vertices u_i and w_i and the edges $\{v_i, u_i\}$ and $\{u_i, w_i\}$. Finally add the vertex z and the edges $\{u_i, z\}$ and $\{v_i, z\}$ for $1 \leq i \leq n$, and set $k' = k + n$. Note that G' is chordal.

We show that there is a dominating set with at most k vertices in G if and only if the geodetic number of G' is less than or equal to k'.

Let D be a dominating set of G with k or fewer vertices. Consider the set $D' = D \cup \{w_i : 1 \leq i \leq n\}$. We need to show that $I[D'] = V(G')$. In order to do this, we need to show that every vertex is in a shortest path between two vertices of D'. First, note that $\{z, u_1, \ldots, u_n\} \subset I[\{w_1, \ldots, w_n\}]$. Hence consider $v_j \in V(G) \setminus D$. Since D is a dominating set of G, there is a vertex $v \in D$ adjacent to v_j. It is easy to see that $v_j \in I[v, w_j]$. Therefore $I[D'] = V(G')$.

Conversely, let D' be a geodetic set of G' with at most k' vertices. Clearly, $w_i \in D'$, for $1 \leq i \leq n$. Hence at most k vertices are in $V(G)$. Suppose that $D = D' \cap V(G)$ is not a dominating set of G. Let v_j be a vertex of $V(G)$ not dominated by D. Note that the shortest path between $v_i \in D$ and w_j is v_i, z, u_j, w_j; and any path from v_i to w_j passing through v_j has at least four edges. Analogously for w_t, $t \neq j$. Further, since the minimum path between two vertices w_j and w_t does not contain a vertex of G and a shortest path between two vertices of D has length at most two, because both are adjacent to z, we conclude that $v_j \notin I[D']$, a contradiction. □

3 Convexity Number

We consider the determination of the convexity number $c(G)$ of a graph G. The following observation leads to a method for finding $c(G)$ for a non-complete graph G.

Lemma 3.1. *Let G be a graph and $k \geq 1$ an integer. Then $c(G) < k$ if and only if $I_h[S] = V(G)$ for every $S \subsetneq V(G)$ with k vertices.*

Proof. If there is a set S such that $|S| = k$ and $I_h[S] \neq V(G)$, since $I_h[S]$ is convex it follows that G contains a proper convex subset of size at least k, a contradiction.

Conversely, if $I_h[S] = V(G)$ for every $S \subseteq V(G)$ with k vertices, it means that there is no proper convex subset of G with k or more vertices, which implies that $c(G) < k$. □

Based on the above lemma we describe a method for verifying if $c(G) \geq k$, for fixed k.

Given an integer k, first check if $I_h[S] = V(G)$, for all subsets S with k vertices. If so, report that $c(G) < k$ and stop. Otherwise, test whether $c(G) < k + 1$. If so then report $c(G) = k$, otherwise report $c(G) > k$, and stop.

Next, we prove that the convexity number problem is NP-complete, restricted to bipartite graphs. We use the NP-completeness of the maximum clique problem [9].

Theorem 3.2. *Given a connected bipartite graph G and an integer $k \geq 3$ it is NP-complete to decide whether the convexity number of G is greater than or equal to k.*

Proof. This problem is in NP because the convex hull of a set can be easily done in polynomial time [11]. The hardness proof is a reduction from the maximum clique problem. Given a graph H and an integer $k \geq 3$, construct a graph G as follows: for every vertex u of H create the vertices w_u, x_u, y_u, z_u in G and the edges $\{w_u, z_u\}, \{x_u, z_u\}, \{y_u, z_u\}$.

For every edge $\{u, v\} \in E(H)$ create $n + 5$ vertices, where $n = |V(H)|$, and the necessary edges to form the subgraph $G_{u,v}$ of G. See Figure 3.1(a). Define the *gadget* graph associated to the edge $\{u, v\}$ as the subgraph of $G_{u,v}$ depicted in Figure 3.1(b).

Finally, add two vertices x and y to G and add the edges $\{x, x_u\}$ and $\{y, y_u\}$, for $u \in V(H)$. Clearly G is bipartite. An example of the entire construction is shown in Figure 3, where $H = P_3$ and $V(H) = \{t, u, v\}$.

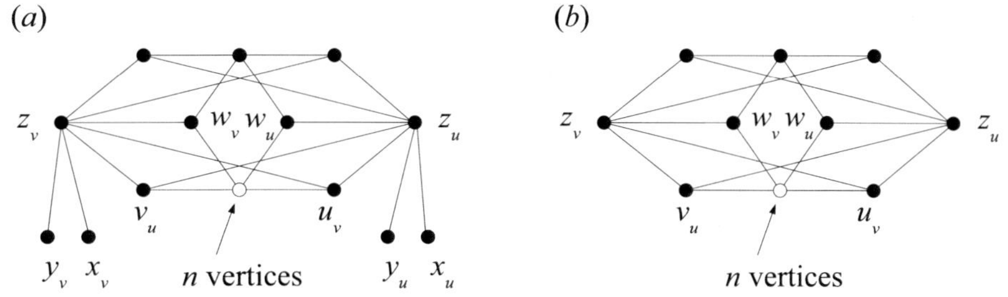

Figure 3.1: (a) The white vertex represents n twin vertices forming an independent set, where each of its vertices has the same neighborhood, consisting of four vertices. (b) Gadget graph associated to edge $\{u, v\}$.

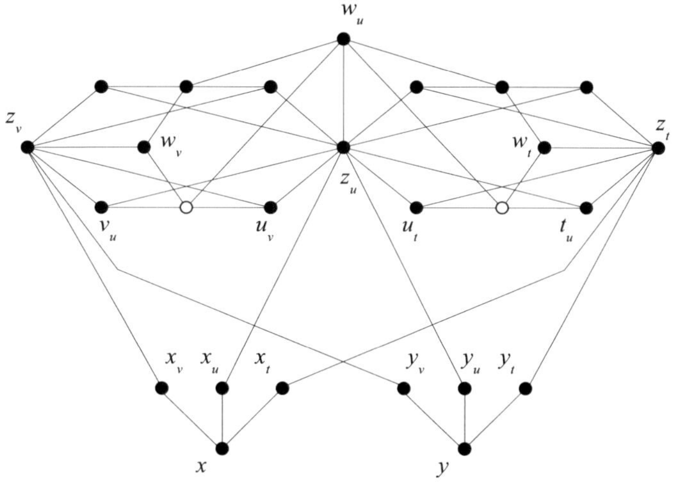

Figure 3.2: The graph G constructed from $H = P_3$.

It is clear that for every edge of H we have a convex set in G with $n + 9$ vertices (the gadget subgraph associated to that edge). We show that H contains a clique with at least k vertices if and only if G contains a proper convex set with at least $3k + (n + 5)\binom{k}{2} + 1$ vertices.

The following claim is straightforward.

Claim I. Let G' be a gadget subgraph, and let $T \subset V(G')$ such that $|T| = 3$ and T contains at most one of the n twin vertices of G'. Then $I_h[T] = V(G')$.

This implies that if a convex set S of G contains a set T as in Claim I, then $V(G') \subseteq S$.

First consider that H contains a clique C with at least k vertices. Construct S as follows: add to S the vertices of each gadget subgraph associated to an edge joining two vertices of C. This implies $|S| \geq 2k + (n + 5)\binom{k}{2}$ vertices. Finally, add to S the vertex x_u for every $u \in C$, and add the vertex x. Observe that S is a proper convex set with at least $3k + (n + 5)\binom{k}{2} + 1$ vertices.

Conversely, consider the partition of $V(G)$ into the sets V_x, V_y, V_z, V_w, V', where $V_x = N[x]$, $V_y = N[y]$, $V_z = \{z_u : u \in V(H)\}$, $V_w = \{w_u : u \in V(H)\}$, and finally $V' = V(G) \setminus \bigcup_{t\in\{x,y,z,w\}} V_t$. Let $S \subsetneq V(G)$ be a convex set with at least $3k + (n+5)\binom{k}{2} + 1$ vertices. Suppose that S contains at most one vertex z_u of V_z. Hence S contains at most one other vertex of each gadget subgraph containing z_u, because if it contains other two for some gadget, by Claim I, S would contain another vertex of V_z. This implies $|S \cap (V_w \cup V_z \cup V')| \leq n$. For the vertices of $V_x \cup V_y$, S contains at most $n+2$ of such vertices, because $I_k[x,y] = V(G)$. Therefore S would contain at most $2n+2$ vertices, a contradiction, since $k \geq 3$. Thus S contains at least two vertices of V_z.

Now suppose that S contains two vertices z_u, z_v of V_z such that $\{u,v\} \notin E(H)$. Since $d_G(z_u, z_v) = 4$, we must have $x, y \in S$. Hence $S = V(G)$, because $I_h[x,y] = V(G)$, which is a contradiction. Then the vertices $u \in V(H)$ such that $z_u \in V_z \cap S$ induce a complete set C in H.

Now, it is sufficient to prove that $|C| \geq k$. Suppose by contradiction that $|C| = t < k$. The gadget subgraphs of the edges joining vertices in C correspond to $2t + (n+5)\binom{t}{2}$ vertices. Denote this subset of vertices by S'.

First we show that there is no other vertex $a \in (V_w \cup V') \setminus S'$ belonging to S. Suppose the contrary. We have three cases.

Case (a). $a \in V_w \setminus S'$. Then $a = w_u$ for some vertex $u \in V(H)$ and $z_u \notin S$. But the closed interval of w_u and any vertex of V_z contains z_u, implying that $z_u \in S$, a contradiction.

Case (b). $a \in V' \setminus S'$ and a belongs to some gadget subgraph G' of an edge $\{u,v\}$ where $v \in C$. Since $a \notin S'$, $z_u \notin S$. But Claim I guarantees that $V(G') \subseteq I_h[a, w_v, z_v]$, which means that $z_u \in S$, a contradiction.

Case (c). $a \in V' \setminus S'$ and a belongs to some gadget subgraph associated to an edge $\{u,u'\}$ where $u,u' \notin C$. It is easy to see that the closed interval of a and any vertex $z_v \in V_z \cap S$ contains z_u or $z_{u'}$, a contradiction.

It remains to check how many vertices of $N[x] \cup N[y]$ belong to S. Observe that if two vertices of $N(x)$ and two of $N(y)$ belong to S, then $S = V(G)$. Therefore, for one of these sets, at most one vertex belongs to S. Thus $|S| \leq 2t + (n+5)\binom{t}{2} + n + 2 \leq 3k + (n+5)\binom{k}{2} + 1$, a contradiction. □

4 Cographs

In this section, we describe polynomial time algorithms for determining the geodetic number and the convexity number of a cograph. In each case, we construct a minimum geodetic set and a maximum proper convex subset, respectively.

Theorem 4.1. *Let G be a connected cograph, and denote by $G_1, \ldots, G_k, G_{k+1}, \ldots, G_t$ the subgraphs of G induced by the vertex sets of the connected components of $\overline{G}$, where $|V(G_i)| \geq 2$ if and only if $i \leq k$. Then:*

$$g(G) = \begin{cases} n, \text{ if } k = 0; \\ g(G_1), \text{ if } k = 1; \\ \min\{4, \min_{1\leq i\leq k}\{g(G_i)\}\}, \text{ if } k \geq 2. \end{cases}$$

Proof. If $k = 0$, then G is a complete graph. If $k = 1$, since G_1 is not a complete graph, a minimum geodetic set of it is not a complete set, and therefore, the closed interval of such set contains all the trivial connected components of $\overline{G}$.

If $k \geq 2$, then there are vertices $v, v' \in V(G_1)$ and $u, u' \in V(G_2)$, such that $\{v, v'\}, \{u, u'\} \notin E(G)$. Clearly $\{v, v', u, u'\}$ is a geodetic set of G. Then $g(G) \leq 4$.

If $g(G_i) = 2$, for some i, since a geodetic set of G_i is also a geodetic set of G, we have $g(G) = 2$. If $g(G) = 2$, any two vertices forming a minimum geodetic set of G are nonadjacent, then they belong to some G_i and $g(G_i) = 2$.

If $\min_{1 \leq i \leq k} \{g(G_i)\} = 3$, by the previous case $g(G) = 3$. If $g(G) = 3$, suppose that there are two vertices $u, v \in V(G_i)$ and one vertex $w \in V(G_j)$, forming a minimum geodetic set such that $i \neq j$. Since $w \in I[u, v]$, $I[u, w] = \{u, w\}$, and $I[v, w] = \{v, w\}$, we have that $\{u, v\}$ is a geodetic set, a contradiction. □

Proposition 4.2. *If G is a disconnected cograph, then $g(G)$ equals the sum of the geodetic numbers of its connected components.*

Proof. Trivial. □

Based on Theorem 4.1 and Proposition 4.2 one can compute the geodetic number of a cograph in linear time, employing modular decomposition [16, 17].

Next, we consider the convexity number. Denote by $\omega(G)$ the clique number of G.

Theorem 4.3. *Let G be a connected cograph, and denote by $G_1, \ldots, G_k$, G_{k+1}, $\ldots, G_t$ the subgraphs of G induced by the vertex sets of the connected components of $\overline{G}$, where $|V(G_i)| \geq 2$ if and only if $i \leq k$. Then:*

$$c(G) = \begin{cases} n - 1, \text{ if } k = 0\,; \\ c(G_1) + t - 1, \text{ if } k = 1\,; \\ \omega(G), \text{ if } k \geq 2\,. \end{cases}$$

Proof. If $k = 0$, then G is a complete graph.

If $k = 1$, note that a vertex is universal if and only if it is a trivial connected component of $\overline{G}$. Since all universal vertices belong to any proper convex set containing two nonadjacent vertices, we have $c(G) = c(G_1) + t - 1$.

If $k \geq 2$, let S be a proper convex set of G. If S contains two nonadjacent vertices, such vertices belong to some G_i, and $I[S]$ contains all the vertices of the other connected components of $\overline{G}$. Since $k \geq 2$ and $S = I[S]$, S contains another pair of nonadjacent vertices, which implies that $S = V(G)$, a contradiction. Then every proper convex set of G is a complete set. □

Proposition 4.4. *Let G be a disconnected cograph. Then $c(G) = n - x$, where $x = \min_{H_i}\{|V(H_i)| - c(H_i)\}$ and H_i is a connected component of G.*

Proof. It follows directly from the fact that a maximum convex set of G contains all but one of the connected componets of G. □

Based on Theorem 4.3 and Proposition 4.4, we can compute the convexity number of a cograph in linear time, using modular decomposition [16, 17].

References

[1] M. Atici, Computational complexity of geodetic set, *Int. J. Comput. Math.*, **79** (2002), pp. 587–591.

[2] K.S. Booth and J.H. Johnson, Dominating sets in chordal graphs, *SIAM J. Comput.*, **11** (1982), pp. 191–199.

[3] J. Cáceres, M.C. Hernando, M. Mora, I.M. Pelayo, M.L. Puertas, and C. Seara, On geodetic sets formed by boundary vertices, *Discrete Math.*, **306(2)** (2006), pp. 188–198.

[4] José Cáceres, Alberto Márquez, Ortrud R. Oellermann and María Luz Puertas, Rebuilding convex sets in graphs, *Discrete Math.*, **297** (2005), pp. 26–37.

[5] S.R. Canoy Jr. and I.J.L. Garces, Convex sets under some graph operations, *Graphs Combin.*, **18(4)** (2002), pp. 787–793.

[6] G. Chartrand, F. Harary and P. Zhang, On the geodetic number of a graph, *Networks*, **39(1)** (2002), pp. 1–6.

[7] G. Chartrand, E.M. Palmer, and P. Zhang, The geodetic number of a graph: A survey, *Congr. Numer.*, **156** (2002), pp. 37–58.

[8] Gary Chartrand, Curtiss E. Wall, and Ping Zhang, The convexity number of a graph, *Graphs Combin.*, **18(2)** (2002), pp. 209–217.

[9] S.A. Cook, The complexity of theorem-proving procedures, in *Proc. 3rd ACM Symp. on Theory of Computing* (1971), pp. 151–158.

[10] A.K. Dewdney, *Fast Turing reductions between problems in NP; Chapter 4; Reductions between NP-complete problems.*, Tech. report, (1981).

[11] M.C. Dourado, J.G. Gimbel, J. Kratochvíl, F. Protti and J.L. Szwarcfiter, On the computation of the hull number of a graph, *Discrete Math.*, (to appear).

[12] P.H. Edelman and R.E. Jamison, The theory of convex geometries, *Geom. Dedicata*, **19** (1985), pp. 247–270.

[13] M. Farber and R.E. Jamison, Convexity in graphs and hypergraphs, *SIAM J. Algebraic Discrete Methods*, **7** (1986), pp. 433–444.

[14] J.G. Gimbel, Some remarks on the convexity number of a graph, *Graphs Combin.*, **19** (2003), pp. 357–361.

[15] F. Harary and J. Nieminen, Convexity in graphs, *J. Differential Geom.*, **16** (1981), pp. 185–190.

[16] R.M. McConnell and J.P. Spinrad, Linear-time modular decomposition and efficient transitive orientation of comparability graphs, in *Proc. of the Fifth Annual ACM-SIAM Symposium on Discrete Algorithms* **5** (1994), pp. 536–545.

[17] R.M. McConnell and J.P. Spinrad, Ordered vertex partitioning, *Discrete Mathe. Theor. Comput. Sci.*, **4** (2000), pp. 45–60.

[18] H. Mueller and A. Brandstädt, The NP-completeness of Steiner tree and dominating set for chordal bipartite graphs, *Theoret. Comput. Sci.*, **53** (1987), pp. 257–265.

Proc. of ICDM 2006, Pages 109–114

Non-binary Codes via the Trace Map

A. Rao[1] and N. Pinnawala[2]
School of Mathematical and Geospatial Sciences, RMIT University, GPO Box 2476V, Melbourne, VIC - 3001, Australia.
Email: [1]asha@rmit.edu.au [2]nimalsiri.pinnawala@rmit.edu.au

Abstract. The uniform distribution of the trace map seems to lend itself very well to the construction of binary and non-binary codes from Galois fields and Galois rings. Here we consider the construction of non-binary codes from Galois Rings using the trace map.

---

1 Introduction

In [7] and [9] the trace map over the Galois rings was used to construct linear codes over $\mathbb{Z}_4$, $\mathbb{Z}_{2^s}$ and $\mathbb{Z}_{p^s}$. During this research it was noticed that the trace map had been used in a similar manner to construct mutually unbiased bases over C_d for d a prime power. An interesting question that arose was whether the equiv-distribution of the trace map would result in different codes depending upon the argument used. In this paper we look at some ways of using the trace map to construct different linear codes over the Galois rings and Galois fields.

A linear code $\mathcal{C}$ of length m over the integers modulo n ($\mathbb{Z}_n = \{0, 1, 2, \ldots, n-1\}$) is an additive subgroup of $\mathbb{Z}_n^m$. An element of $\mathcal{C}$ is called a codeword and a generator matrix of $\mathcal{C}$ is a matrix whose rows generate $\mathcal{C}$. The Hamming weight $W_H(x)$ of an m-tuple $x = (x_1, x_2, \ldots, x_m)$ in $\mathbb{Z}_n^m$ is the number of nonzero components of x and the Lee weight $W_L(x)$ of x is $\sum_{i=1}^{m} \min \{x_i, n - x_i\}$. The Euclidean weight $W_E(x)$ of x is $\sum_{i=1}^{m} \min \{x_i^2, (n-x_i)^2\}$ and the Chinese Euclidean weight $W_{CE}(x)$ of x is $\sum_{i=1}^{m} \{2 - 2\cos\left(\frac{2\pi x_i}{n}\right)\}$. The Hamming, Lee, Euclidean and Chinese Euclidean distances between $x, y \in \mathbb{Z}_n^m$ are defined and denoted as $d_H(x,y) = W_H(x-y)$, $d_L(x,y) = W_L(x-y)$, $d_E(x,y) = W_E(x-y)$ and $d_{CE}(x,y) = W_{CE}(x-y)$ respectively.

A code C over Z_p, p a prime, is called a simplex code if every pair of codewords are the same Hamming distance apart. In [4] Gupta introduced the simplex code of type α and β over Z_4 and Z_{2^s} and in [5] Gupta et. al. constructed senary simplex codes of type α, β and γ. A major distinguishing characteristic of a simplex code of type α over either Z_4, Z_{2^s} or Z_6 is that each row of its generator matrix contains every element of the alphabet equally often (see [4], [5], etc.).

In this paper we construct codes over $\mathbb{Z}_{p^s}$ for any prime p and extend this construction via the weighted trace map to n for all integers n, obtaining non-binary codes over $\mathbb{Z}_n$. Some of these codes were found to be simplex codes.

In the next section we give details of the trace map over Galois rings. In Section 3 we look at the construction of codes over Galois rings via the trace map. The general case of codes over $\mathbb{Z}_n$ for a composite integer n is considered in Section 4. The final section outlines work currently in progress and future directions.

2 The Trace map over Galois Rings

The starting point of this work is to study the properties of the trace map. Here we only consider the trace map over the Galois Rings.

Let $p > 2$ be a prime and s a positive integer. The ring of integers modulo p^s is the set $\mathbb{Z}_{p^s} = \{0, 1, 2, \ldots, p^s - 1\}$. Let $h(x) \in \mathbb{Z}_{p^s}[x]$ be a monic basic irreducible polynomial of degree m that divides $(x^{p^m-1} - 1)$. The Galois ring of characteristic p^s and dimension m is defined to be the quotient ring $\mathbb{Z}_{p^s}[x]/(h(x))$ and is denoted by $GR(p^s, m)$. The element $\zeta = x + (h(x))$ is a root of $h(x)$ and hence ζ is a primitive $(p^m - 1)$th root of unity. Consequently ζ is a primitive element of $GR(p^s, m)$ and $GR(p^s, m) = \mathbb{Z}_{p^s}[\zeta]$. Thus $GR(p^s, m) =< 1, \zeta, \zeta^2, \ldots, \zeta^{m-1} >$ and hence $|GR(p^s, m)| = p^{sm}$.

Every element $u \in GR(p^s, m)$ has a unique representation as $u = \sum_{i=0}^{s-1} p^i u_i$, where $u_i \in \mathcal{T} = \{0, 1, \zeta, \zeta^2, \ldots, \zeta^{p^m-2}\}$. This representation is called the p-adic representation of the elements of $GR(p^s, m)$ and the set $\mathcal{T}$ is called the Teichmuller set. Note that u is invertible if and only if $u_0 \neq 0$. Hence every non-invertible element of $GR(p^s, m)$ can be written as $u = \sum_{i=k}^{s-1} p^i u_i, \;\; k = 1, 2, \ldots, s-1$. By using the p - adic representation of elements of $GR(p^s, m)$, the Frobenius automorphism f is defined in [1, 2, 10], etc. as

$$f : GR(p^s, m) \to GR(p^s, m)$$
$$f(u) = \sum_{i=0}^{s-1} p^i u_i^p.$$

Note that when $s = 1$, f is the usual Frobenius automorphism for the Galois field $GF(p, m)$ (see [6]).

The relative trace map over $GR(p^s, m)$ is defined as

$$T : GR(p^s, m) \to \mathbb{Z}_{p^s}$$
$$T(u) = u + f(u) + f^2(u) + \ldots + f^{m-1}(u).$$

In addition to being a surjective linear transformation, the trace map also satisfies the following property:

Lemma 2.1. *For any $b \in GR(p^s, m)$, as x ranges over $GR(p^s, m)$, $T(xb)$ takes elements in $D_k = \{p^k t \mid t = 0, 1, 2, \ldots, p^{s-k} - 1\}$ equally often, i.e., $p^{s(m-1)+k}$ times, where $k = 0, 1, 2, \ldots, s-1$.*

Proof. For any $x \in GR(p^s, m)$, consider the m-tuple $V_x = (T(x), T(\zeta x), \ldots, T(\zeta^{m-1} x))$ over $\mathbb{Z}_{p^s} = D_0$. Let $V = \{V_x | x \in GR(p^s, m)\}$ and consider the following correspondence:

$$\alpha : GR(p^s, m) \to V.$$

It is easy to see that α sets up a one to one correspondence between the elements of $GR(p^s, m)$ and the m - tuples of V over D_0. Thus as x ranges over $GR(p^s, m)$, each component $T(x\zeta^i)$, for $i = 0, 1, 2, \ldots, m-1$, must take each element of D_0 equally often, i.e., $\frac{p^{sm}}{p^s} = p^{s(m-1)}$ times. In general, for an invertible element $b \in GR(p^s, m)$ (i.e., $b = \sum_{i=0}^{s-1} p^i u_i; \;\; u_i \in \mathcal{T}$ and $u_0 \neq 0$), as x ranges over $GR(p^s, m)$, $T(xb)$ must also assume each element of D_0 equally often, i.e., $p^{s(m-1)}$ times. If b is not invertible then $b = \sum_{i=k}^{s-1} p^i u_i, \;\; k = 1, 2, \ldots, s-1$. Now from the expansion of $T(xb)$ and induction on k, as x ranges over $GR(p^s, m)$, $T(xb)$ must take each element of D_k equally often, i.e., $\frac{p^{sm}}{p^{s-k}} = p^{sm-(s-k)} = p^{s(m-1)+k}$ times. This completes the proof.

3 Linear Codes over $\mathbb{Z}_{p^s}$

Using the trace map we can construct linear codes over $\mathbb{Z}_{p^s}$. As noted in [7, 9] the codes obtained are similar in structure to the simplex codes of type α over $\mathbb{Z}_4$, $\mathbb{Z}_{2^s}$ and $\mathbb{Z}_6$ found by [4, 5], but are not simplex codes in the case $p > 2$ and $s > 1$.

Proposition 3.1. *Let p be a prime, $p > 2$. Let $GR(p^s, m)$ be the Galois ring of characteristic p^s and C_{p^s} be the multiplicative group of all complex (p^s)th roots of unity.*

The rows of the matrix $A = [T(c_i c_j)]_{\forall c_i, c_j \in GR(p^s,m)}$ form a linear code over $\mathbb{Z}_{p^s}$ with parameters
$[n, k, d_L, d_H] = \left[p^{sm}, m, p^{s(m-1)}\left(\frac{p^{2s}-p^{2(s-1)}}{4}\right), p^{sm-1}(p-1)\right]$.

Proof. Consider the matrix A:

$$A = [T(c_i c_j)]_{\forall c_i, c_j \in GR(p^s,m)}.$$

Since $T(c_i c_j) \in \mathbb{Z}_{p^s}$, the rows of A are codewords over $\mathbb{Z}_{p^s}$.

Next construct the matrix G_A as follows:

$$G_A = \begin{bmatrix} T(c_i), & i = 1, 2, \ldots, p^{sm} \\ T(\zeta c_i), & i = 1, 2, \ldots, p^{sm} \\ \vdots & \vdots \\ T(\zeta^{m-1} c_i), & i = 1, 2, \ldots, p^{sm} \end{bmatrix}_{m \times p^{sm}},$$

where $c_i \in GR(p^s, m)$. Since the ζ^i are invertible in $GR(p^s, m)$, from Lemma 2.1, each row of G_A contains each element of $\mathbb{Z}_{p^s}$ equally often, i.e., $p^{s(m-1)}$ times. Further the rows of G_A are linearly independent. Therefore the code generated by G_A is a linear code over $\mathbb{Z}_{p^s}$. In addition the structure of G_A is very similar to the generator matrices of the simplex codes in [4, 5, 7] over $\mathbb{Z}_4$, $\mathbb{Z}_{2^s}$ and $\mathbb{Z}_6$.

Taking all linear combinations of the rows of G_A we obtain

$$A = [T(c_i c_j)]_{\forall c_i, c_j \in GR(p^s,m)}.$$

Therefore G_A is a generator matrix of the code A and hence A is a linear code over $\mathbb{Z}_{p^s}$ with the p-dimension $k = m$. Let x be any nonzero codeword in A. Then x can be written as $x = (x_1, x_2, \ldots, x_{p^{sm}})$, where $x_i \in D_k$. From Lemma 2.1, each element in D_k will appear in x equally often, i.e., $p^{s(m-1)+k}$ times. Therefore the Lee weight of x is $W_L(x) = p^{s(m-1)}\left(\frac{p^{2s}-p^{2k}}{4}\right)$ and the Hamming weight is $W_H(x) = p^{s(m-1)+k}\left(p^{s-k} - 1\right)$. The minimum Lee and Hamming weights of the codewords of A are obtained when $k = s - 1$. Thus $\min W_L(x) = p^{s(m-1)}\left(\frac{p^{2s}-p^{2(s-1)}}{4}\right)$ and $\min W_H(x) = p^{sm-1}(p-1)$. Therefore the parameters of the code A are given by

$$[n, k, d_L, d_H] = \left[p^{sm}, m, p^{s(m-1)}\left(\frac{p^{2s} - p^{2(s-1)}}{4}\right), p^{sm-1}(p-1)\right]$$

Note that the code A is not equidistant with respect to either the Lee or Hamming distances. Therefore this is not a simplex code over $\mathbb{Z}_{p^s}$.

Remark.

1. When $p > 2$ and $s = 1$, $GF(p,m)$ is the Galois field of order p^m. Hence the rows of the matrix $A = [T(c_i c_j)]_{\forall c_i, c_j \in GF(p,m)}$, form a $\mathbb{Z}_p$ - linear code with parameters $[n,k,d_H] = [p^m, m, p^{m-1}(p-1)]$, where d_H is the minimum Hamming distance. Also every nonzero codeword has constant Hamming weight $W_H = p^{m-1}(p-1)$ and constant Lee weight $W_L = \frac{p^{m-1}}{4}(p^2-1)$. Thus the code A is a simplex code over $\mathbb{Z}_p$.

2. In the case $p = 2$, $A = [T(c_i c_j)]_{\forall c_i, c_j \in GF(2,m)}$ is a binary linear code with parameters $[n,k,d_L] = [2^m, m, 2^{m-1}]$.
   In addition the rows of the matrix A^* obtained by deleting the all zero column of A form an $\mathbb{Z}_2$ - simplex code $[2^m - 1, m, 2^{m-1}]$.

4 Codes over $\mathbb{Z}_n$

We can now consider codes over the integer ring $\mathbb{Z}_n$, for a composite integer n. To be able to extend the construction outlined in Section 3, we need to extend the trace map first. We do this by defining the weighted trace map.

In the first instance let $n = p_1^{e_1} p_2^{e_2}$, where $p_1 \neq p_2 \geq 2$ are primes and e_1, e_2 are positive integers. It is well known that $\mathbb{Z}_n \simeq \mathbb{Z}_{p_1^{e_1}} \times \mathbb{Z}_{p_2^{e_2}}$ and moreover for any positive integer m, $\mathbb{Z}_n^m \simeq (\mathbb{Z}_{p_1^{e_1}} \times \mathbb{Z}_{p_2^{e_2}})^m$. For more details on these results see for example [3].

Let $f_1(x)$ and $f_2(x)$ be basic monic irreducible polynomials of degree m over $\mathbb{Z}_{p_1^{e_1}}$ and $\mathbb{Z}_{p_2^{e_2}}$ respectively. As in Section 3 the Galois rings of characteristics $p_1^{e_1}$ and $p_2^{e_2}$ and common dimension m are defined as the quotient rings $\mathbb{Z}_{p_1^{e_1}}[x]/(f_1(x))$ and $\mathbb{Z}_{p_2^{e_2}}[x]/(f_2(x))$ respectively. These rings are denoted by $GR(p_1^{e_1}, m)$ and $GR(p_2^{e_2}, m)$. If ζ_1 and ζ_2 are defined to be $\zeta_1 = x + (f_1(x))$ and $\zeta_2 = x + (f_2(x))$, the two rings can then be expressed as $GR(p_1^{e_1}, m) = \langle 1, \zeta_1, \zeta_1^2, \ldots, \zeta_1^{m-1} \rangle$ and $GR(p_2^{e_2}, m) = \langle 1, \zeta_2, \zeta_2^2, \ldots, \zeta_2^{m-1} \rangle$. This tells us that $GR(p_1^{e_1}, m) = \mathbb{Z}_{p_1^{e_1}}[\zeta_1]$ and $GR(p_2^{e_2}, m) = \mathbb{Z}_{p_2^{e_2}}[\zeta_2]$. Hence any element $a \in GR(p_1^{e_1}, m)$ can be considered as an m-tuple $a = (a_0, a_1, \ldots, a_{m-1})$ over $\mathbb{Z}_{p_1^{e_1}}$ while $b \in GR(p_2^{e_2}, m)$ as $b = (b_0, b_1, \ldots, b_{m-1})$ over $\mathbb{Z}_{p_2^{e_2}}$.

Let $R(n,m) = GR(p_1^{e_1}, m) \times GR(p_2^{e_2}, m)$. Any element $c \in R(n,m)$ can be written as $c = (a, b)$, where $a \in GR(p_1^{e_1}, m)$ and $b \in GR(p_2^{e_2}, m)$ and further as $c = (a_0, a_1, \ldots a_{m-1}, b_0, b_1, \ldots, b_{m-1})$. Since $\mathbb{Z}_n \simeq \mathbb{Z}_{p_1^{e_1}} \times \mathbb{Z}_{p_2^{e_2}}$, c can be given as an m-tuple $c = (c_0, c_1, \ldots, c_{m-1})$ over $\mathbb{Z}_n$, where $c_i = (a_i, b_i) i = 1, 2, \ldots, m-1$. Let any $c, c' \in R(n,m)$. It is easy to see that $R(n,m)$ is a ring under the addition $c + c' = ((c_0 + c_0'), (c_1 + c_1'), \ldots, (c_{m-1} + c_{m-1}'))$ and the multiplication $cc' = (c_0 c_0', c_1 c_1', \ldots, c_{m-1} c_{m-1}')$. Obviously $|R(n,m)| = n^m = (p_1^{e_1} p_2^{e_2})^m = p_1^{e_1 m} p_2^{e_2 m} = |GR(p_1^{e_1}, m)||GR(p_2^{e_2}, m)|$.

Using the fact that given a prime p and any integer a, $a^p \equiv a (\text{mod } p)$, we can define the weighted trace map as follows:

Definition 4.1 (Weighted trace map). *Let T_1 and T_2 be the trace maps over $GR(p_1^{e_1}, m)$ and $GR(p_2^{e_2}, m)$ respectively (see Section 3). For any $c = (c_1, c_2) \in R(n,m)$, the weighted trace map T over $R(n,m)$ is defined by*

$$T : R(n,m) \rightarrow \mathbb{Z}_n$$
$$T(c) = p_2^{e_2} T_1(c_1) + p_1^{e_1} T_2(c_2)$$

It is easy to show that this weighted trace map satisfies the following properties: For any $c, c' \in R(n,m)$ and $\alpha \in \mathbb{Z}_n$

(i) $T(c+c') = T(c) + T(c')$

(ii) $T(\alpha c) = \alpha T(c)$

(iii) T is surjective

(iv) For any $c \in R(n,m)$, as x ranges over $R(n,m)$, $T(cx)$ takes each element in

$$S_{i,j} = \left\{ p_1^i p_2^j t \,|\, t = 0, 1, 2, \ldots, \frac{n}{p_1^i p_2^j} - 1 \right\} \tag{1}$$

equally often $p_1^i p_2^j n^{m-1}$ times, where $i = 0, 1, 2, \ldots, e_1$ and $j = 0, 1, 2, \ldots, e_2$.

Since the map T satisfies properties similar to those satisfied by the trace map, we call it the weighted trace map. This weighted trace map can now be used to obtain codes over $\mathbb{Z}_n$.

Theorem 4.2. *Let $R(n,m) = GR(p_1^{e_1}, m) \times GR(p_2^{e_2}, m)$. Let T_1 and T_2 be the trace maps over $GR(p_1^{e_1}, m)$ and $GR(p_2^{e_2}, m)$ respectively. Let T be the weighted trace map as defined above. Then the rows of the matrix $A = [T(ab)]_{\forall a,b \in R(n,m)}$ form a linear code over $\mathbb{Z}_n$ with parameters $[n, k, d_H] = \left[n^m, m, (n - p_1^{e_1} p_2^{e_2} \ldots p_k^{e_k - 1}) n^{m-1}\right]$*

Proof. Let $B = [T_1(c_{1i}c_{1j})]_{\forall c_{1i}, c_{1j} \in GR(p_1^{e_1}, m)}$ and $D = [T_2(c_{2i}c_{2j})]_{\forall c_{2i}, c_{2j} \in GR(p_2^{e_2}, m)}$ be the codes over $\mathbb{Z}_{p_1^{e_1}}$ and $\mathbb{Z}_{p_2^{e_2}}$ respectively as described in Theorem 3.1. Let G_B and G_D be the generator matrices of the codes B and D and consider the $m \times n^m$ matrix

$$G_A = p_2^{e_2} \, [p_2^{e_2 m} \text{ copies of } G_B] + p_1^{e_1} \, [p_1^{e_1 m} \text{ copies of } G_D] \tag{2}$$

i.e.,

$$G_A = p_2^{e_2} \begin{bmatrix} T_{1l}(c_{1i}), & 1 \le i \le p_1^{e_1 m}, 1 \le l \le p_2^{e_2 m} \\ T_{1l}(\zeta_1 c_{1i}), & 1 \le i \le p_1^{e_1 m}, 1 \le l \le p_2^{e_2 m} \\ \vdots & \vdots \\ T_{1l}(\zeta_1^{m-1} c_{1i}), & 1 \le i \le p_1^{e_1 m}, 1 \le l \le p_2^{e_2 m} \end{bmatrix}$$

$$+ p_1^{e_1} \begin{bmatrix} T_{2t}(c_{2j}), & 1 \le j \le p_2^{e_2 m}, 1 \le t \le p_1^{e_1 m} \\ T_{2t}(\zeta_2 c_{2j}), & 1 \le j \le p_2^{e_2 m}, 1 \le t \le p_1^{e_1 m} \\ \vdots & \vdots \\ T_{2t}(\zeta_2^{m-1} c_{2j}), & 1 \le j \le p_2^{e_2 m}, 1 \le t \le p_1^{e_1 m} \end{bmatrix}$$

After some calculations we can show that by taking linear combinations of all the rows of G_A we can generate the rows of the matrix A. The Hamming weight of each non-zero codeword (which is just a row of A) is given by $(n - p_1^i p_2^j) n^{m-1}$, where $i = 0, 1, 2, \ldots, e_1$ and $j = 0, 1, 2, \ldots, e_2$. (Note that i and j do not take the values e_1 and e_2 respectively simultaneously). If $p_1 > p_2$ and $e_1 \ge e_2$, the minimum Hamming weight is given by $(n - p_1^{e_1} p_2^{e_2 - 1}) n^{m-1}$. Since A is a linear code the minimum Hamming distance $d_H = (n - p_1^{e_1} p_2^{e_2-1}) n^{m-1}$. Thus $[n, k, d_H] = \left[n^m, m, (n - p_1^{e_1} p_2^{e_2 - 1}) n^{m-1}\right]$.

It is fairly straight-forward to extend these results to the case $n = \prod_{i=1}^{k} p_i^{e_i}$. The only onerous task being calculating the Lee, Euclidean and Chinese Euclidean weights of the codewords. See [8] for details.

5 Conclusion

This paper outlined the use of the trace map over Galois Rings to obtain non-binary codes over $\mathbb{Z}_n$ for any integer n.

It is possible to get different codes when the argument of the trace map is varied. Current research includes cataloguing self-orthogonal codes constructed over the Galois Fields $GF(p, 2)$ as well as constructing mutually unbiased bases of non-prime power dimensions, using the weighted trace map.

References

[1] J.T. Blackford and D.K. Ray-Chaudhuri, A transform approach to permutation group of cyclic codes over Galois rings, *IEEE Transactions on Information Theory*, **46(7)** (2000), pp. 2350–2358.

[2] A.B. Calderbank and N.J.A. Sloane, Modular and p - adic cyclic codes, *Designs Codes and Crypto*, **6** (1995), pp. 21–35.

[3] A.J. Gareth and J.J. Mary, *Elementary number theory*, Springer-Verlag, (1998).

[4] M.K. Gupta, *On some linear codes over* Z_{2^s},. Ph.D, Department of Mathematics, Indian institute of Technology, Kanpur, India (Dec. 1999).

[5] M.K. Gupta, D.G. Glynn and T.A. Gulliver, On some quaternary self orthogonal codes, In *S. Boztas and I.E. Shparlinski, editors, Applied Algebra, Algebraic Algorithms and Error-Correcting Codes; AAECC-14, volume LNCS 2227 of Lecture Notes in Computer Science*, pp. 112–121. Springer, (2001).

[6] R. Lidl and H. Niederreiter, *Finite Fields*, Cambridge University press, (1997).

[7] N. Pinnawala and A. Rao, Cocyclic simplex codes of type α over $\mathbb{Z}_4$ and $\mathbb{Z}_{2^s}$, *IEEE Trans. Info. Theory*, **50(9)** (2004), pp. 2165–2169.

[8] N. Pinnawala and A. Rao, Finite Fields and Applications, *Chapter Cocyclic Butson Hadamard matrices and codes over Zn via the trace map. Number CONM/461 in Contemporary Mathematics Series*, AMS, (2008), (To appear).

[9] A. Rao and N. Pinnawala, New linear codes over Z_{p^s} via the trace map, in *IEEE International Symposium on Information Theory, Adelaide, Australia, 4–9 September 2005* (2005), pp. 124–126.

[10] Z. X. Wan, *Lectures on finite fields and Galois rings*, World Scientific, New Jersey (2003).

Proc. of ICDM 2006, Pages 115–124

Vertex Identifying Codes for Fault Isolation in Communication Networks

KRISHNAIYAN THULASIRAMAN*
School of Computer Science, University of Oklahoma, Norman, Oklahoma, 73019, USA
MIN XU†
School of Mathematical Sciences, Beijing Normal University, Beijing, 100875, China
Also at *School of Computer Science, University of Oklahoma, Norman, Oklahoma, 73019, USA*
YING XIAO
Packet Design Inc., Palo Alto, California, 94304, USA
XIAO-DONG HU‡
Institute of Applied Mathematics, Chinese Academy of Sciences, Beijing, 100080, China

Abstract. Rapid advances in semiconductor technology have made possible the design of systems with a large number of components. It is difficult to build such systems without defects. Thus fault diagnosis, testing and tolerance have become issues of great interest in the design and analysis of computer and communication systems. It was in this context that Preparata, Metze and Chien [1] at the University of Illinois proposed a model, now well known as the PMC model, and a framework for the diagnosis of large scale digital systems. Recently, Karpovsky et al. [5] proposed the concept of identifying codes and its application in the detection/isolation of faults in communication systems. In this paper we give a broad survey of certain important results in this area along with our recent work on a generalization of the identifying codes concept and an approximation algorithm.

1 Introduction

Rapid advances in semiconductor technology have made possible the design of systems with a large number of components. It is difficult to build such systems without defects. Thus fault diagnosis, testing and tolerance have become issues of great interest in the design and analysis of computer and communication systems. It was in this context that Preparata, Metze and Chien [1] at the University of Illinois proposed a model, now well known as the PMC model, and a framework for the diagnosis of large scale digital systems.

*Krishnaiyan Thulasiraman thanks for the support of the US National Science Foundation under the ITR grant ECS 0426831.

†Min Xu thanks for the support of NNSF of China No. 10701074, No. 10626053 and China Postdoctoral Science Foundation.

‡Xiao-Dong Hu thanks for the support of NNSF of China No. 70221001 and No. 10531070.

In the PMC model for the diagnosis of multiprocessor systems, the problem is formulated by what is called a test graph. Each vertex in the test graph represents a processor. It is assumed that each processor tests a subset of the remaining processors. A link directed from the vertex representing processor p_i to the vertex representing processor p_j indicates that processor p_i has tested p_j. If p_i tests p_j as fault free then a bit 0 is associated with the corresponding link $(p_i,\ p_j)$; otherwise a 1 bit is associated with this link. The test result arising from a faulty testing processor could be arbitrary, that is, unreliable. The collection of all test results (0 or 1) is called a syndrome. The problem is to identify all faulty processors subject to certain assumptions. In the last forty years or so since the introduction of this approach to diagnosis, several fundamental results have been reported on various generalizations and related diagnosis algorithms. For examples, see [2]-[4] and the references therein.

Recently Karpovsky et al. [5] introduced the concept of identifying codes and studied several issues relating to the construction of these codes. Consider an undirected graph G with vertex set V and edge set E. A ball of radius $t \geq 1$ centered at a vertex v is defined as the set of all vertices that are at distance t or less from v. The vertex v is said to cover itself and all the vertices in the ball with v as the center. The identifying codes problem defined by Karpovsky et al. [5] is to find a minimum set D such that every vertex in G belongs to a unique set of balls of radius $t \geq 1$ centered at the vertices in D. The set D may be viewed as a code identifying the vertices and is called an identifying code. Two important applications have triggered considerable research on the identifying codes problem. One of these is the problem of diagnosing faulty processors in a multiprocessor system [5]. Another application is robust location detection in emergency sensor networks [6]. Next we briefly describe the application of identifying codes in fault diagnosis.

Consider a communication network modeled as an undirected graph G. Each vertex in the graph represents a processor and each edge represents the communication link connecting the processors represented by its end vertices. Some of the processors could become faulty. To simplify the presentation let us assume that at most one processor could become faulty at any given time. Assume that a processor, when it becomes faulty, can trigger an alarm placed on an adjacent processor. We would like to place alarms on certain processors that will facilitate unique identification of the faulty processors. We would also like to place alarms on as few processors as possible. If D is a minimum identifying code for the case $t = 1$, then placing alarms on the processors represented by the vertices in the set D will help us to uniquely identify the faulty processor. Notice that we only need to consider $t = 1$ because if $t > 1$ is desired, we can proceed with G^t, the tth power of G.

Karpovsky et al. [5] have studied the identifying codes selection problem extensively and have established bounds on the cardinality of the identifying codes. They have shown how to construct the identifying codes for specific topologies such as binary cubes and trees. Several problems closely related to the identifying codes problem have been studied in the literature. Some of these may be found in [7], [8].

This paper is organized as follows. In Section 2, a formal definition of the identifying codes problem is first given. This is followed by a review of results on this problem for certain specific topologies, namely, cycles, trees and hypercubes. In Section 3, a generalization of the identifying codes problem and an approximation algorithm for this problem are discussed.

2 Identifying Codes Problem and Construction of Codes

Let $G = (V, E)$ be a simple, connected, undirected graph and $r \geq 1$ be an integer. Given a vertex $x \in V$, we define $B_r(x) = \{y : d(x, y) \leq r\}$ where $d(x, y)$ denotes the distance of the shortest path between x and y in G. For a subset C of V and any vertex set $X \subseteq V$, we define $I_r(X) = I_r(C; X) = (\cup_{x \in X} B_r(x)) \cap C$ and we say that C r-covers X if $I_r(X) \neq \emptyset$. We say that a subset C, r-separates two distinct vertex sets X and Y if and only if $I_r(X) \neq I_r(Y)$. For a graph G, let S be a collection of all the vertex sets of $V(G)$ with cardinality at most l, then an $(r, \leq l)$-identifying code of G is a set $C \subseteq V$ which r-covers all the vertex set in S and r-separates any pair of distinct vertex set in S.

The identifying codes problem is to find the minimum cardinality of such a code, which we denote by $M_r^{(\leq l)}(G)$. For convenience, we use $M_r(G)$ for $l = 1$.

The problem of determining an identifying code with minimum cardinality in a graph has been proved to be NP-Complete [9]. Many researchers have focused on the study of identifying codes in some restricted classes of networks, for example [10]–[14]. In this section, we survey results on this problem for the case of cycles, trees and hypercubes.

2.1 Identifying Codes for Cycles

A cycle C_n for $n \geq 3$ denotes a graph $(V(C_n), E(C_n))$ where $V(C_n) = \{v_i : 0 \leq i \leq n-1\}$ and $E(C_n) = \{v_i v_{i+1} : 0 \leq i \leq n-1\}$, and the subscripts are taken modulo n. We will present some known results about the r-identifying code of C_n.

When n is even, Bertrand *et al.* give the following theorem in [11].

Theorem 2.1 (Bertrand *et al.* [11]). *For all $r \geq 1$, we have $M_r(C_{2r+2}) = 2r+1$ and $M_r(C_n) = \frac{n}{2}$ for $n \geq 2r+4$ even. Furthermore, the identifying code can be selected as $C' = V(C_n) \setminus \{v_0\}$ and $C'' = \{v_i : i \text{ is odd}\}$ for $n = 2r+2$ and $n \geq 2r+4$, respectively.* □

In [13], Gravier *et al.* consider the problem with odd n. They get the following result.

Theorem 2.2 (Gravier *et al.* [13]). *For all $r \geq 1$ and $n \geq 2r+3$ odd, we have*

$$\frac{n+1}{2} + \frac{\gcd(2r+1, n) - 1}{2} \leq M_r(C_n) \leq \frac{n+1}{2} + r.$$

□

For some subcases, they get the theorems below.

Theorem 2.3 (Gravier *et al.* [13]). *Let $r \geq 1$, n be an odd integer such that $3r+2 \leq n \leq 4r+1$, then $M_r(C_n) = \frac{n+1}{2} + \frac{gcd(2r+1,n)-1}{2}$.* □

Theorem 2.4 (Gravier *et al.* [13]). *Let $r \geq 1$, n be an odd integer such that $gcd(2r+1, n) = 1$, and $4r+5 \leq n \leq 8r+1$. Then $M_r(C_n) = \frac{n+1}{2}$.* □

Theorem 2.5 (Gravier *et al.* [13]). *Let $r \geq 1$, n be an odd integer such that $n \geq 3r+2$ and $gcd(2r+1, n) \neq 1$. Then $M_r(C_n) = \frac{n+1}{2} + \frac{gcd(2r+1,n)-1}{2}$.* □

Theorem 2.6 (Gravier *et al.* [13]). *$M_1(C_5) = 3$; $M_1(C_n) = \frac{n+1}{2}$ for all $n \geq 7$, n odd; $M_r(C_n) = \lfloor \frac{2n}{3} \rfloor$ for all $r \geq 1$, $n = 2r + 3$; $M_r(C_{4r+3}) = 2r + 3$.* □

Recently, we studied in [14] the r-identifying code about the cycles C_n for odd n with $\gcd(2r + 1, n) = 1$ and established the following results.

Theorem 2.7 (Xu *et al.* [14]). *Let $r \geq 1$, n be an odd integer such that $n \geq 3r + 2$, and $\gcd(2r + 1, n) = 1$. If $n = 2m(2r + 1) + x$ for $m \geq 1$ and $2 \leq x \leq 2r$ or $n = (2m + 1)(2r + 1) + y$ for $m \geq 1$, $1 \leq y \leq 2r - 1$ or $m = 0$, $r + 1 \leq y \leq 2r - 1$, then $M_r(C_n) = \frac{n+1}{2}$.* □

Theorem 2.8 (Xu *et al.* [14]). *Let $r \geq 1$, n be an odd integer such that $n \geq 3r + 2$, and $\gcd(2r + 1, n) = 1$. If $n = 2m(2r + 1) + x$ for $m \geq 1$ and $x = 1$ or $n = (2m + 1)(2r + 1) + y$ for $m \geq 1$, $y = 2r$, then $M_r(C_n) = \frac{n+1}{2} + 1$.* □

Remark 2.9 In Theorem 2.3, 2.4, 2.5, 2.7, we can define an r-identifying code with minimum cardinality as $C = \bigcup_{j=0}^{a-1} \bigcup_{i=0}^{\frac{n}{a}-1} \{v_{i(2r+1)+j} : i \text{ is even}\}$, where $a = \gcd(2r + 1, n)$.

Table 1 shows all of the results about the r-identifying code of cycle C_n.

Table 1:

	$n =$ $2r+3$	$2r+5 \leq n$ $< 3r+2$	$3r+2 \leq n$ $\leq 8r+1$	$n > 8r+1$ $\gcd(2r+1, n) \neq 1$	$n > 8r+1$ $\gcd(2r+1, n) = 1$
odd n	√	?	√	√	√
even n	√				

From Table 1, we can see that, when $2r + 5 \leq n \leq 3r + 1$, the r-identifying code of C_n with minimum cardinality for odd n is not determined yet.

2.2 Identifying Codes for Trees

Tree is a connected network without cycle. We select a vertex in the tree as root. The tree with root is called a rooted-tree. In a rooted-tree any vertex of one degree, unless it is the root, is called a leaf. Each node is either a leaf or an internal vertex in a rooted-tree. The maximum length from root to all leaves is called the height of the rooted-tree. For a rooted-tree, if its root is of degree k and all other internal vertices are of degree at most $k + 1$, then we call this rooted-tree a k-ary tree.

In [10], Bertrand *et al.* give a lower bound about the tree with n vertices.

Theorem 2.10 (Bertrand *et al.* [10]). *For a tree T with n vertices, $M_1(T) \geq \frac{3(n+1)}{7}$.* □

Karpovsky *et al.* consider the p-ary tree in [5].

Theorem 2.11 (Karpovsky *et al.* [5]). *For a p-ary tree T with height l ($l \geq 3$), we have*

$$p^{l-3}(p^2 + 1) \leq M_1(T) \leq \begin{cases} \frac{p^{l+1}-1}{p^2-1}, & \text{if } l \text{ is odd} \\ \frac{p^{l+1}-p}{p^2-1}, & \text{if } l \text{ is even.} \end{cases}$$ □

Corollary 2.12 (Karpovsky *et al.* [5]). *For p-ary trees T with height $l = 3$, $M_1(T) = p^2 + 1$, while for a p-ary tree T' with height $l = 4$, $M_1(T') = p(p^2 + 1)$.* □

They also consider the case when $r > 1$. But they proved that the vertices in a tree are not identifiable if $r > 1$.

Theorem 2.13 (Karpovsky *et al.* [5]). *It is not possible to uniquely identify the vertices of a p-ary l-height tree for $r > 1$.* □

For the complete binary trees, Bertrand *et al.* get the following result in [10].

Theorem 2.14 (Bertrand *et al.* [10]). *For a complete binary tree T with $2^h - 1$ vertices, $M_1(T) = \lceil 20(2^h - 1)/31 \rceil$.* □

A path P_n is a special tree where $V(P_n) = \{v_1, v_2, \ldots, v_n\}$ and $E(P_n) = \{(v_i, v_{i+1}) : i = 1, 2, \ldots, n-1\}$. In [11], Bertrand *et al.* get the following results.

Theorem 2.15 (Bertrand *et al.* [11]). *For a path P_n, $M_1(P_n) = \frac{n+1}{2}$ if $n \geq 1$ is odd and $M_1(P_n) = \frac{n}{2} + 1$ if $n \geq 4$ is even. Furthermore, the identifying code can be selected as $C' = \{1, 3, \ldots, n\}$ and $C'' = \{1, 3, \ldots, n-7, n-5, n-3, n-2, n-1\}$ for odd n and even n, respectively.* □

Theorem 2.16 (Bertrand *et al.* [11]). *For $r \geq 2$, $n \geq 2r + 1$, $M_r(P_n) \geq \lceil \frac{n+1}{2} \rceil$.* □

Theorem 2.17 (Bertrand *et al.* [11]). *Let k be a nonnegative integer. For any fixed $r \geq 2$, and for $n = (4r+2)k + 1$, $M_r(P_n) \leq \frac{n+1}{2}$.* □

2.3 Identifying Codes for Hypercubes

The hypercube Q_n is a popular interconnection network. In Q_n, $V(Q_n) = \{(x_1, x_2, \ldots, x_n) : x_i \in \{0, 1\},\ i = 1, 2, \ldots, n\}$ and two vertices $x = (x_1, x_2, \ldots, x_n)$ and $y = (y_1, y_2, \ldots, y_n) \in V(Q_n)$ are linked by an edge if and only if $x - y = (0, 0, \ldots, \pm 1, 0, \ldots, 0)$. Karpovsky *et al.* [5] first obtained some lower bounds on $M_1(Q_n)$.

Theorem 2.18 (Karpovsky *et al.* [5]). *For an n-dimensional hypercube Q_n, $n \geq 3$, $M_1(Q_n) \geq \frac{n \times 2^{n+1}}{n(n+1)+2}$.* □

Let $K(n, q)$ be the size of an optimal code C^* in $V(Q_n)$ satisfying that every vertex in $V(Q_n)$ is at distance at most q from a codeword of C^*. An upper bound on $M_1(Q_n)$ follows from the theorem below.

Theorem 2.19 (Karpovsky *et al.* [5]). *Let C^* be an optimal code of Q_n and covering radius 2, i.e., C^* has $K(n, 2)$ codewords. Then, for $r = 1$, a code C identifying vertices in the n-dimensional hypercube can be selected as $C = \{w : \exists v \in C^*,\ d(v, w) = 1\}$ ($d(v, w)$ is Hamming distance between v and w).* □

Corollary 2.20 (Karpovsky *et al.* [5]). *The number of codewords in an optimal identifying code with $r = 1$ for a n-dimension hypercube Q_n ($n \geq 3$) is upper-bounded by $M_1(Q_n) \leq nK(n, 2)$.* □

Theorem 2.19 can be extended to the case $r > 1$ as follows.

Theorem 2.21 (Karpovsky *et al.* [5]). *For any given $r < \frac{n}{2}$, a code C for identifying vertices in the n-dimensional hypercube ($n > 2$) can be obtained by selecting as codewords all vertices at distance exactly r from the codewords of a optimal code C^* which has covering radius $2r$, i.e., $C = \{x : \exists u \in C^*,\ d(x, u) = r\}$.* □

Corollary 2.22 (Karpovsky *et al.* [5]). *For $r < \frac{n}{2}$, the number of codewords required for identifying vertices in a hypercube is upper bounded by $M_r(Q_n) \leq K(n, 2r)\binom{n}{r}$.* □

Corollary 2.23 (Karpovsky *et al.* [5]). *The number of codewords required for a hypercube with $(4s+1)t$ dimensions using balls of radius st is upper-bounded by $M_{st}(Q_{(4s+1)t}) \leq \binom{(4s+1)t}{st} 2^t$.* □

A code $C \subseteq V(Q_n)$ is called a u-fold r-covering if every vertex in $V(Q_n)$ is r-covered by at least u codewords of C. Let us denote by $K(n, r, u)$ the smallest size of a u-fold r-covering in hypercube Q_n.

Theorem 2.24 (Laihonen [15]). *Let $l \geq 3$. A code $C \subseteq V(Q_n)$ is $(1, \leq l)$ identifying code of Q_n if and only if it is a $(2l-1)$-fold 1-covering. In particular $M_1^{\leq l}(Q_n) = K(n, 1, 2l-1) \geq \lceil (2l-1)\frac{2^n}{n+1} \rceil$.* □

Corollary 2.25 (Laihonen [15]). *Let $l \geq 3$. Then $M_1^{\leq l}(Q_n) = (2l-1)\frac{2^n}{n+1}$ if and only if there are integers $i \geq 0$, $u_0 > 0$ such that $u_0|(2l-1)$ and $2l-1 \leq 2^i u_0$ and $n = u_0 2^i - 1$.* □

3 *d*-Identifying Codes Problem

Karpovsky et al. [5] have shown that unique identification of vertices may not always be possible for certain topologies. In other words, triggering of alarms on a set of processors could mean that one of several candidate processors could be faulty. Once such a set of possible faulty processors has been identified then testing each processor in this set will identify the faulty processor. This motivates the generalization of the identifying codes problem to d-identifying codes problem defined below [16]. This generalization is similar to the introduction of t/s diagnosable systems that generalize the t-diagnosable systems introduced by Preparata, Metze and Chien [1]. An introduction to t-diagnosable systems and their generalization may be found in [2], [3].

3.1 Definition of the *d*-Identifying Codes Problem

Consider an undirected graph $G(V, E)$ with each vertex $v \in V$ associated with an integer cost $c(v) > 0$ and an integer weight $w(v) > 0$.

Let $N[v]$ be the set of vertices containing v and all its neighbors. For a subset of vertices $S \subseteq V$, define the cost and weight of S as

$$c(S) = \sum_{v \in S} c(v) \text{ and } w(S) = \sum_{v \in S} w(v).$$

Two vertices $u, v \in V$ are distinguished by vertex w iff $|N[w] \cap \{u, v\}| = 1$. A set of vertices $D \subseteq V$ is called an identifying set if (1) every unordered vertex pair (u, v) is distinguished by some vertex in D and (2) D is a dominating set of G, i.e., each vertex in G is adjacent to at least one vertex in D (we will relax this requirement later).

Given $D \subseteq V$, define $I_D(v) = N[v] \cap D$ and an equivalence relation $u \equiv v$ iff $I_D(u) = I_D(v)$. The equivalence relation partitions V into equivalence classes $V_D = \{S_1, S_2 \ldots, S_l\}$ such that $u, v \in S_i \iff I_D(u) = I_D(v)$.

For any $D \subseteq V$, let V_D be the equivalence classes induced by D. If D is a dominating set of G and $d \geq \max\{w(S_1), w(S_2) \ldots, w(S_l)\}$, then D is called a d-identifying code of G. The d-identifying codes problem is to find a d-identifying code $D \subseteq V$ with minimum cost.

Note that if $d = 1$ then the d-identifying codes problem reduces to the identifying codes problem if the vertex costs and weights are equal to unity. Also, whereas the cost of the d-identifying code is a measure of the cost of installing alarms, the value of d is a measure of the degree of uncertainty in the identification of faulty processors. Since the value of d is also a measure of the expenses involved in testing each processor in an equivalence class, d has to be set at a small value.

The identifying code must be a dominating set. However we can drop this requirement after a simple transformation of the graph, i.e., adding a new isolated vertex with weight d and a very big cost such that any cost aware algorithm will not include this vertex in the solution set. Thus it will be the only vertex not adjacent to the identifying set. So we will ignore the dominating set condition for the simplicity of presentation.

3.2 An Approximation Algorithm for the *d*-Identifying Codes Problem

We now present a greedy approximation algorithm for the d-identifying codes problem. This algorithm, presented in [16], was inspired by a heuristic for the minimum probe selection problem [17] based on ideas from information theory. In the following $H(V_s)$ is called the entropy defined on V_s which is the set of equivalence classes induced by S. Similarly, $I(V_s, v) = H(V_s) - H(V_{S+v})$ is called the information content of $v \in V - S$ with respect to S. The definition of a required entropy function will be presented later. The framework of this greedy algorithm without specific entropy function is applicable to any class of d-identifying codes problems whose detailed specifications can be hidden in the definition of the entropy. Based on the framework of this greedy algorithm, one only needs to focus on the design of the entropy for other variations of the d-identifying codes problem, e.g., the strong identification codes problem [8].

Algorithm 1 Greedy Algorithm

```
Initialize D = ∅
while H(D) > 0 do
    Select vertex v* = argmax_{v∈V−D} I(V_D; v)/c(v)
    D ← D ∪ {v*}.
end while
```

The time complexity of the above greedy algorithm is $O(n^2 T_H(n))$, where T_H is the time complexity function of the algorithm computing $H(\cdot)$. The following theorem is the main result on the approximation ratio of the greedy algorithm.

Theorem 3.1 (Xiao *et al.* [16]). *Denote V_D as the set of equivalence classes induced by $D \subseteq V$. Suppose an entropy function $H(\cdot)$ satisfies the following conditions:*

(a) $H(V_D) = 0$ *for any d-identifying set D,*

(b) *If $H(V_S) \neq 0$, then $H(V_S) \geq 1$, and*

(c) $I(V_S; v) \geq I(V_{S+u}; v)$ *for all $u \neq v, S \subseteq V$,* *then the greedy algorithm returns a d-identifying code D such that* $c(D)/c(D^*) < \ln[H(V_\emptyset)] + 1$ *(recall that by definition* $V_\emptyset = V$*), where* $D^* = \{v_1^*, v_2^*, \ldots, v_{|D^*|}^*\}$ *is the minimum d-identifying code.* □

Corollary 3.2 (Xiao *et al.* [16]). *Let* $G(V, E)$ *be a graph with n vertices with equal cost which are labeled such that* $I(V_\emptyset; v_1) \geq I(V_\emptyset; v_2) \geq \cdots \geq I(V_\emptyset; v_n)$*. Then the optimal cost of the minimum d-identifying set,* $OPT_d(G) \geq K$*, where K is the smallest integer such that* $\sum_{i=1}^{K} I(V_\emptyset; v_i) \geq H(V_\emptyset)$*.* □

Given $d \geq 2$, let the function $f_d(n)$ defined be as follows.

$$f_d(n) = \begin{cases} n \lg(n/d), & n \geq d \\ 0, & \text{otherwise.} \end{cases} \tag{1}$$

Theorem 3.3 (Xiao *et al.* [16]). *Using the entropy* $H_d(V_D) = \sum_{S \in V_D} f_d(w(S))$ *with* $f_d(\cdot)$ *as defined above, the greedy algorithm guarantees the approximation ratio of* $1 + \ln d + \ln(|V| \lg |V|)$*.* □

Corollary 3.4 (Xiao *et al.* [16]). *For the identifying codes problem, the greedy algorithm guarantees the approximation ratio of* $1 + \ln |V| + \ln(\lg |V|)$*.* □

In fact, we can show that the function $f_d(n)$ defined in (1) is optimal in asymptotic sense, i.e., the approximation ratio based on Theorem 3.1 cannot be improved by finding better entropy.

Lemma 3.5 (Xiao *et al.* [16]). $f(n) \geq \Theta(n \lg n)$. □

3.3 Hardness of the *d*-Identifying Codes Problem

Lemma 3.6 (Xiao *et al.* [16]). *Assume that the identifying codes problem is feasible. For any fixed* $d \geq 2$*, if there exists a polynomial time* ϕ*-approximation algorithm for the d-identifying codes problem, there also exists a polynomial time* ϕ*-approximation algorithm for the identifying codes problem.* □

Lemma 3.6 means that for any fixed *d*, the *d*-identifying codes problem is at least as hard as the identifying codes problem in term of approximability. Thus, with an application of the results in [7], we have the following theorem.

Theorem 3.7 (Xiao *et al.* [16]). *For any given* $d \geq 1$*, the d-identifying codes problem with unit vertex costs and weights is not approximable within* $(1 - \epsilon) \ln |V|$ *unless* $NP \in DTIME(n^{\lg \lg n})$*.* □

In view of Corollary 3.4, we can see that the approximation ratio of the greedy algorithm is quite tight for the *d*-identifying codes problem where the vertex costs and weights are 1. Furthermore, we can expect that the approximation ratio is also very tight for general *d*-identifying codes problem as in the special case.

In [16] the authors reported a probabilistic analysis on random graphs assuming that vertex costs and weights are all equal. It has been shown that a *d*-identifying set of certain cardinality exists with very high probability. It has also been shown that a *d*-identifying code of cardinality smaller than this number does not exist with a high probability.

4 Summary

System level diagnosis proposed in [1] and the fault diagnosis framework based on identifying codes [5] are two powerful approaches of importance in fault diagnosis, testing and tolerance studies of computer and communication systems. References [2]–[4] point to several advances in the system level diagnosis area. In this paper we have given a broad survey of certain key results on the identifying codes problem, its generalization and related algorithms.

There are two main differences in the application of the above two frameworks in fault diagnosis. In the application of the system level diagnosis we are required to construct a test graph that satisfies certain measures of diagnosis as desired by the application. Constructing such a graph is not always easy. Also, we often need sophisticated algorithms for diagnosis using the test graph. In applying the identifying codes technique to diagnosis, we need to find a identifying code of the given graph. In certain cases an identifying code may not even exist. If such a code does not exist, we can decompose the graph into simple subgraphs for which identifying codes exist. Then we can place monitors on each of the subgraphs and use them to isolate faults within those subgraphs. Since identifying codes can be constructed for graphs such as trees, cycles and hypercubes as discussed in this paper, the diagnosis approach based on identifying codes presents an interesting and alternative approach to the system level diagnosis approach.

References

[1] F. Preparata, G. Metze and R. Chien, On the connection assignment problem of diagnosiable systems, *IEEE Trans. on Electronic Computers*, **16** (1967), 848–854.

[2] A. Das and K. Thulasiraman, Diagnosis of t/s-diagnosable systems, In *International Workshop on Graph-Theoretic Concepts in Computer Science*, (1990), 193–205.

[3] A. Das, K. Thulasiraman and V. Agarwal, Diagnosis of t/(t+1)-diagnosable systems, *SIAM J. Comput.*, **23(5)** (1994), 895–905.

[4] A. Das, K. Thulasiraman, V.K. Agarwal and K.B. Lakshmanan, Multiprocessor Fault Diagnosis Under Local Constraints, *IEEE Trans. on Computers*, **42(8)** (1993), 984–988.

[5] M. Karpovsky, K. Chakrabarty and L. Levitin, On a new class of codes for identifying vertices in graphs, *IEEE Trans. on Information Theory*, **44(2)** (1998), 599–611.

[6] S. Ray, R. Ungrangsi, F. Pellegrini, A. Trachtenberg and D. Starobinski, Robust location detection in emergency sensor networks, *INFOCOM*, 2003.

[7] Bjarni V. Halldórsson, Magnús M. Halldórsson and R. Ravi, On the approximability of the minimum test collection problem, In *ESA*, (2001), 158–169.

[8] Tero Laihonen, Optimal codes for strong identification, *European Journal of Combinatorics*, **23(3)** (2002), 307–313.

[9] I. Charon, O. Hudry and A. Lobstein, Minimizing the size of an identifying or locating-dominating code in a graph is NP-hard, *Theoretical Computer Science,* **290(3)** (2003), 2109–2120.

[10] N. Bertrand, I. Charon, O. Hudry and A. Lobstein, 1-Identifying codes on trees, *Australasian Journal of Combinatorics*, **31** (2005), 21–35.

[11] N. Bertrand, I. Charon, O. Hudry and A. Lobstein, Identifying and locating-dominating codes on chains and cycles, *European Journal of Combinatorics,* **25(7)** (2004), 969–987.

[12] G. Cohen, I. Honkala, S. Litsyn and A. Lobstein, *Covering codes*, Amsterdam, The Netherlands: Elsevier, 1997.

[13] S. Gravier, J. Moncel and A. Semri, Identifying codes of cycles, *European Journal of Combinatorics*, **27(5)** (2006), 767–776.

[14] M. Xu, K. Thulasiraman and X.-D. Hu, The identifying codes of cycles with odd order, Accepted for publication in European Journal of Combinatorics.

[15] T.K. Laihonen, Sequences of optimal identifying codes, *IEEE Transaction on Information Theory*, **48(3)** (2002), 774–776.

[16] Y. Xiao, C. Hadjicostis and K. Thulasiraman, The d-Identifying codes problem for vertex identification in graphs: probablistic analysis and an approximation algorithm, *Lecture Notes in Computer Science*, **4112** (2006), 284–298.

[17] M. Brodie, I. Rish and S. Ma, Optimizing probe selection for fault localization, In *International Workshop on Distributed Systems: Operations and Management*, 2001.

Proc. of ICDM 2006, Pages 125–130

On the Berge's k-Optimal Path Partition Conjecture

Sriraman Sridharan
Département de Mathématiques et d'Informatique,
Université de Perpignan Via Domitia, 52 Avenue Paul Alduy,
66100 Perpignan France.
Email: sriraman@univ-perp.fr

Dedicated to the memory of Professor Claude Berge

Abstract. In this paper, the Berge's **weak** path partition conjecture is proved for few simple families of graphs.

1 Introduction

We follow the notation and terminology of Berge [1]. Let G be a directed graph with vertex set X and arc set U. The number of vertices is denoted by n and the number of arcs(directed edges) by m. Unless stated otherwise, our graphs are directed graphs. A directed *elementary path* will be simply referred to as *path*. A *stable set* of G is a set of vertices of G no two of which are joined by an arc.
A *partial k-coloring* of G is a family of k pairwise disjoint (possibly empty) stable sets. A partial k coloring is sometimes denoted by $(S_1, S_2, \ldots, S_k)$ where S_i is a stable set for $i = 1, 2, \ldots, k$. The vertices belonging to S_i are said to be colored with the color i. The vertices not belonging to $\cup_{i=1}^{k} S_i$ are not colored at all by the partial k coloring $(S_1, S_2, \ldots, S_k)$.
A partial k coloring $(S_1, S_2, \ldots, S_k)$ is *optimal* if $|\cup_{i=1}^{k} S_i| = \alpha_k$, the maximum number of vertices of an induced subgraph of G which can be colored in k colors. In particular, α_1, denoted simply by α, is the *stability number* or the *independence number* of G.
If P is a path, we also denote by P the set of vertices of the path P. The following is the classic theorem of Dilworth [3] concerning the partially ordered set(poset).

Theorem 1.1. *In the graph of a partially ordered set, the minimum number of paths partitioning the vertex set is equal to the cardinality of a maximum stable set.*

The graph of a partially ordered set is transitive. If the condition of transitivity is dropped, we have the following theorem due to Gallai and Milgram [5].

Theorem 1.2. *In a directed graph, the minimum number of paths partitioning the vertex set is less than or equal to the stability number.*

This theorem is the best possible in the sense that, equality is always attained if we *reorient* all the arcs of G incident to a maximum stable set S either towards exterior of S or towards interior of S [1].
This observation gives the following formula:
In any *undirected* graph G we have:

$$\text{The stability number} \quad \alpha = \max_{\vec{G}} \big(\min(|\mathcal{P}| \mid \mathcal{P} \text{ is a path partition of an orientation of the edges of } G) \big)$$

where $\mathcal{P}$ is a partition into directed paths of the vertex set X of $\vec{G}$, an orientation of the edges of G, and $|\mathcal{P}|$ is the number of paths in the path partition $\mathcal{P}$. Here, the maximum is taken over all possible orientation of the edges of the graph G.

If the graph is a transitive graph, every path induces a clique and inversely a complete graph contains a hamilton path (see Berge [1]).
The following is the dual of the Dilworth's theorem (see Berge [1]).

Theorem 1.3. *In the graph of a partially ordered set, the minimum number of stable sets partitioning the vertex set is equal to the number of vertices in a longest path.*

If the condition of transitivity is dropped from the graph, we have the following theorem proved independently by Gallai [4] and Roy [9].

Theorem 1.4. *In any graph, the chromatic number γ is equal to or less than the number of vertices in a longest path.*

This theorem is the best possible in the following sense.
We can always find an orientation of the edges of an undirected graph such that the chromatic number of the graph is equal to the number of vertices in a longest path (see Berge [1]).
The above observation leads to the following formula:
The chromatic number of an undirected graph

$$\gamma = \min_{\vec{G}} \big(\max(|P| \mid P \text{ is a directed path of an orientation of the edges of } G) \big).$$

Here, $\vec{G}$ is an orientation of the edges of G, $|P|$ is the number of vertices of the path P and the minimum is taken over all possible orientation $\vec{G}$ of the edges of G.

In order to unify the theorems of Gallai-Milgram and Gallai-Roy, Berge [2] posed the following two conjectures. To state the two conjectures, we need a few definitions.

Definition 1.5. *Let P be a path in a directed graph and k an integer such that $1 \leq k \leq \max_P |P|$, the maximum is taken over all paths P of G where $|P|$ is the number of vertices of the path P. Consider now a partial k coloring of G. The path P is said to be strongly colored or well colored or the partial k coloring is strong for P, if the number of different colors that the path P meets is exactly $\min\{k, |P|\}$. That is, if $|P| \leq k$, then, all the vertices of the path P are distinctly colored. If on the other hand, if, $|P| > k$, then P meets all the k colors.*

Definition 1.6. *Consider a directed graph and a path partition $\mathcal{P} = (P_1, P_2, \ldots, P_s)$ of the vertex set X. Let k be any integer satisfying $1 \le k \le \max_P |P|$, the maximum is taken over all paths P of G. Then consider the sum*

$$B_k(\mathcal{P}) = \sum_{i=1}^{s} \min(k, |P_i|).$$

A path partition $\mathcal{P}'$ is k-finer than $\mathcal{P}$ if $B_k(\mathcal{P}') < B_k(\mathcal{P})$. A path partition $\mathcal{P}''$ is k-finest or k-optimal or k-saturated if there is no other k-finer partition of the vertex set. In other words, the partition $\mathcal{P}''$ minimizes the sum $B_k(\mathcal{P}'')$.

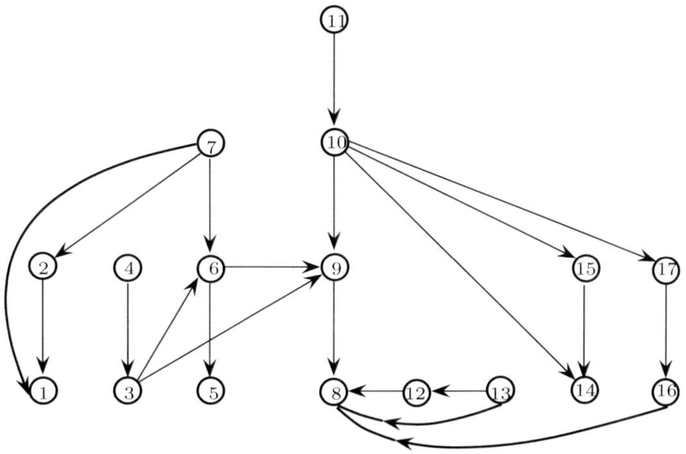

Figure 1.1:

Example. Let us refer to the graph G of Figure 1.1. Consider the path partition $\mathcal{P} = (P_1, P_2, \ldots, P_8)$ where $P_1 = (2,1), P_2 = (4,3)$, $P_3 = (7,6,5)$, $P_4 = (11,10,9,8)$, $P_5 = (12)$, $P_6 = (13)$, $P_7 = (15,14)$, $P_8 = (17,16)$.

$B_2(\mathcal{P}) = 14$ and $\mathcal{P}$ is not a 2-optimal partition of the vertex set of G as the path partition $\mathcal{P}' = (P'_1, P'_2, \ldots, P'_8)$ where $P'_1 = (7,2,1), P'_2 = (3,6,9)$, $P'_3 = (13,12,8)$, $P'_4 = (10,15,14)$, $P'_5 = (17,16)$, $P'_6 = (4)$, $P'_7 = (5)$, $P'_8 = (11)$ verifies $B_2(\mathcal{P}') = 13$ and hence $\mathcal{P}'$ is 2-$finer$ than the partition $\mathcal{P}$.

The partition $\mathcal{P}'' = (P''_1, P''_2, \ldots, P''_6)$ where $P''_1 = (4,3,6,9)$, $P''_2 = (11,10,17,16)$, $P''_3 = (13,12,8)$, $P''_4 = (7,2,1)$, $P''_5 = (15,14)$, $P''_6 = (5)$ can be seen to be a 2-$optimal$ partition of the graph G of Figure 1.1 and $B_2(\mathcal{P}'') = 11$.

We are now ready to state the conjectures of Berge [2].

Conjecture 1.7 (Strong Path Partition Conjecture). For every k-optimal path partition of a directed graph, there is a partial k-coloring which colors all paths of the partition strongly.

For $k = 1$, this conjecture is equivalent to a slightly stronger result of the Gallai-Milgram [5] theorem, proved by Linial [7].
If k is equal to the number of vertices of a longest path, then we get a stronger result of the Gallai-Roy Theorem proved by Berge [2].
In fact, the above conjecture can be considered as an extension of the following celebrated generalization of the Dilworth's theorem due to Greene-Kleiteman [6].

Theorem 1.8. *Let G be the graph of a partially ordered set and let $\mathcal{P}$be a k-optimal path partition of the vertex set of G. Then*

$$B_k(\mathcal{P}) = \alpha_k.$$

We now state the weak path partition conjecture.

Conjecture 1.9 (The Weak Path Partition Conjecture). Every graph G admits a path partition $\mathcal{P}$ such that $B_k(\mathcal{P}) \leq \alpha_k$ where α_k is the maximum number of vertices of an induced subgraph which can be colored in k colors.

It can be seen that the Strong Path Partition Conjecture (SPPC) implies The Weak Path Partition Conjecture (WPPC). The SPPC gives a structural approach to the WPPC.

Definition 1.10 (Level Sets Associated with a Path Partition). *Consider a path partition $\mathcal{P} = (P_1, P_2, \ldots, P_s)$ of the vertex set X. A vertex $x \in P_i$, is said to be at level k of the partition $\mathcal{P}$, if the section of the path P_i starting from the vertex x to the terminal vertex of P_i contains exactly k vertices. The set of vertices at level k of all the paths of the partition $\mathcal{P}$ is denoted by $L_i(\mathcal{P})$.*

It is not difficult to see that

$$\sum_{i=1}^{s} \min\{k, |P_i|\} = |L_1(\mathcal{P})| + |L_2(\mathcal{P})| + \cdots + |L_k(\mathcal{P})|.$$

Set $|P_i| = r_i$ for $i = 1, 2, \ldots, s$. Let r_k^* be the number of integers of the sequence (r_i) which are greater than or equal to the integer $k(\geq 1)$.
The sequence (r_i^*) is called the *dual* of the sequence (r_i). Let us note that $r_k^* = 0$ if $k > \max_i r_i$ and that $|L_i(\mathcal{P})| = r_i^*$.
The number of paths in the partition $\mathcal{P}$ is denoted by $|\mathcal{P}|$.

2 The Result

Theorem 2.1. *Let G be a graph with a $(k+1)$-optimal path partition $\mathcal{P} = (P_1, P_2, \ldots, P_s)$. Let this path partition satisfy the following property: The vertices of the paths of the partition $\mathcal{P}$ from the level 1 to the the level k (including the level k) can be colored strongly using k colors. Then the paths of the partition $\mathcal{P}$ can be colored strongly with $k+1$ colors. In particular, the WPPC is true for $k+1$.*

Proof. Consider the graph G' induced by the vertices $X \setminus \{L_1(\mathcal{P}) \cup L_2(\mathcal{P}) \cup \cdots \cup L_k(\mathcal{P})\}$. Let $\mathcal{P}' = (P_1 \setminus \cup_{i=1}^k L_i(\mathcal{P}), P_2 \setminus \cup_{i=1}^k L_i(\mathcal{P}), \ldots, P_s \setminus \cup_{i=1}^k L_i(\mathcal{P}))$ be the path partition of graph G' induced by the partition $\mathcal{P}$. Then $\mathcal{P}'$ is a path partition of G' such that the set of terminal vertices of $\mathcal{P}'$, $L_1(\mathcal{P}')$, is minimal with respect to the inclusion. Otherwise, there is a path partition $\mathcal{P}''$ of G' such that the set of terminal vertices of $\mathcal{P}''$ is properly contained in the set of terminal vertices of $\mathcal{P}'$ and that the set $L_1(\mathcal{P}'')$ is minimal with respect to this inclusion. In particular, the number of paths in $\mathcal{P}''$ is strictly less than the number of paths in the partition $\mathcal{P}'$. This means that the path partition $\mathcal{P}''$ of G' together with the paths from the level 1 to the level k(including the level k) of the partition $\mathcal{P}$ constitute a new path partition $\mathcal{P}'''$ of the graph G. But then, $B_{k+1}(\mathcal{P}''') = \sum_{i=1}^k r_i^* + |\mathcal{P}''| < \sum_{i=1}^k r_i^* + |\mathcal{P}'| = B_{k+1}(\mathcal{P})$, a contradiction to the fact that $\mathcal{P}$ is a $(k+1)$- optimal path partition of G. This contradiction proves that the path partition $\mathcal{P}'$ partitions the vertex set of G' with the property that the set of terminal vertices of $\mathcal{P}'$ is minimal relative to the inclusion. By applying the theorem of Linial [7] to the partition $\mathcal{P}'$ of G', we have a stable set S of G' which intersects exactly once each paths of $\mathcal{P}'$. Now let us color the vertices from the level 1 to the level k(including the level k) of the partition $\mathcal{P}$ strongly with the first k colors. Then finally, color the vertices of the stable set S with the last color $k+1$. This gives us a strong $k+1$ coloring of the paths of the partition $\mathcal{P}$. Hence the theorem. □

Example. Let G be a bipartite graph and consider *any* path partition $\mathcal{P}$ of the vertex set X. Then the vertices from the level 1 to the level k (including the level k, k even) can be strongly colored with k colors. That is, $\mathcal{P}$ satisfies the property of the above theorem for all *even* integer k (see Berge [2]).
The above result motivates the following conjecture:

Conjecture 2.2 (Semi-Strong Path Partition Conjecture (SSPPC)). For any graph, there is a path partition whose paths can be strongly colored with k colors.

Evidently, SPPC implies SSPPC which in turn implies WPPC.

Corollary 2.3. *Let $\mathcal{P}$ be a 2-optimal path partition of the vertex set X of the graph G. Let us suppose the terminal vertices of the paths of $\mathcal{P}$ form a stable set. Then,* **SSPPC** *is true for $k=2$.*

Proof. If the terminal vertices form a stable set, then the corollary follows by applying the above theorem for $k=1$. □

Now we consider the class of graphs satisfying the following property:

Property (k). The *m*aximum number of vertex disjoint k-chromatic induced subgraphs is equal to the *m*inimum number of vertices intersecting every k-chromatic induced subgraph.

Fact 2.3.1. *Every* cactus *(a graph in which every block is an elementary cycle or K_2) satisfies the above property for $k=3$.*

Proof. Let us observe that the only induced blocks of chromatic number three are odd cycles. The proof is by induction on the number of odd cycles. If there is no odd cycle or there is exactly one odd cycle the property verifies trivially.
Suppose the property for all cactus with $\leq c$ odd cycles and consider a cactus G with $c+1$ odd cycles. Let us take any odd cycle and an articulation point a(cut-point) on it. Then the property is true for the graph $G-a$ and hence true for G. □

On the other hand, the *E*scher wall of height h (See B. Reed [8]) contains neither 2 vertex disjoint odd cycles nor *a*n odd cycle cover (a set of vertices hitting every odd cycle) of less than h vertices.

Theorem 2.4. *Let G be a graph satisfying the above property(k). Then* **WPPC** *is true for $k-1$.*

Proof. Let $\{x_1, x_2, \ldots, x_s\}$ be a minimum set of vertices meeting all induced subgraphs of chromatic number k. Consider a family of s induced subgraphs $G_i = (X_i, U_i)$ for $i=1,2,\ldots,s$ where the the chromatic number of G_i is k and $X_i \cap X_j = \emptyset$ for $i \neq j$. Extract a path P_i of k vertices (length $k-1$) from each of the induced subgraph G_i for $i=1,2,\ldots,s$. Such a path exists by virtue of the theorem of Gallai-Roy.

Now let us consider a path partition $\mathcal{P} = (P_1, P_2, \ldots, P_s, P_{s+1}, \ldots, P_q)$ of the vertex set X of G where each path P_j represents a singleton vertex of the graph generated by $X - \{P_1, P_2, \ldots, P_s\}$ for $j = s+1, s+2, \ldots, q$.
Now it can be seen that $B_{k-1}(\mathcal{P}) = |X - \{x_1, x_2, \ldots, x_s\}| = n - s$ and the chromatic number of the induced subgraph $G - x_1 - x_2 - \cdots - x_s$ is $\leq k-1$ because of the property P_k. Hence $B_{k-1}(\mathcal{P}) \leq \alpha_{k-1}$. □

The SPPC is proved for graphs in which no two cycles intersect at a vertex (see [10]).
Now, we prove the

Corollary 2.5. *The WPPC is true for cactus.*

Proof. If there is no odd cycle, then the graph is bipartite and the conjecture is true (see Berge [2]). If not, the chromatic number of the cactus is 3 and the conjecture is trivially true for $k \geq 3$. For $k = 2$, the conjecture follows from the above theorem because the cactus satisfies the property(k) with $k = 3$. For $k = 1$, it follows from the theorem of Gallai-milgram [5]. □

References

[1] C. Berge, *Graphes*, Gauthier-Villars, Paris (1983).

[2] C. Berge, k-optimal partitions in directed graphs, *European J. of Combinatorics*, **3** (1982), pp. 97–101.

[3] R.P. Dilworth, A decomposition theorem for partially ordered sets, *Ann. of Math.*, **51** (1950),.

[4] T. Gallai, In *Directed paths and circuits, Theory of Graphs, (P.Erdös and G.O.H. Katona, eds)*, pp. 115–118. Academic Press, New York (1968).

[5] T. Gallai and A.N. Milgram, Verallgemeinerung eines Graphentheoretischen Satzes Von Rédei, *Acta Sci, Math.*, **21** (1960), pp. 181–186.

[6] C. Greene and D.J. Kleitman, The structure of Sperner k-families, *J. Combin. Theory Ser. A*, **20** (1976), pp. 41–68.

[7] N. Linial, Covering digraphs by paths, *Discrete Math.*,**23** (1978), pp. 257–272.

[8] B. Reed, Mangoes and Blueberries, *Combinatorica*, **19(2)** (1999), pp. 267–296.

[9] B. Roy, Nombre chromatic et plus longs chemins, *Rev; Fr. Automat. Informat.*, **1** (1967), pp. 127–132.

[10] Sriraman Sridharan, On the strong path partition conjecture of Berge, *Discrete Mathematics*, **117** (1993), pp. 265–270.

Proc. of ICDM 2006, Pages 131–140

Circular Perfect Graphs

XUDING ZHU
Department of Applied Mathematics, National Sun Yat-sen University, Kaohsiung, Taiwan 80424
and
National Center for Theoretical Sciences
Email: zhu@math.nsysu.edu.tw

Abstract. For $1 \leq d \leq k$, let $K_{k/d}$ be the graph with vertices $0, 1, \cdots, k-1$, in which $i \sim j$ if $d \leq |i-j| \leq k-d$. The circular chromatic number $\chi_c(G)$ of a graph G is the minimum of those k/d for which G admits a homomorphism to $K_{k/d}$. The circular clique number $\omega_c(G)$ of G is the maximum of those k/d for which $K_{k/d}$ admits a homomorphism to G. A graph G is circular perfect if for every induced subgraph H of G we have $\chi_c(H) = \omega_c(H)$. This paper surveys results on circular perfect graphs.

1 Introduction

Suppose $G = (V, E)$ and $G' = (V', E')$ are graphs. A *homomorphism* from G to G' is mapping $f : V \to V'$ such that $f(x)f(y) \in E'$ whenever $xy \in E$. Homomorphism of graphs is a generalization of graph colourings. An n-colouring of a graph G is equivalent to a homomorphism of G to K_n. We write $G \preceq G'$ if there exists a homomorphism from G to G'. Two graphs G and G' are *homomorphically equivalent*, written as $G \approx G'$, if $G \preceq G'$ and $G' \preceq G$. The relation $\approx$ is an equivalence relation. Let $\mathcal{F}$ be the set of finite graphs. Then $\preceq$ defines a partial order on $\mathcal{F}/\approx$.

Let $\mathcal{Z}_{\mathcal{G}}$ be the set of complete graphs, i.e., $\mathcal{Z}_{\mathcal{G}} = \{K_1, K_2, \cdots, \}$. Then $\mathcal{Z}_{\mathcal{G}}$ forms an infinite increasing chain in the partial order $(\mathcal{F}/\approx, \preceq)$. Every graph $G \in \mathcal{F}$ admits a homomorphism to some members of the set $\mathcal{Z}_{\mathcal{G}}$, and contains some member of $\mathcal{Z}_{\mathcal{G}}$ as its subgraphs. The *chromatic number* $\chi(G)$ is the minimum n such that $G \preceq K_n$. The *clique number* $\omega(G)$ is the maximum n such that $K_n \preceq G$. We may view the set $\mathcal{Z}_{\mathcal{G}}$ as a representation of natural numbers, with K_n be a representation of the integer n. Then $\chi(G)$ is the least element of the set $\mathcal{Z}_{\mathcal{G}}$ which is "above" G in the order $\preceq$, and $\omega(G)$ is the maximum element of $\mathcal{Z}_{\mathcal{G}}$ which is "below" G in the order $\preceq$. In this sense, we may view the set $\mathcal{Z}_{\mathcal{G}}$ as a scale that measures a dimension of graphs.

Just as the set of natural numbers is extended to the set of rational numbers, we can "extend" the set $\mathcal{Z}_{\mathcal{G}}$ into a larger set. For a fraction k/d with $k \geq 2d$, let $K_{k/d}$ be the graph with vertex set $\{0, 1, \cdots, k-1\}$ and edge set $\{ij : d \leq |i-j| \leq k-d\}$. We shall denote by $\mathcal{Q}_{\mathcal{G}}$ the set $\{K_{k/d} : k \geq 2d\} \cup \{K_1\}$. Note that $K_{k/1} = K_k$. So $\mathcal{Q}_{\mathcal{G}}$ is an extension of $\mathcal{Z}_{\mathcal{G}}$. Moreover, the set $\mathcal{Q}_{\mathcal{G}}$ is also linearly ordered. It was shown in [1, 10] that if $k'/d' \geq 2$ and $k/d \geq 2$, then $k'/d' \leq k/d$ if and only if $K_{k'/d'} \preceq K_{k/d}$. In particular, if $k/d = k'/d'$, then $K_{k/d} \approx K_{k'/d'}$. We call the graphs $K_{k/d}$ *circular cliques* or *circular complete graphs*.

The set $\mathcal{Q}_\mathcal{G}$ together with the order $\preceq$ may be viewed as a representation of those rational numbers $r \geq 2$ or $r = 1$ [18]. The *circular chromatic number* $\chi_c(G)$ of a graph is defined as

$$\chi_c(G) = \inf\{k/d : G \preceq K_{k/d}\},$$

and the *circular clique number* $\omega_c(G)$ of G is defined as

$$\omega_c(G) = \sup\{k/d : K_{k/d} \preceq G\}.$$

For any finite graph G, the infimum in this definition of $\chi_c(G)$ and the supremum in the definition of $\omega_c(G)$ are always attained [10, 16]. Hence, $\chi_c(G)$ is the smallest member of $\mathcal{Q}_\mathcal{G}$ which is above G, and $\omega_c(G)$ is the largest member of $\mathcal{Q}_\mathcal{G}$ which is below G in the order $\preceq$.

We view both $\mathcal{Z}_\mathcal{G}$ and $\mathcal{Q}_\mathcal{G}$ as scales that measures a dimension of graphs. As $\mathcal{Z}_\mathcal{G}$ is a subset of $\mathcal{Q}_\mathcal{G}$, the set $\mathcal{Q}_\mathcal{G}$ is a finer scale. The circular chromatic number $\chi_c(G)$ of G is a refinement of its chromatic number $\chi(G)$, and the circular clique number $\omega_c(G)$ of G is a refinement of its clique number $\omega(G)$. Indeed, it follows easily from the definitions (and the fact the infimum and the supremum are attained in the definitions) that for any finite graph G,

$$\omega(G) \leq \omega_c(G) \leq \chi_c(G) \leq \chi(G), \; \omega(G) = \lfloor \omega_c(G) \rfloor, \text{ and } \chi(G) = \lceil \chi_c(G) \rceil.$$

A graph G is *perfect* if for every induced subgraph H of G we have $\chi(H) = \omega(H)$. Perfect graphs have been studied extensively in the literature since it was introduced by Berge in 1961, culminating in the proof by Chudnovsky, Robertson, Seymour, and Thomas [2] of the strong perfect graph theorem. The following definition is a natural generalization of this concept to circular colouring.

Definition 1.1. *A graph G is called* circular perfect *if for every induced subgraph H of G we have $\chi_c(H) = \omega_c(H)$.*

The circular chromatic number of a graph was first introduced by Vince [10] under the name "star chromatic number". An equivalent definition of the parameter, using circular colouring, was given in [12]. Circular chromatic number of graphs has attracted considerable attention. Many interesting results are obtained and techniques are developed. Surveys on this topics are given in [13, 19]. The circular clique number of a graph and the concept of circular perfect graphs were introduced in [16]. In this article, we survey results obtained concerning circular perfect graphs. Similar to the strong perfect graph theorem, one might expect a characterization of circular perfect graphs parallel to the characterization of perfect graphs. However, it is unlikely that there is a simple characterization of circular perfect graphs, as we shall explain. If restricted to special classes of graphs, the characterization might be possible. We present some necessary and some sufficient conditions for a graph to be circular perfect, discuss circular perfectness of graphs for special classes of graphs, and present examples of circular perfect graphs with given properties.

2 Basic Facts

Since $\omega(G) \leq \omega_c(G) \leq \chi_c(G) \leq \chi(G)$ for any graph G, if G is perfect, then G is circular perfect. So the class of circular perfect graphs contains the class of perfect graphs as a subclass. To study circular perfect graphs, we concentrate on circular perfect graphs that are not perfect.

By definition, if G contains $K_{k/d}$ as a subgraph, then $\omega_c(G) \geq k/d$. The converse is also true, although it is not obvious from the definition.

Lemma 2.1. *If G is a graph with $\omega_c(G) = k/d$ and $gcd(k,d) = 1$, then $K_{k/d}$ is an induced subgraph of G.*

It follows from Lemma 2.1 that determining the circular clique number of a graph can be accomplished by testing all its induced subgraphs. This is certainly not an efficient way, but most likely there is no efficient way as the problem of determining the circular clique number of a graph is NP-hard. Indeed, once the circular clique number of a graph is determined then its clique number is determined. As the latter problem is NP-hard, so is the former one.

Observe that the induced subgraphs of circular complete graphs are usually not circular complete graphs. Thus it is not obvious that circular complete graphs are circular perfect. Nevertheless, we have the following result:

Lemma 2.2. *If G is an induced subgraph of a circular complete graph $K_{k/d}$, then $\chi_c(G) = \omega_c(G)$.*

Corollary 2.3. *Suppose $k \geq 2d$. The circular complete graph $K_{k/d}$ is circular perfect.*

In particular, for $k \geq 2$, the circular complete graphs $K_{(2k+1)/k}$ and $K_{(2k+1)/2}$, which are odd holes and odd anitholes, are circular perfect. As odd holes and odd antiholes are not perfect, the class of perfect graphs is a proper subclass of the class of circular perfect graphs. Although a circular perfect graph need not be perfect, the neighbourhood of any vertex in such a graph must be a perfect graph. For a vertex x of a graph G, let $N_G[x]$ (or $N[x]$) denote the close neighbourhood of x.

Theorem 2.4. *If G is circular perfect then for every vertex x of G, $N_G[x]$ induces a perfect graph.*

3 Some Sufficient Conditions for a Graph to be Circular Perfect

The necessary condition in Theorem 2.4 for a graph to be circular perfect is not sufficient. By adding some further requirements, we can obtain a sufficient condition.

Theorem 3.1. *Suppose G is a graph such that for every vertex x of G, $N[x]$ is a perfect graph and $G - N[x]$ is a bipartite graph with no induced P_5 (i.e., a path with 5 vertices). Then G is circular perfect.*

It is easy to see that if G satisfies the condition of Theorem 3.1, then any induced subgraph of G satisfies that condition. Therefore to prove Theorem 3.1, it suffices to prove the following:

Theorem 3.2. *Suppose G is a graph and for every vertex x of G, $N[x]$ is a perfect graph and $V - N[x]$ is a bipartite graph which contains no induced P_5. Then $\chi_c(G) = \omega_c(G)$.*

The proof of Theorem 3.2 is quite long. The following example shows that these conditions are quite tight in some sense. Let G be the graph as depicted in Figure 3 below. It is easy to verify that $\chi_c(G) = 8/3$ and $\omega_c(G) = 5/2$. So G is not circular perfect. But G "almost satisfies" the condition of Theorems 3.2. For each vertex x of G, $N[x]$ is a star, and $G - N[x]$ is either a P_4 or a P_5.

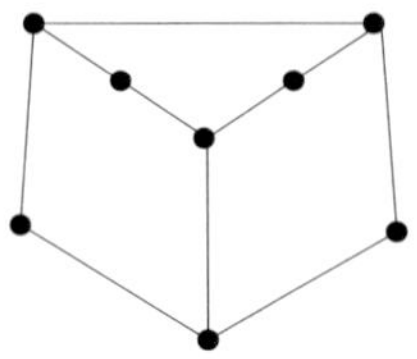

Figure 3.1: A non-circular perfect graph G

On the other hand, it is easy to construct circular perfect graphs G which contain vertices x such that $G - N[x]$ is not bipartite, or $G - N[x]$ is bipartite but contains induced path length greater than 5. Indeed, if $k = 2d + 1$ and $d \geq 4$, then the circular complete graph $K_{k/d}$ does not satisfy the condition of Theorem 3.1.

The next sufficient condition for a graph to be circular perfect concerns triangle free graphs. Suppose $G = (V, E)$ is a graph and x is a vertex of G. For two vertices $u, v \in V - N[x]$,

- $u \leq^x v$ means $N(u) \cap N(x) \subseteq N(v) \cap N(x)$;
- $u <^x v$ means $N(u) \cap N(x) \subset N(v) \cap N(x)$;
- $u =^x v$ means $N(u) \cap N(x) = N(v) \cap N(x)$.

Here $A \subset B$ means that A is a proper subset of B.

Definition 3.3. *Given an induced path $P_n = (p_0, p_1, \cdots, p_n)$ of $G - N[x]$, we say P_n is* badly-linked *with respect to x if one of the following holds:*

1. *There are three indices $i < j < k$ of the same parity such that $p_i \not\leq^x p_j$ and $p_k \not\leq^x p_j$.*

2. *There are three indices $i < j < k$ of the same parity such that $p_j \not\leq^x p_i$ and $p_j \not\leq^x p_k$.*

3. *There are two even indices $i < j$ and two odd indices $i' < j'$ such that $p_i \not\leq^x p_j$ and $p_{i'} \not\leq^x p_{j'}$.*

An induced path P of $G - N[x]$ is called well-linked *with respect to x if it is not badly-linked with respect to x.*

Theorem 3.4. *Suppose G is a triangle free graph such that for every vertex x of G, $G - N[x]$ is a bipartite graph with no induced C_n for $n \geq 6$, and any induced path of $G - N[x]$ is well-linked. Then G is circular perfect.*

It is easy to see that if G satisfies the condition of Theorem 3.4, then any induced subgraph of G satisfies that condition. Therefore to prove Theorem 3.4, it suffices to prove the following:

Theorem 3.5. *Suppose G is a triangle free graph. If for every vertex x of G, $V - N[x]$ is a bipartite graph which contains no induced C_n for $n \geq 6$, and every induced path of $G - N[x]$ is well-linked, then $\chi_c(G) = \omega_c(G)$.*

The proof of Theorem 3.5 is again quite complicated. Theorems 3.1 and 3.4 are used to prove an analogue of Hajós Theorem for circular chromatic number. In [14], three graph operations are given to play the role of the Hajós sum in original Hajós Theorem. Namely it was proved in [14] that if $k/d \geq 3$, then all the graphs G with $\chi_c(G) \geq k/d$ can be constructed from $K_{k/d}$ by adding edges and vertices, by identifying non-adjacent vertices and by applying the three graph operations that replace the Hajós sum. Moreover all such constructed graphs G have $\chi_c(G) \geq k/d$. The same result is proved for $2 \leq k/d < 3$ in [15], where four new graph operations are introduced in place of the Hajós sum.

4 Minimal Circular Imperfect Graphs

A graph G is *minimal imperfect* if G is not perfect, but every induced subgraph of G is perfect. Similarly, we call a graph G *circular imperfect* if G is not circular perfect, and *minimal circular imperfect* if G is circular imperfect, but every proper induced subgraph of G is circular perfect. The strong perfect graph theorem characterizes the family of perfect graphs by determining all minimal imperfect graphs. It is natural that we would like to study minimal circular imperfect graphs. The problem of characterizing all minimal circular imperfect graphs is difficult. Thus we restricted to special classes of graphs.

First we consider norms of circular cliques. Suppose G is a graph. The *norm* of G, denoted by norm(G), is the spanning subgraph H of G that consists of all the edges e that are contained in a maximum clique of G. The following result, proved in [8], characterizes all circular cliques whose norms are circular imperfect, and those whose norms are minimal circular imperfect.

Theorem 4.1. *Suppose $gcd(k, d) = 1$. Then*

- norm$(K_{k/d})$ *is circular imperfect if and only if* $k \not\equiv -1 \pmod d$ *and* $\lfloor k/d \rfloor \geq 3$.
- norm$(K_{k/d})$ *is minimal circular imperfect if and only if* $k = 3d+1$ *and* $d \geq 3$.

A graph G is a *(p, q)-partitionable graph* if $|V(G)| = pq + 1$, and for each vertex v of G, $G \setminus \{v\}$ admits a partition into p cliques of cardinality q as well as a partition into q stable sets of cardinality p. A graph is *partitionable* if it is a (p, q)-partitionable graph for some $p, q \geq 2$. Partitionable graphs were introduced by Lovász [4] and Padberg [5] as a tool in the study of perfect graphs. Before the proof of the strong perfect graph theorem, it was shown by Lovász [4] and Padberg [5] that every minimal imperfect graph is a partitionable graph. Partitional graphs satisfy $\omega(G) + 1 = \chi(G)$. As circular perfect graphs satisfies $\omega(G) + 1 \geq \chi(G)$, a natural question is which partitionable graphs are circular perfect. The following theorem, proved in [8], characterizes all partitionable graphs that are circular imperfect.

Theorem 4.2. *All partitionable graphs apart from circular complete graphs are circular imperfect.*

It is likely most of the partitionable graphs are not minimal circular imperfect. However, this is not proved yet.

It is not difficult to show that every outerplanar graph is circular perfect.

Theorem 4.3. *All outerplanar graphs are circular perfect.*

Outerplanar graphs is a subclass of series-parallel graphs. If we move from the class of outerplanar graphs to the the class of series-parallel graphs, then we can easily find circular imperfect graphs. It would be nice to characterize all the minimal circular imperfect series-parallel graphs. However, the task is not finished yet. It is easy to see that odd cycles and K_2 are the only circular cliques that are series-parallel graphs. Thus if a series-parallel graph G is circular perfect then the core of G is an odd cycle or K_2. It is known [6] that for every rational $r \in [2, 8/3]$, there is a series-parallel graph G with $\chi_(G) = r$. If $r \neq 2 + 1/k$, then such a series-parallel graph is circular imperfect. It remains an open problem as to determine which rational is the circular chromatic number of a minimal circular imperfect series-parallel graph. The following result can be proved by using the construction method given in [6].

Theorem 4.4. *If k/d is a rational number and there is an integer $q \geq 2$ such that $(2q+1)d = kq - 1$, then there is a minimal circular imperfect series-parallel graph G with $\chi_c(G) = k/d$.*

Figure 4.1 is a minimal circular imperfect series-parallel graph with $\chi_c(G) = 13/5$.

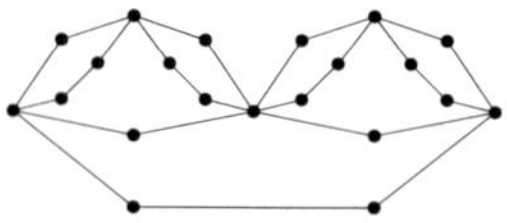

Figure 4.1: A minimal circular imperfect series-parallel graph G with $\chi_c(G) = 13/5$

For planar graphs, the problem of characterizing circular perfect planar graphs is more difficult. However, planar circular cliques are easy to identify: K_4, K_3 and $K_{(2k+1)/k}$ for some $k \geq 2$.

Theorem 4.5. *If a planar graph G is circular perfect then either $\chi_c(G) = 4$ or $\chi_c(G) \in \{2 + \frac{1}{k} : k = 1, 2, \cdots\}$.*

It follows from Theorem 4.5 that if G is a planar graph with $\chi(G) = 4$ but $\omega(G) = 3$, then G is circular imperfect; if $\chi(G) = 3$ and the core of G is not an odd cycle, then G is circular imperfect.

The following are two classes of minimal circular imperfect planar graphs.

Example 4.1. Let G be the graph with vertex set

$$V = \{x_1, x_2, \cdots, x_k, u_1, u_2, \cdots, u_{k-1}, v_1, v_2, \cdots, v_{k-1}\}$$

and edge set

$$\{x_i v_i, x_i, u_i, u_i v_i, x_{i+1} v_i, x_{i+1} u_i : i = 1, 2, \cdots, k-1\} \cup \{x_1 x_k\}.$$

Let $1 = i_1 < i_2 < \cdots < i_t = k+1$. For $j = 1, 2, \cdots, t-1$, identify u_{i_j}, u_{i_j+1}, ..., $u_{i_{j+1}-1}$ into a single vertex. The resulting graph is a minimal circular imperfect graph, provided that it contains no K_4.

Figure 4.2 is an example of such a graph.

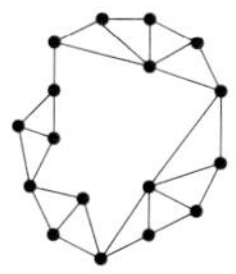

Figure 4.2: A minimal circular imperfect planar graph

Another class of planar graphs that are minimal circular imperfect is defined as follows:

Example 4.2. Let G be obtained from an odd wheel by subdividing each edge into a path so that each finite face is has length $2k+1$ for some $k \geq 1$, and the infinite face has length at least $2k+1$. Then G is minimal circular imperfect.

Figure 4.3 is an example of such a graph.

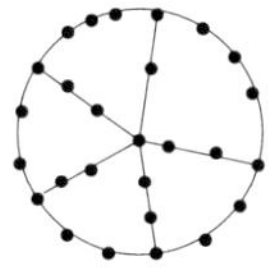

Figure 4.3: A minimal circular imperfect planar graph

Suppose G is a series-parallel graph which is circular imperfect. Suppose u, x, v are three consecutive vertices of a face of G. Let H be obtained from G by identifying u, v into a single vertex. If H has the same odd girth as G, then H is circular imperfect. Usually, if G is minimal circular imperfect, H is also minimal circular imperfect. Such minimal circular imperfect graphs may not be included in the two classes of minimal circular imperfect graphs presented above. So it seems not easy to characterize all minimal circular imperfect planar graphs.

The family of minimal circular imperfect line graphs was studied by Xu. The following result is proved in [11].

Theorem 4.6. *Suppose $G = L(H)$ is the line graph of H. Then G is minimal circular imperfect if and only if H is 3-critical, i.e., $\Delta(H) = 3$ and $\chi'(H) = 4$, and for any edge e of H, $\chi'(H - e) = 3$.*

It is a difficult problem to determine 3-critical graphs, and most likely, there is no simple characterization of such graphs. Thus this theorem suggests that there is no easy characterization of minimal circular imperfect graphs.

Line graphs is a subclass of claw-free graphs. Minimal circular imperfect claw-free graphs is studied in [9], where the following result is proved.

Theorem 4.7. *If G is a minimal circular imperfect graph and G is claw-free, then either $\alpha(G) \leq 3$ or $\omega(G) \leq 3$.*

Note that minimal imperfect graphs G have $\min\{\alpha(G), \omega(G)\} = 2$. A natural question is whether $\min\{\alpha(G), \omega(G)\} \leq c$ for some constant c for minimal circular imperfect graphs. This question was answered in the negative in [7].

Theorem 4.8. *For any integer k, there is a minimal circular imperfect graph G with $\min\{\alpha(G), \omega(G)\} > k$.*

Figure 4.4 below is a minimal circular imperfect graph G with $\min\{\alpha(G), \omega(G)\} = 5$. The subset $V = \{v_0, v_1, \cdots, v_{14}\}$ of $V(G)$ induces a copy of $K_{15/2}$, i.e., $v_i v_j$ is an edge if $2 \le |i-j| \le 13$. For simplicity, these edges are not shown in the figure.

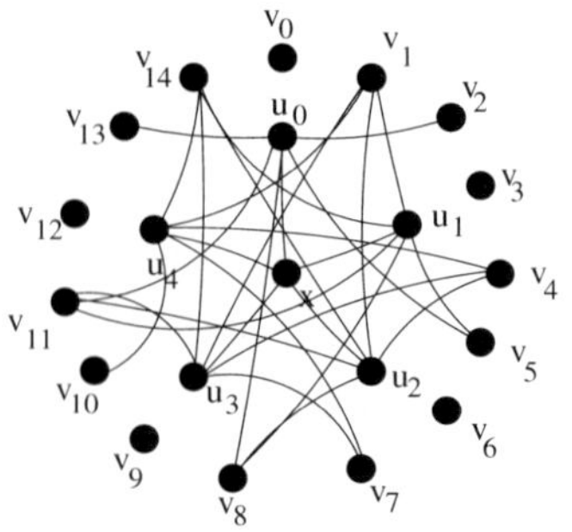

Figure 4.4: A minimal circular imperfect graph G with $\min\{\alpha(G), \omega(G)\} = 5$, with edges between the v_i's missing.

5 Strongly Circular Perfect Graphs

A nice property of perfect graphs is that if G is perfect then its complement $\overline{G}$ is also perfect. This is not true for circular perfect graphs. For example, the complement of circular complete graphs are usually not circular perfect. For example, $\overline{K_{8/3}}$ is not circular perfect. Indeed, for most circular cliques $K_{k/d}$, their complements $\overline{K_{k/d}}$ are not circular perfect.

Theorem 5.1 ([3]). *Suppose $k \ge 2d$. Then $\overline{K_{k/d}}$ is circular perfect if and only if $d \le 2$ or $k = 2d+1$ or $d = 3$ and $k \equiv 0 \pmod 3$.*

A graph is called *strongly circular perfect* if both G and $\overline{G}$ are circular perfect. It follows from Theorem 5.1 that if G is a strongly circular perfect graph, then $\chi_c(G) = k/d$ for some $d \le 2$, or $\chi_c(G) = 2 + 1/d$ for some integer d. Perfect graphs are strongly circular perfect. Odd cycles and the complements of odd cycles are also strongly circular perfect. The family of triangle free strongly circular perfect graphs is characterized by Coulnges, Pecher and Wagler [3].

A graph G is called an *interlaced odd hole* if $V(G)$ is partitioned into $(\cup_{i=0}^{2k} A_i) \cup (\cup_{i=0}^{2k} B_i)$ such that the following hold:

- $A_i \ne \emptyset$ for each i.
- If $|A_i| \ge 2$, then $|A_{i-1}| = |A_{i+1}| = 1$, where summation in the index is modulo $2k+1$.
- If $B_i \ne \emptyset$, then $|A_i| = 1$.
- For any i, for any $v \in A_i$, $N(v) = A_{i-1} \cup A_{i+1} \cup B_i$.
- For any i, for any $v \in B_i$, $N(v) = A_i$.

Figure 5.1 is an example of an interlaced odd hole.

Theorem 5.2 ([3]). *A triangle free graph G is strongly circular perfect if and only if either G is bipartite or G is an interlaced odd hole.*

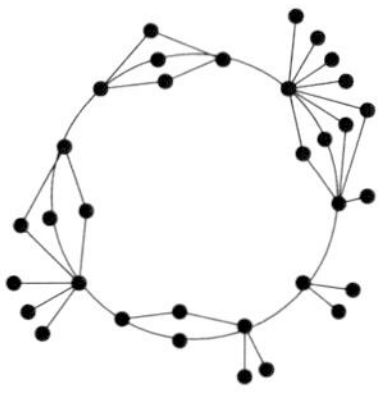

Figure 5.1: An interlaced odd hole

6 Some Open Questions

Question 6.1. *Is it true that if G is minimal circular imperfect then $\chi(G) = \omega(G) + 1$?*

Question 6.2. *For which rational number r, there is a minimal circular imperfect series-parallel graph G with $\chi_c(G) = r$?*

Question 6.3. *Characterize minimal circular imperfect series-parallel graphs.*

Question 6.4. *Suppose G is a series-parallel graph. Assume that the core of G is not an odd cycle, and for any vertex x of G, the core of $G - x$ is an odd cycle. Is it true that G is a minimal circular imperfect graphs?*

Z. Pan said the answer is no. Denote by GsH the series join of G and H, and by GpH the parallel join of G and H. Let $H = ((((P_5pP_6)s(P_5pP_6))pP_4)sP_2)pP_5$. Let $H' = P_4pP_5$. Then $L_{9,4}(H) = \{2, 7\} = L_{9,4}(H')$. Let $G = (H'sHsH')pP_1$. Then $L_{9,4}(G) = \emptyset$. So $\chi_c(G) > 9/4$. For any vertex x, $\chi_c(G - x) = 9/4 = \omega_c(G - x)$. However, G is not minimal circular imperfect because H is not circular perfect. The graph is shown in Figure 6.1 below.

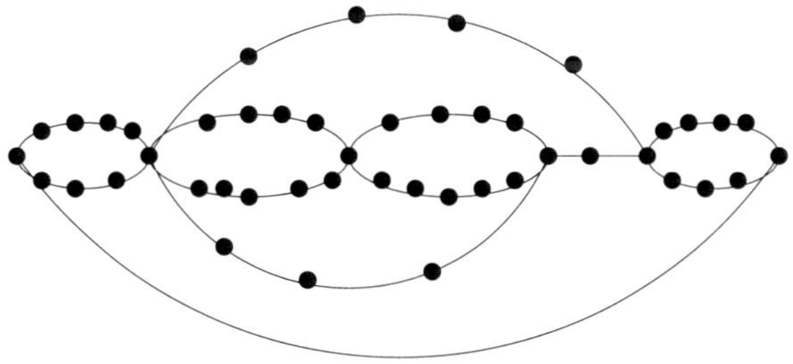

Figure 6.1: A counterexample to Question 6.4

Question 6.5. *Characterize minimal circular imperfect planar graphs.*

References

[1] J.A. Bondy and P. Hell, A note on the star chromatic number, *J. Graph Theory*, **14** (1990), pp. 479–482.

[2] M. Chudnovsky, N. Robertson, P. Seymour and R. Thomas, The Strong Perfect Graph Theorem, *Annals of Mathematics*, **164(1)**(2006), pp. 51–229.

[3] S. Coulonges, A. Pêcher and A.K. Wagler, Triangle-free strongly circular perfect graphs,, *LaBRI Research Report RR-1360-05*, 2005.

[4] L. Lovász, A characterization of perfect graphs,, *J. Combinatorial Theory Ser. B*, **13** (1972), pp. 95–98.

[5] M.W. Padberg, Perfect zero-one matrices, *Math. Programming*, **6**(1974), pp. 180–196.

[6] Z. Pan and X. Zhu, Density of the circular chromatic numbers of series-parallel graphs, *J. Graph Theory*, **46(1)**(2004), pp. 57–68.

[7] Z. Pan and X. Zhu, Minimal circular imperfect graphs of large clique number and large independence number, *European J. Combinatorics*, To appear.

[8] A. Pêcher and A. Wagler, Three classes of minimal circular-imperfect graphs, manuscript, 2005.

[9] A. Pêcher and X. Zhu, Claw-free graphs that are minimal circular imperfect, in preparation.

[10] A. Vince, Star chromatic number, *J. Graph Theory*, **12** (1988), pp. 551–559.

[11] B. Xu, On minimally circular imperfect graphs, manuscript, 2006.

[12] X. Zhu, Star chromatic numbers and products of graphs, *J. Graph Theory*, **16** (1992), pp. 557–569.

[13] X. Zhu, The circular chromatic number, a survey, *Discrete Math.*, **229(1–3)** (2001), pp. 371–410.

[14] X. Zhu, An analogue of Hajós theorem for the circular chromatic number, *Proceedings of the American Mathematical Society*, **129** (2001), pp. 2845–2852.

[15] X. Zhu, An analogue of Hajós' theorem for the circular chromatic number (II), *Graphs and Combinatorics*, **19** (2003), pp. 419–432.

[16] X. Zhu, Circular perfect graphs, *J. Graph Theory*, **48(3)** (2005), pp. 186–209.

[17] X. Zhu, Perfect graphs for generalized colouring - circular perfect graphs, in *AMS, DIMACS series: Proceedings of the DIMACS/DIMATIA Workshop on Graphs, Morphisms and Statistical Physics* **63** (2004), pp. 177–193.

[18] X. Zhu, Circular coloring and graph homomorphism, *Bulletin of the Australian Mathematical Society*, **59** (1999), pp. 83–97.

[19] X. Zhu, Recent developments in circular colouring of graphs, *Topics in Discrete Mathematics: Dedicated to Jarik Nesetril on the occasion of his 60th birthday, Algorithms and Combinatorics Series*, **28** pp. 497–550.

Proc. of ICDM 2006, Pages 141–148

Generalized Colourings and Matrix Partitions

PAVOL HELL
School of Computing Science, Simon Fraser University, Burnaby, B.C., Canada V5A 1S6
Email: pavol@cs.sfu.ca

Abstract. Very few graph colourability problems can be characterized by finitely many forbidden subgraphs. On the other hand, certain natural generalizations of graph colouring admit such characterizations somewhat more frequently. In these notes, we survey recent results on this topic. We also discuss related questions on the descriptive and computational complexity of such generalized colouring problems.

1 Introduction

Let S be a set of *colours*. An *S-colouring* of a graph G is a mapping f of $V(G)$ to S in which each pair of adjacent vertices u, v of G has $f(u) \neq f(v)$ in S.

For instance, if $S = \{1\}$, the constraint on the colouring can be illustrated as in Figure 1.1 (a). Deciding whether or not a given graph G has such a colouring (that is, a colouring with one colour) is trivial - a graph G is S-colourable if and only if it has no edge, i.e., no induced K_2. When $S = \{1, 2\}$, the constraints are illustrated in Figure 1.1 (b), and a graph G is S-colourable if and only if it is bipartite, i.e., if and only if in has no induced odd cycle. Thus in this case there are infinitely many forbidden induced subgraphs; however, there is still a polynomial time algorithm to detect colourability. (On the other hand, when G is restricted to be a perfect graph, then G is bipartite if and only if it contains no triangle, so we again have a characterization by finitely many forbidden induced subgraphs.) When $S = \{1, 2, 3\}$, the colourability problem becomes NP-complete. Thus for classical colourings there is only the case $S = \{1\}$ (and the trivial case $S = \emptyset$), in which a characterization by finitely many forbidden subgraphs is possible (unless we restrict the input graphs G). The first step in generalizing colourings is to consider *graph homomorphisms* [33]. If H is a graph with vertex set S, a homomorphism of G to H is a mapping f of $V(G)$ to S in which each pair of adjacent vertices u, v of G has $f(u)$ adjacent to $f(v)$ in H. Such a mapping f is also called an *H-colouring* of G.

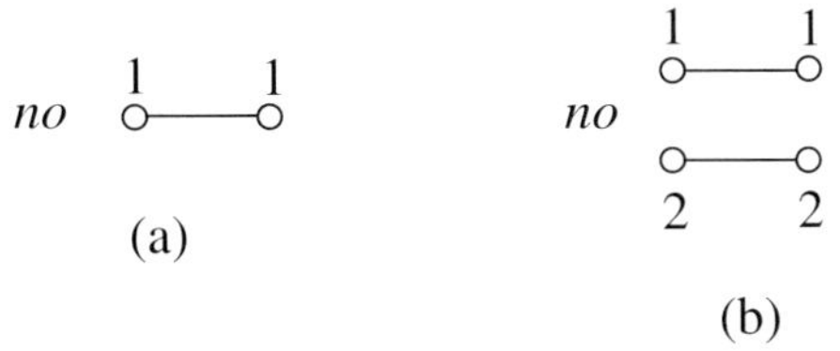

Figure 1.1:

For instance, when H has the vertices $S = \{1, 2, 3\}$ and the edges $22, 33, 12, 13$, then an H-colouring of G satisfies the constraints illustrated in Figure 1.2 (a). In other words, there can be no edge amongst the vertices coloured 1, and between a vertex coloured 2 and a vertex coloured 3. (Thus the set of vertices coloured 1 is an independent set, and there are no edges joining the sets of vertices coloured 2 and 3.) We have enriched the syntax of describing our colouring problems. Unfortunately, no new non-trivial problems arise which can be characterized by finitely many forbidden induced subgraphs [33]. (The situation is different for digraphs, where several such characterizations are possible [33, 35].)

We may further enrich the syntax of describing generalized colouring problems by allowing constraints for colours on non-adjacent pairs of vertices. This is illustrated in Figure 1.2 (b) (dotted lines depict non-adjacency). In this problem we again have colours $S = \{1, 2, 3\}$ and again the set of vertices coloured 1 must be an independent set, and there can be no edges between the sets coloured 2 and 3. Additionally, the set of vertices coloured 2 must be a clique, and there must be all edges between the vertices coloured 1 and 3. These problems are best described

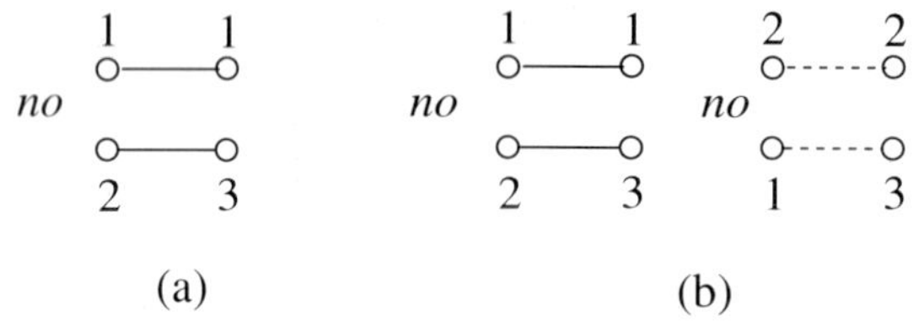

Figure 1.2:

as *matrix partition problems* (sometimes also presented as *trigraph homomorphism problems*) [23, 33].

Let M be a symmetric m by m matrix over $\{0, 1, *\}$. An *M-partition* of a graph G is a partition of $V(G)$ into parts $V_1, V_2, \ldots, V_m$ such that for distinct vertices $u \in V_i$, $v \in V_j$, we have $uv \in E(G)$ if $M(i, j) = 1$, and $uv \notin E(G)$ if $M(i, j) = 0$. Note that we admit $i = j$; in particular, if $M(i, i) = 0$, the set V_i is independent in G, and if $M(i, i) = 1$, it is a clique. Also note that $*$ means no restriction. For each fixed symmetric matrix M we obtain the graph *M-partition problem* - to decide whether or not a given graph G admits an M-partition. For instance, if C_m is the matrix in which the diagonal entries are 0 and all other entries are $*$, then the C_m-partition problem asks whether or not the graph G is m-colourable, in the usual sense. The example given in Figure 2 (b) leads to the M-partition problem with the matrix M

$$\begin{pmatrix} 0 & * & 1 \\ * & 1 & 0 \\ 1 & 0 & * \end{pmatrix}$$

We shall focus on matrix partitions as the right generalization f colourings. We have observed that matrix partition problems include the classical colouring problems, as well as H-colouring problems (i.e., graph homomorphism problems). Specifically, suppose that M has no 1's: let M' be obtained from M by replacing each $*$ with a 1, and let H be the graph whose adjacency matrix is M'. Then M-partitions are precisely H-colourings. Many other graph partition problems, especially those arising in the study of perfect graphs [7, 8, 20]), can be formulated as matrix partition problems; these include problems such as deciding the existence of a clique cutset, or a skew cutset [8, 37], and so on [7].

Take, for instance, the problem of recognizing split graphs [28]. This is precisely the M-partition problem in which M is the two by two matrix

$$\begin{pmatrix} 0 & * \\ * & 1 \end{pmatrix}$$

For this problem there is a well known characterization by forbidden induced subgraphs - G is partitionable (is a split graph) if and only if it does not have an induced $2K_2, C_4, C_5$ [28]. This suggests we have been successful in generalizing colourings to a context where forbidden induced subgraph characterizations may be more common. Thus we ask for other finite forbidden subgraph characterizations of partition problems.

More generally, we have three basic questions we may ask, for a given matrix M:

- Can M-partitionable graphs be recognized in polynomial time?
- Can M-partitionable graphs be described by a first-order sentence?
- Can M-partitionable graphs be characterized by a finite set of forbidden induced subgraphs?

Note that a characterization by finitely many forbidden induced subgraphs is a description in first-order logic, and a description by a first-order sentence yields a polynomial time algorithm. Thus the questions have been ordered so that a positive answer to a later question implies a positive answer to an earlier question.

There are several variants of the basic M-partition problem. In the *digraph M-partition problem*, the matrix M is not necessarily symmetric, and we apply the above definition of M-partition to a digraph G.

In the *list M-partition problem* (and the *digraph list M-partition problem*), each vertex v of the input graph (respectively digraph) G is equipped with a list $L(v)$ (a set of parts), and vertices are restricted to be placed in parts from their lists.

Other variants include extensions to *full* constraint satisfaction problems [16], and surjective M-partition problems in which each part is required to be nonempty [11, 38].

2 Computational Complexity

In this section we briefly review what is known about the first of our basic questions above.

Historically, the most studied case is that of homomorphism problems. Thus let us first focus on the case when **the matrix M has no 1's**. For graphs, the complexity of H-colouring problems has been classified in [32], as polynomial time solvable, when H has a loop or is bipartite, or NP-complete (otherwise). However, for digraphs (when M is not necessarily symmetric), it is not even known whether all digraph H-colouring problems are polynomial time solvable or NP-complete; such a *dichotomy* has been conjectured by Feder and Vardi [26], and is a subject of intense investigation cf. [3, 4, 29, 30, 33, 34].

In some cases, dichotomy is known, but there is no simple classification of which cases are tractable and which are not. This is, for instance, the case of oriented cycles [15]: for any oriented cycle H, the H-colouring problem is polynomial time solvable or NP-complete, but no classification in known.

Much changes with the introduction of lists. For graphs, all list H-colouring problems have been classified in [19] as polynomial time solvable, when H is a bi-arc graph, or NP-complete (otherwise). Dichotomy has then been established by Bulatov [4] in the more general context of constraint satisfaction problems with lists (which includes both graphs and digraphs).

We next consider *general matrices* M (with entries $0, 1, *$); of course, here even less is known. A kind of dichotomy has been established in [16] - namely each list M-partition problem is NP-complete, or is solvable in *quasi-polynomial time*, $n^{O(\log n)}$. Surprizingly, there are concrete small matrices M for which the list M-partition problem (and the M-partition problem) is solvable in quasi-polynomial time, but no polynomial time algorithms are known [5, 16, 22]. The simplest problem of this sort is actually a "full" constraint satisfaction problem [16], but it has a simple combinatorial formulation:

Given a complete graph in which the edges are coloured $1, 2, 3$, decide if there is a colouring of the vertices by $1, 2, 3$, so that there is no monochromatic edge (an edge of colour i with both endpoints coloured i).

This problem has received much attention [16], yet the best algorithm currently known has complexity $n^{O(\log n/ \log\log n)}$ [22].

Restrictions of (list) M-partition problems to classes of perfect graphs G have been studied in [12, 13, 17, 21, 24, 30, 31]. For instance it turns out that there are matrices M for which the M-partition problem (without lists) is NP-complete even when restricted to chordal graphs. This was unexpected, as chordal graphs generally admit simple algorithms for various colouring problems; and, in fact, there are polynomial time algorithms for (list) M-partition problems in chordal graphs for many important classes of matrices M [21].

3 Descriptive Complexity, Forbidden Subgraphs

We don't have much to offer as answer to our second basic question, above. When the matrix M has no 1's (for homomorphism problems), it follows from [1, 36], that the last two basic questions have the same answer, i.e., that if (list) M-partitionable graphs (digraphs) can be described in first order logic, then they can be characterized by finitely many forbidden subgraphs. This is not known for general matrices M (that may include 1's). In the remainder of this section we focus on the third basic question, of characterizations of M-partitionability by finite sets of forbidden induced subgraphs.

We shall discuss the basic M-partition problem without lists; in this situation it makes sense to assume that all $M(i, i) \neq *$. Indeed, if some $M(i, i) = *$, then every graph G is M-partitionable. Thus we may assume that $M(1, 1) = \cdots = M(k, k) = 0, M(k+1, k+1) = \cdots = M(m, m) = 1$. Let A denote the submatrix of M with rows $1, \ldots, k$ and columns $1, \ldots, k$; let B denote the submatrix with rows $k+1, \ldots, m$ and columns $k+1, \ldots, m$; and let C denote the submatrix with rows $1, \ldots, k$ and columns $k+1, \ldots, m$. A matrix which has no $*$ in the submatrices A and B is called *friendly*.

Theorem 3.1 ([25, 39]). *If M is not a friendly matrix, then M-partitionable graphs cannot be characterized by a finite set of forbidden induced subgraphs.*

This is shown by exhibiting infinitely many *minimal obstructions to M-partition*, i.e., graphs G which are not M-partitionable, but such that for each vertex v of G, the graph $G - v$ is M-partitionable.

We have several classes of friendly matrices M for which M-partitionable graphs are known to have a characterization by finitely many forbidden subgraphs. The simplest case occurs when M has no $*$ entries at all. In [18] we have shown the following fact.

Theorem 3.2 ([18]). *If M has no $*$ entries, then all minimal obstructions to M-partition have at most $(k+1)(m-k+1)$ vertices. Moreover, there are at most two minimal obstructions with precisely $(k+1)(m-k+1)$ vertices.*

The fact that there are only finitely many minimal obstructions in this case also follows from [2].

We can extend the validity of this result by applying the so-called sparse-dense technique from [17]. We formulate it here specifically for the application at hand; its derivation from the general sparse-dense technique is easy to see (by setting $\mathcal{S}$ to be all A-partitionable graphs and $\mathcal{D}$ to be all B-partitionable graphs).

We claim that for any fixed M there exists an integer r such that any graph G that is both A-partitionable and B-partitionable has at most r vertices. Indeed, such a graph G is, in particular, k-colourable and hence contains no clique of size $k+1$, and its complement $\overline{G}$ is $(m-k)$-colourable and hence contains no clique of size $m-k+1$. Thus the existence of such an integer r follows from Ramsey's theorem.

A *labeled* graph G is a graph in which each vertex has a label, either A or B. Label A means that the vertex has to be placed in parts $V_1, \ldots, V_k$, label B means the vertex has to be placed to parts $V_{k+1}, \ldots, V_m$. (Thus the labels are lists of a restricted kind.) A *labeled M-partition* of G is an M-partition of G that satisfies these constraints. A *minimal labeled obstruction* to M-partition is a labeled graph G which has no labeled M-partition, such that for each vertex $v \in V(G)$ the graph $G - v$ (with the inherited labels) has a labeled M-partition.

Theorem 3.3 ([17]). *Suppose each minimal labeled obstruction to M-partition has at most p vertices. Then each minimal obstruction to M-partition has at most $2p^{2r+1}$ vertices.*

Now suppose M is a matrix in which C has only $*$ entries. In this case a labeled graph G is M-partitionable if and only if its subgraph on the vertices labeled A is A-partitionable and its subgraph on the vertices labeled B is B-partitionable. In particular, we obtain the following result.

Corollary 3.4. *If M is a friendly matrix in which C has only $*$ entries, then M-partitionable graphs can be characterized by a finite set of forbidden induced subgraphs.*

This result extends Theorem 3.2, and has many other interesting applications. It implies, for instance, a characterization by finitely many forbidden induced subgraphs for the class of split graphs [28], (k,ℓ)-polar graphs [12], and so on.

It may seem that all friendly matrices admit characterizations by finite sets of forbidden induced subgraphs. In fact, it has been shown in [25, 39], that this is so for all small friendly matrices M.

Corollary 3.5. *If m is a friendly matrix of size $m \leq 5$, then M-partitionable graphs can be characterized by a finite set of forbidden induced subgraphs.*

However, for $m > 5$ we have a counterexample. Let M_6 be the matrix

$$\begin{pmatrix} 0 & 1 & 0 & * & 0 & 0 \\ 1 & 0 & 1 & * & 0 & 0 \\ 0 & 1 & 0 & 0 & * & * \\ * & * & 0 & 1 & 0 & 1 \\ 0 & 0 & * & 0 & 1 & 0 \\ 0 & 0 & * & 1 & 0 & 1 \end{pmatrix}$$

Theorem 3.6. *We have the following facts.*

- *The matrix M_6 is friendly.*
- *The M_6-partition problem cannot be characterized by a finite set of forbidden induced subgraphs.*

As we have remarked in the introduction, characterizations by finitely many forbidden induced subgraphs become more numerous if we restrict the problems to classes of perfect graphs. For instance, for cographs, we have the following result.

Theorem 3.7 ([24]). *For any symmetric matrix M with m rows, there exists a characterization of M-partitionable cographs by a set of forbidden induced subgraphs with at most 36^m vertices.*

The fact that there are only finitely many forbidden induced subgraphs for any M-partition problem on cographs also follows from [10].

4 $\mathcal{M}$-Partitions

Let $\mathcal{M}$ be a set of matrices M. The *$\mathcal{M}$-partition problem* asks whether or not an input graph G admits an M-partition for some $M \in \mathcal{M}$. In some cases, the answer is trivially yes, and we may want the smallest size (number of rows) of a matrix $M \in \mathcal{M}$ such that G admits an M-partition.

It should be clear that this allows us to formulate problems such as the chromatic number, and the cochromatic number [14]. We close the notes with a discussion of *polar* graphs. A graph is polar, if it has an M-partition with a matrix M in which A has all off-diagonal entries 1, B has all off-diagonal entries 0, and C consist of $*$'s. Thus a graph is polar if and only if it can be partitioned into a complete multipartite graph and a complement of a complete multipartite graph. Polar graphs generalize bipartite, cobipartite, and split graphs, but are NP-complete to recognize [6]. Recognition of polar graphs is clearly an $\mathcal{M}$ -partition problem. We contrast two recent results on polarity.

Theorem 4.1 ([13]). *Polar cographs can be characterized by finitely many forbidden induced subgraphs (and hence can be recognized in polynomial time).*

Theorem 4.2 ([12]). *Polar chordal graphs cannot be characterized by finitely many forbidden induced subgraphs, but can be recognized in polynomial time.*

References

[1] A. Atserias, On Digraph coloring problems and treewidth duality, in *20th IEEE Symposium on Logic in Computer Science (LICS)* (2005), pp. 106–115.

[2] R.N. Ball, J. Nešetřil and A. Pultr, Dualities in full homomorphisms, Mansucript 2006.

[3] A.A. Bulatov, Tractable conservative constraint satisfaction problems, in *18th IEEE SOLCS* (2003), pp. 321–330.

[4] A.A. Bulatov, P.G. Jeavons and A.A. Krokhin, Constraint satisfaction problems and finite algebras, in *27th ICALP*(2000), pp. 272–282.

[5] K. Cameron, E.E. Eschen, C.T. Hoang and R. Sritharan, The list partition problem for graphs, in *Proc. 15th Annual ACM-SIAM Symposium on Discrete Algorithms (SODA)* (2004), pp. 391–399.

[6] Z.A. Chernyak and A.A. Chernyak, About recognizing (a, b)-classes of polar graphs, *Discrete Math.*, **62** (1986), pp. 133-138.

[7] M. Chudnovsky, N. Robertson, P. Seymour and R. Thomas, The strong perfect graph theorem, *Annals of Mathematics*, **164** (2006), pp. 51–229.

[8] V. Chvátal, Star-cutsets and perfect graphs, *J. Combinatorial Theory B*, **39** (1985), pp. 189–199.

[9] B. Courcelle, Graph Operations and Monadic Second-Order Logic: A Survey, in *LPAR* (2000), pp. 20–24.

[10] P. Damaschke, Induced subgraphs and well quasi-ordering, *J. Graph Theory*, **14** (1990),

[11] S. Dantas, C.M.H. de Figueiredo, S. Gravier and S. Klein, Finding H-partitions efficiently, *Theoret. Informatics Appl.*, **39**(2005), pp. 133-144.

[12] T. Ekim, P. Hell, J. Stacho and D. de Werra, Polarity of chordal graphs, *Discrete Applied Mathematics*, in press.

[13] T. Ekim, N.V.R. Mahadev and D. de Werra, Polar cographs, *Electronic Notes in Discrete Mathematics*, **28** (2007), pp. 317–323.

[14] P. Erdos and J. Gimbel, Some problems and results in cochromatic theory, in: *Quo Vadis, Graph Theory?, Annals of Discrete Mathematics*, **55**(1993), pp. 261–264.

[15] T. Feder, Classification of homomorphisms to oriented cycles and of k-partite satisfiability, *SIAM J. on Discrete Math.*, **14** (2001), pp. 471–480.

[16] T. Feder and P. Hell, Full constraint satisfaction problems, *SIAM J. on Computing*, **36** (2006), 230–246.

[17] T. Feder and P. Hell, Matrix partitions of perfect graphs, *Discrete Math.*, **306** (2006), pp. 2450–2460.

[18] T. Feder and P. Hell, On realizations of point determining graphs, and obstructions to full homomorphisms, *Discrete Math.*, **308** (2008), pp. 1639–1652.

[19] T. Feder, P. Hell and J. Huang, Bi-arc graphs and the complexity of list homomorphisms, *J. Graph Theory*, **42** (2003), 61–80.

[20] T. Feder, P. Hell, S. Klein and R. Motwani, Complexity of list partitions, *SIAM J. Discrete Mathematics*, **16** (2003), 449–478.

[21] T. Feder, P. Hell, S. Klein, L. Nogueira and F. Protti, List matrix partitions of chordal graphs, *Theoretical Computer Science*, **349** (2005), 52–66.

[22] T. Feder, P. Hell, D. Kral and J. Sgall, Two algorithms for list matrix partitions, in *Proc. 16th Annual ACM–SIAM Symposium on Discrete Algorithms (SODA)* (2005), pp. 870–876.

[23] T. Feder, P. Hell and K. Tucker-Nally, Digraph matrix partitions and trigraph homomorphisms, *Discrete Applied Math.*, **154** (2006), pp. 2458–2469.

[24] T. Feder, P. Hell and W. Hochstättler, Generalized colourings (matrix partitions) of cographs, *Graph Theory, (J. Ramirez ed.) Volume in Memory of Claude Berge, Trend in Mathematics, Birkhäuser*, Verlag, Basel pp. 149–167. (2006).

[25] T. Feder, P. Hell and W. Xie, Matrix partitions with finitely many obstructions, *Electronic Journal of Combinatorics*, **R58** **14** (2007).

[26] T. Feder and M.Y. Vardi, The computational structure of monotone monadic SNP and constraint satisfaction: a study through Datalog and group theory, *SIAM J. Comput.*, **28** (1998), 236–250 (ALSO in STOC 25 (1993) 612–622).

[27] C.M.H. de Figueiredo, S. Klein, Y. Kohayakawa and B. Reed, Finding skew partitions efficiently, *J. Algorithms*, **37** (2000), pp. 505–521.

[28] S. Foldes and P. Hammer, Split graphs, *Congr. Numer.*, **19** (1977), 311–315.

[29] P. Hell, Algorithmic aspects of graph homomorphisms, in in *Surveys in Combinatorics 2003 (C.D. Wensley ed.), London Math. Soc. Lecture Note Series* **307** (2003), pp. 239–276.

[30] P. Hell, From graph colouring to constraint satisfaction: there and back again, Invited chapter in *Topics in Discrete Mathematics: Springer Verlag Algorithms and Combinatorics Series*, **26** (2006), pp. 407–432.

[31] P. Hell, S. Klein, L. Tito Nogueira and F. Protti, Partitioning chordal graphs into independent sets and cliques, *Discrete Applied Math.*, **141** (2004), pp. 185–194.

[32] P. Hell and J. Nešetřil, On the complexity of H-coloring, *J. Comb. Theory, Series B*, **48** (1990), pp. 92–110.

[33] P. Hell and J. Nešetřil, *Graphs and Homomorphisms*, Oxford University Press, (2004).

[34] P.G. Jeavons, On the algebraic structure of combinatorial problems, *Theoret. Comput. Sci.*, **200** (1998), pp. 185–204.

[35] A. Pultr and J. Nešetřil, On classes of relations and graphs determined by subobjects and factorobjects, *Discrete Math.*, **22** (1978), pp. 287–300.

[36] B. Rossman, Existential positive types and preservation under homomorphisms, in *20th IEEE LICS* (2005),

[37] R.E. Tarjan, Decomposition by clique separators, *Discrete Math.*, **55** (1985), pp. 221–232.

[38] N. Vikas, Computational complexity of compaction to reflexive cycles, *SIAM J. on Computing*, **32** (2003), pp. 253–280.

[39] W. Xie, Forbidden subgraph characterizations of matrix partitions, M.Sc. thesis, Simon Fraser University (2006).

Proc. of ICDM 2006, Pages 149–164

Applications of Orthogonal Arrays to Computer Science

K. Gopalakrishnan
Department of Computer Science, East Carolina University, Greenville, NC 27858, U.S.A.
Douglas R. Stinson
David R. Cheriton School of Computer Science, University of Waterloo, Waterloo, Ontario–N2L 3G1, Canada.

Abstract. Orthogonal arrays (OAs) are basic combinatorial structures, originally studied by statisticians motivated by their applications to design of experiments. In recent years, they have found numerous applications in computer science. Among their applications are **derandomization of algorithms, random pattern testing of VLSI chips, authentication codes, universal hash functions, threshold schemes,** and **perfect local randomizers**. In this article, we survey the applications of orthogonal arrays to computer science.
keywords: Orthogonal Arrays, Computer Science, Cryptography, Applications.

---

1 Introduction

The field of combinatorial design theory has gained a lot with the advent of computers. Many questions, which remained open for several years, about the existence of certain combinatorial designs have been settled by computer searches. A classic example is the proof of the non-existence of projective planes of order 10. However, more often than not, the contributions of combinatorial designs to computer science have been overlooked and not appreciated.

Recent years have seen numerous interesting applications of combinatorics to computer science in general and to cryptography in particular. This is exemplified by the following surveys: applications of error-correcting codes to cryptography (Sloane [45]), applications of combinatorial designs to computer science (Colbourn and Van Oorschot [14]), applications of finite geometry to cryptography (Beutelspacher [8]), applications of combinatorial designs to cryptography (Stinson [50]) and applications of combinatorial designs to communications, cryptography and networking (Colbourn et al., [15]). Thus the interaction between combinatorial designs and computer science is symbiotic in nature rather than parasitic.

There are a wide variety of combinatorial designs and in this article, we will focus exclusively on the applications of **orthogonal arrays**. The applications of orthogonal arrays to statistical design of experiments are well known and can even be found in textbooks (see Hedayat et al., [24] for example). Hence, we will focus exclusively on the applications of orthogonal arrays to computer science. Among their applications are **derandomization of algorithms, random pattern testing of VLSI chips, authentication codes, universal hash functions, threshold schemes,** and **perfect local randomizers**. The above collection of applications coming from diverse areas attests to the usefulness of orthogonal arrays.

In Section 3, applications of OAs to some sampling strategies associated with randomized algorithms are presented. Section 4 explains how the same techniques could be exploited in random pattern testing of VLSI chips in an effective manner. Under appropriate circumstances, a randomized algorithm can be transformed into a deterministic polynomial time algorithm using orthogonal arrays. This is known as total derandomization and it is the topic of Section 5.

Next we focus on applications of OAs to cryptography. Sections 6 and 7 discuss the applications of OAs to authentication codes and universal hash functions respectively. In Section 8, the construction of ideal secret sharing schemes for threshold access structures using orthogonal arrays is illustrated. Finally, perfect local randomizers, which are really orthogonal arrays in disguise, are discussed in Section 9.

2 Definitions

An *orthogonal array* $OA_\lambda(t, k, v)$ is a $\lambda v^t \times k$ array of v symbols, such that in any t columns of the array every one of the possible v^t ordered t-tuples (not necessarily distinct) of symbols occurs in exactly λ rows. Usually t is referred to as the *strength* of the orthogonal array, k is called the number of *factors*, v is called the number of *levels* and λ is called the *index* of the orthogonal array. If $\lambda = 1$, then we write $OA(t, k, v)$. An orthogonal array is said to be *simple* if no two rows are identical. Of course, an array with $\lambda = 1$ is simple. In this article, we consider only simple arrays. As an example, an $OA(3, 4, 2)$ is shown in Table 1. We refer the reader to the excellent book by Hedayat et al., [24] for the basic theory of orthogonal arrays including methods of constructions and bounds. In this survey, our primary focus will be on their applications.

Table 1: An $OA(3, 4, 2)$

0	0	0	0
1	1	0	0
1	0	1	0
0	1	1	0
1	0	0	1
0	1	0	1
0	0	1	1
1	1	1	1

3 Partial Derandomization of Monte Carlo Algorithms

3.1 Basics

There are two basic types of randomized algorithms viz., Las Vegas algorithms and Monte Carlo algorithms. Here we will be concerned primarily with the Monte Carlo algorithms. For any problem instance I, a Monte Carlo algorithm always gives an answer, but the answer may be incorrect with some probability ϵ. Monte Carlo algorithms are usually assumed to be algorithms for decision problems, and we will also make that assumption here.

A Monte Carlo algorithm is defined to be *yes-biased* if the following conditions are satisfied:

1. If the instance I is a no-instance, then the algorithm answers "no".

2. If the instance I is a yes-instance, then the probability that the algorithm answers "yes" is at least $1 - \epsilon$, where $\epsilon \geq 0$ is some fixed constant.

Hence, if the algorithm answers "yes", then we know that the answer is correct. However, if the algorithm answers "no", then there is the possibility that the answer may be incorrect. A *no-biased* Monte Carlo algorithm is defined in the obvious way and there is no loss of generality in considering only yes-biased Monte Carlo algorithms.

A yes-biased Monte Carlo algorithm can be viewed as a two-stage procedure in which first a "sample point" is chosen at random and next a deterministic procedure is applied to the sample point. This view is usually known as the *witness paradigm*. In the generic case the sample point is an integer $r \in \{0, 1, \ldots, n-1\}$ and the deterministic procedure can be viewed as computing a zero-one valued function $f(I, r)$. If $f(I, r) = 0$ then the algorithm answers "no" and if $f(I, r) = 1$ then the algorithm answers "yes". The function f should satisfy the following: If I is a no-instance, then for all $r \in \{0, 1, \ldots, n-1\}$, $f(I, r) = 0$ and If I is a yes-instance, then the fraction of the values of r for which $f(I, r) = 1$ is at least $1 - \epsilon$.

There are many problems for which no fast deterministic algorithm is known, but which can be solved efficiently using randomized algorithms. Notable examples are the massive use of randomness in computational number theory [4] and in parallel algorithms [32]. A good survey of randomization techniques used in sequential and distributed algorithms is presented in [22].

3.2 Two-Point Based Sampling

The reason why a biased algorithm is useful is that the error probability can be made as small as desired by repeated application of the algorithm. Suppose we have a yes-biased algorithm, and we apply the algorithm k times. If we get at least one "yes" answer we can quit, since the instance x must be a yes-instance. On the other hand, if x is a yes instance, then the probability of getting k "no" answers is at most ϵ^k, which can be made arbitrarily small by taking k large enough.

However, the conclusion that the error probability can be made exponentially small is based on the assumption that the sequence of random numbers used in the successive trials are independent and uniformly distributed. When a randomized algorithm is implemented in actual practice, one always uses a pseudo-random number generator which produces a sequence whose later values come from the previous values in a *deterministic* fashion. This invalidates the assumption of independence and might cause one to regard results about randomized algorithms with suspicion.

In view of this objection, many researchers attempted to construct pseudo-random number generators for which provable bounds could be obtained on the probability that none of the k successive values produced is a witness. A simple and effective method known as *two-point sampling* is developed by Chor and Goldreich.

For the sake of simplicity, we assume that $U = Z_p$, the set of residues modulo a prime p. If $n = |U|$ is not a prime, then we can use the smallest prime number p such that $n \leq p \leq 2n$; that such a prime exists is proved by Chebyshev and later by Paul Erdös. The construction, developed by Chor and Goldreich [12] building on an observation by Joffe [27], proceeds as follows: Choose two random independent elements a and b in Z_p with uniform probability. Compute the residues $r_i = a + i \times b \mod p$, for $1 \leq i \leq k$. These residues are then used as the random numbers in the k trials of the Monte Carlo algorithm. It is not too hard to see that the r_i's are pairwise-independent random variables. By using Chebyshev's inequality, Chor and Goldreich prove that the probability that none of the r_i's is a witness is at most $\epsilon/(1-\epsilon)k$.

If the successive random numbers used are independent of each other, then the error probability is ϵ^k and the number of true random bits needed is $k \log n$. In the above technique, the number of random bits used is $2 \log n$ and the error probability achieved is $\epsilon/(1-\epsilon)k$. This is significant as random bits are a scarce resource and hence need to be used sparingly.

3.3 Relevance of Orthogonal Arrays

Orthogonal arrays could be used in two-point sampling strategies as follows. Consider an $OA(2, k, n)$. This array has n^2 rows. Generate $2 \log n$ random bits and use them to index a specific row of the OA. Now test the Monte Carlo algorithm on the k points which are elements of the row selected.

A bound on the error probability can be calculated using elementary combinatorial properties of orthogonal arrays. We proceed to do this now. Let U denote the universe of sample points and let $|U| = n$. Let $S \subseteq U$ be the set of witnesses or "good" sample points. Then $|S| = (1-\epsilon)n = m$ (say).

Let x_i denote the number of rows of the orthogonal array in which there are exactly i occurrences of the elements from S. Call a row of the matrix a *bad row* if none of the elements in the row is a witness. Then the error probability is simply the probability that the randomly selected row is a bad row. Hence, for a given S, the error probability is

$$err(S) = \frac{x_0}{n^2}.$$

Note that we actually do not know the set S explicitly. The only thing that we know about S is the fact that $|S| = m$. So the actual error probability is bounded above by

$$err = \max\{err(S) : S \subseteq U, |S| = m\}.$$

We first derive three simple equations from the elementary properties of orthogonal arrays. Since an $OA(2, k, n)$ has n^2 rows, we have

$$\sum_{i=0}^{n} x_i = n^2.$$

We can count the number of occurrences of good points in the array in two ways. There are exactly x_i rows in which there are i occurrences of good points. An $OA(2, k, n)$ is also an $OA_n(1, k, n)$. Hence in any column any point occurs exactly n times. So, the number of occurrences of good points in a column is nm. As there are k columns, the total number of occurrences of good points in the whole array is knm. This yields the following equation:

$$\sum_{i=0}^{n} i x_i = knm.$$

Once again, we can count the number of occurrences of pairs of good points which lie in the same row in two ways. In any row in which there are i occurrences of good points, there will be $i(i-1)$ occurrences of pairs of good points. On the other hand, if we look at any two columns, the number of occurrences of pairs of good points which lie in the same row is simply m^2. This is because any particular pair of good points occurs only once in any given pair of columns since the array is an $OA(2,k,n)$, and there are m^2 different such pairs of good points. Now the two columns can be selected in $k(k-1)$ ways and so the total number of occurrences is $k(k-1)m^2$. This yields the following equation:

$$\sum_{i=0}^{n} i(i-1)x_i = k(k-1)m^2.$$

To compute the error probability, we need an upper bound on x_0. Equivalently, we need a lower bound on $\sum_{i=1}^{n} x_i$, subject to the constraints

$$\begin{aligned} \sum_{i=1}^{n} ix_i &= knm \\ \sum_{i=1}^{n} i(i-1)x_i &= k(k-1)m^2. \end{aligned}$$

Now, let z be any real number. Then clearly

$$\begin{aligned} 0 &\leq \sum_{i=1}^{n} (i-z)^2 x_i \\ &= \sum_{i=1}^{n} (i^2 - 2zi + z^2)x_i \\ &= \sum_{i=1}^{n} i^2 x_i - 2z\sum_{i=1}^{n} ix_i + z^2 \sum_{i=1}^{n} x_i \\ &= k(k-1)m^2 + knm - 2zknm + z^2 \sum_{i=1}^{n} x_i \end{aligned}$$

It follows that

$$\sum_{i=1}^{n} x_i \geq \frac{2knmz - knm - k(k-1)m^2}{z^2}.$$

The right hand side of the above equation is maximized when

$$z = \frac{n+(k-1)m}{n}.$$

Hence we get

$$\sum_{i=1}^{n} x_i \geq \frac{kmn^2}{n+(k-1)m}.$$

Equivalently, we have

$$x_0 \leq n^2 - \frac{kmn^2}{n+(k-1)m}.$$

Hence we get the following bound on the error probability:

$$err \leq 1 - \frac{km}{n + (k-1)m}.$$

Since $m = n(1-\epsilon)$, we have that

$$err \leq \frac{\epsilon}{1 + (k-1)(1-\epsilon)}.$$

Chor and Goldreich's construction of sequence of a two-point based random numbers is essentially the construction of a specific $OA(2, k, n)$. Our technique presented above is based on an *arbitrary* $OA(2, k, n)$ and hence is more general. On the other hand, despite the generality of the technique, our bound of

$$\frac{\epsilon}{1 + (k-1)(1-\epsilon)}$$

is *always* less than or equal to their bound of

$$\frac{\epsilon}{(1-\epsilon)k}.$$

Also, while the proof of our bound uses only elementary combinatorial properties, the proof of their bound uses sophisticated machinery like Chebyshev's inequality. Thus we have a simpler proof of a better bound which is applicable in a more general context. This is due to the fact that two-point sampling is simply an application of orthogonal arrays.

In a slightly different context, Spencer (see [46] and Section 4) developed a bound on the error probability using the Bonferroni inequalities. His bound is

$$\frac{1}{1 + k(1-\epsilon)}.$$

Once again, despite the use of complex techniques, this bound is inferior to the above bound developed using orthogonal arrays.

Further details on the analysis of two-point based sampling can be found in [21]. An extension of the same technique to the analysis of t-point based sampling can be found in [17].

4 Testing VLSI Chips

4.1 Background

The basic testing problem is to determine an optimal testing procedure to identify digital systems which have become faulty because of physical failures. In most cases involving VLSI chips, one is not usually concerned with detecting the exact physical failure; what is desired is merely to determine the existence of (or absence of) any physical failure.

The large number and complex nature of physical failures dictates that a practical approach to testing should avoid working directly with physical failures. Often the problem is attacked by describing the effects of physical failures at some higher level. This description is called a fault model [35]. If the fault model accurately describes all the physical failures of interest, then one only needs to derive tests to detect all the faults in the fault model.

In what follows, the digital system chosen is combinational circuits. This is because, in spite of their simplicity, combinational circuits capture the complexity of the testing process. Further, the widespread use of design for testability techniques, such as Level-Sensitive Scan Design or LSSD [18], that transform a sequential circuit into a combinational one for testing purposes, has lent increasing importance to specialized methods for testing combinational circuits.

In what follows, the fault model selected is the classical *Single, permanent Stuck-at Fault (SSF)* model. In this model, testing is done to detect the capability of every line in the circuit to carry both a 0 and a 1 signal. A Single Stuck-at Fault in a circuit is a good model of various physical failures, such as a break in a diffusion line caused by an unintentional scratch on IC chip. Further the high coverage of multiple faults by Single Fault test sets has made the SSF model even more important [1, 35].

The basic testing problem, in the specific context of combinational circuits and SSF model, can be described as follows.

Let C be a well formed combinational circuit [23] with l lines and F be the set of $2l$ Single Stuck-at Faults in C. Let $x_1, x_2, \ldots, x_n$ be the set of all primary inputs and $y_1, y_2, \ldots, y_m$ be the set of all primary outputs of C. The symbol $Z[X]$ is used to denote the output vector $Y = (y_1, y_2, \ldots, y_m)$, when the input vector $X = (x_1, x_2, \ldots, x_n)$ is applied to the circuit. Let $Z_f[X]$ represent the output vector of the circuit in the presence of the fault $f \in F$. It is clear that any input vector X_t which produces different output vectors $Z[X_t]$ and $Z_f[X_t]$ is a test vector for f. The set of all test vectors which can detect f will be denoted by $T(f)$. Thus

$$T(f) = \{X_t \mid Z[X_t] \neq Z_f[X_t]\}.$$

The basic testing problem is to construct a set of test vectors $T(F)$ such that for every fault $f \in F$, there exists at least one test vector in $T(F)$ which will detect f i.e,

$$\forall\ f \in F\ \exists\ X_t \in T(F) \text{ such that } X_t \in T(f).$$

In practice, it is difficult to test for all the faults and hence what is attempted is merely a high amount of fault coverage (95% to 99%). The term *fault coverage* refers to the fraction of the modeled faults detected by the test vectors. Further the test costs may be broken down into two categories viz., cost of test generation and cost of test application. The former is a one-time cost and is measured by the computational effort required to generate the vectors. The later is a recurrent cost and is measured by the time taken to apply the vectors to the circuit under test.

4.2 Built-in Self Test

The traditional testing strategy is to compute a set $T(F)$ of input vectors such that for each possible fault there is an input vector in $T(F)$ that distinguishes the correct circuit from the circuit with the fault. If this strategy is used, the time required to fetch the patterns, from where they are stored to the circuit, severely limits the rate at which the patterns can be applied. Built-in self test (BIST) has been suggested as a way to overcome this bottleneck. Simply speaking, the idea is to have the chip test itself. The test vectors are generated by special purpose hardware which itself is a part of the chip. As a consequence the test vectors can be generated and applied as fast as one pattern per machine cycle. The speed gained is normally worth the additional investment in chip area.

Random pattern testing strategy is typically used in BIST schemes. In this approach, a random pattern generator is used to generate the test vectors which are applied to the circuit under test. A comparator compares the actual response from the circuit under the test and the expected response to determine whether or not an error has been observed. The random test may stop either on the first observed error or after a sufficient number of random tests (test length) yield a correct response. Several random self-test architectures have been proposed in the literature [5]. Basically, all these versions use a linear feedback shift register (LFSR) to generate pseudo-random inputs that are fed to the circuit.

The fault coverage desired essentially dictates the test length in a random pattern test generator. So, let us focus on the estimation of test length [39] to achieve a prespecified amount of fault coverage. The detection probability q_i, of a fault f_i, is the probability that a randomly selected input vector will detect the fault. In other words

$$q_i = \frac{|T(f_i)|}{2^n}.$$

The problem of exact computation of detectability of a fault is NP-hard. However, algorithms which work well in practice for exact and approximate computations of detectability of each fault in a circuit are available [10, 40, 41]. The escape probability e_i, of a fault f_i, is the probability that the fault will go undetected after the application of t random input vectors. The escape probability is given by

$$e_i = (1 - q_i)^t.$$

Similarly, the escape probability e of a fault set $\{f_1, f_2, \ldots f_m\}$ is the probability that at least one member of the fault set will be left undetected after the application of t random input vectors. The escape probability of a fault set can be computed using the Markov chain model (using the escape probabilities of individual faults). The minimum random pattern test length is then merely the smallest t that satisfies $e \leq e_{th}$, where e_{th} is the acceptable error probabaility. Finally, observe that this analysis has assumed independence of the random vectors generated.

4.3 Provably Good Pattern Generators

As with randomized algorithms, the primary objection to the random pattern testability is that the analysis of test length and escape probability assumed complete independence of the vectors generated and randomness of each vector. In practice we only use pseudo-random generators where the successive vectors produced are completely determined by the initial seed vector. Thus the analysis does not reflect reality and hence we cannot be really sure of the actual fault coverage provided.

In view of this objection, T. Spenser [46] developed a pseudo-random pattern generator for use in VLSI testing. This generator is essentially based on the idea of two-point sampling method for randomized algorithms developed by Chor and Goldreich [12]. The generator is implemented as a modification of the classical LFSR such that the increases in the delay and the required hardware are only by at most a small constant factor. Thus, it can be used in practice.

The input vectors can be considered as elements of the finite field $GF(2^n)$ with 2^n elements. Let z be a generator of the cyclic multiplicative group of this field. The seed for the generator consists of an ordered pair (a, b) of elements of $GF(2^n)$, such that $a \neq 0$. The ith element produced by the generator is $az^{i-1} + b$. The escape probability of a fault f_i using this pattern generator is shown to be $1/(1+q_i t)$ provided $q_i t$ is an integer. The analysis of the escape probability is realistic and reflects the actual generator that is being used. Once again, this generator can be interpreted as an orthogonal array $OA(2, t, 2^n)$ and hence the analysis developed in Section 3 is applicable here as well.

5 Total Derandomization

Any randomized algorithm typically uses random bits in the course of its computation. So, we somehow need to generate random bits to actually implement randomized algorithms. In this regard John Von Neumann says that "*anyone who attempts to generate random numbers by an arithmetical process is in a state of sin*". On the other hand, there exists chaotic processes in nature, such as radioactive decay and the thermal noise in a transistor, that allow the construction of a random number generator whose behavior is for all practical purposes equivalent to that of a true random source. Even in these cases whether the chaotic processes involved are truly random is a philosophical question related to the question of whether the universe is deterministic or not, and seems to be impossible to answer to everyone's satisfaction. For a philosophical discussion of randomness the reader is referred to the book by Von Mises [31] and the one by Popper [34]. In any case, it is clear that randomness should be considered as a scarce resource which should be used as meagerly as possible. Thus we could talk of *randomness complexity* like we do of time and space complexity. The randomness complexity of an algorithm could be defined as the function that describes the growth of the number of random bits used by the algorithm with the size of the input. Under this view we would like to make sure that the randomness complexity of our randomized algorithms is fairly small.

Hence many researchers have developed techniques to reduce the number of random bits needed in computation and to eliminate completely, if possible, the use of random bits. When we completely eliminate the random bits needed in the randomized algorithm, we are in effect converting the randomized algorithm to a deterministic algorithm. This process is called (total) derandomization. For an excellent survey of some derandomization techniques the reader is referred to the recent book by Alon and Spencer [3].

Our emphasis here will be on a derandomization technique known as the t-wise independence method. This method is a general technique for converting randomized algorithms, whose analysis depends only on t-wise rather than fully independent random choices, into deterministic ones. This method was first developed by M. Luby for the case $t = 2$ in the context of the maximal independent set problem [29] and later generalized to arbitrary constant t by Alon, Babai and Itai [2].

The basic idea is to replace an exponentially large sample space by one of polynomial size. If a random variable on such a space takes a certain value with positive probability, then we can find a point in the sample space in which this happens in polynomial time simply by deterministically checking all the points. Moreover, the checking can be parallelized, if desired, using a polynomial number of parallel processors.

For the sake of simplicity, consider a randomized algorithm which requires n random bits, where n is also the size of the input. Then the size of the sample space is 2^n. Suppose that the analysis of the randomized algorithm does not really require the complete independence of the random bits but only t-wise independence. Then, one could use any $OA_\lambda(t, n, 2)$ as the sample space instead of the above sample space with 2^n points.

If we can construct an $OA_\lambda(t, n, 2)$ with polynomial (in n) number of rows, then we have a polynomial sample space with t-wise independent random variables.

Such polynomial sample spaces exist for appropriate parameters is the content of the following theorem.

Theorem 5.1. *Suppose $n = 2^m - 1$ and $t = 2e$. Then there exists a symmetric probability space Ω of size $(n+1)^e$ and t-wise independent random variables $y_1, y_2, \ldots, y_n$ over Ω, each of which takes the values 0 and 1 with equal probability.*

The above theorem can be proved by constructing an orthogonal array $OA_{2^{e(m-2)}}(2e, n, 2)$. Orthogonal arrays with these parameters can be constructed from BCH codes with designed distance $2e+1$ as explained in [2].

While this method is simple and powerful, it does not extend to the cases where $f(n)$-wise independence of the random variables is needed, where $f(n)$ is an increasing (however small the growth rate may be) function of n, as shown by the following theorem. This fact was first observed by Chor *et al* in [13].

Theorem 5.2 ([13]). *The size of the sample space constructed in Theorem 5.1 is asymptotically optimal and hence the size of a sample space with n t-wise independent non-trivial random variables can be polynomial in n only when t is a constant.*

The proof is a straightforward application of Rao bounds (see [36]) for orthogonal arrays and can be found in [2] and [13].

6 Authentication Without Secrecy

Authentication codes were invented in 1974 by Gilbert, MacWilliams and Sloane [19] and have been extensively studied in recent years. The general theory of unconditional authentication was developed by Simmons (see e.g. [43]). In this section, we will give a brief review of some relevant known results and explain the relevance of orthogonal arrays in this context. For a general survey on authentication, the reader is referred to [44].

In the usual model for authentication, there are three participants: Alice, Bob, and an opponent (Oscar). Alice wants to communicate some information to Bob using a public communication channel. The purpose of authentication code is to protect the integrity of the information. That is, when Bob receives a message from Alice, he wants to be sure that the message was really sent by Alice and was not tampered with. However, Oscar has the ability to introduce messages into the channel and/or to modify existing messages. When Oscar places a (new) message into the channel, this is called *impersonation*. When Oscar sees a message m and changes it to a message $m' \neq m$, this is called *substitution*.

We will use the following notation. An *authentication code without secrecy* is a four-tuple $(\mathcal{S}, \mathcal{A}, \mathcal{K}, \mathcal{E})$, defined as follows:

1. $\mathcal{S}$ is a finite set of *source states* (a source state is analogous to a plaintext). Let $\mid \mathcal{S} \mid = k$.

2. $\mathcal{A}$ is a finite set of *authenticators*. Let $\mid \mathcal{A} \mid = l$.

3. $\mathcal{K}$ is a finite set of *keys*.

4. For each $K \in \mathcal{K}$, there is an *authentication rule* $e_K \in \mathcal{E}$. Each $e_K : \mathcal{S} \to \mathcal{A}$.

Alice and Bob will follow the following protocol. First, they jointly choose a secret key $K \in \mathcal{K}$. At a later time, Suppose Alice wants to communicate a source state $s \in \mathcal{S}$ to Bob. Alice uses the authentication rule e_K to produce the authenticator $a = e_K(s)$. The *message* $m = (s, a)$ is sent over the channel. When Bob receives m, he verifies that $a = e_K(s)$ to authenticate the source state s. Since each authentication rule is a function from $\mathcal{S}$ to $\mathcal{A}$, we can represent an authentication code by an $|\mathcal{E}| \times |\mathcal{S}|$ matrix, where the rows are indexed by authentication rules, the columns are indexed by source states, and then entry in row e and column s is $e(s)$. We call this matrix the *authentication matrix*

When Oscar performs impersonation or substitution, his goal is to have his bogus message $m' = (s', a')$ accepted as authentic by Bob, thus misleading Bob as to the state of the source. That is, if e is the authentication rule being used (which is *not* known to Oscar), then he is hoping that $a' = e(s')$.

We assume that there is some probability distribution on $\mathcal{S}$, which is known to all the participants. Given this probability distribution, Alice and Bob will choose a probability distribution for $\mathcal{E}$, called an *authentication strategy.* Once Alice and Bob have chosen their authentication strategy, it is possible to compute, for $i = 0, 1$, a *deception probability* denoted by P_{d_i}, which is the probability that Oscar can deceive Bob by impersonation and substitution, respectively. In computing the deception probabilities, it is assumed that Oscar is using an optimal strategy.

It is not difficult to show that $P_{d_0} \geq 1/|\mathcal{A}|$ [43] and $P_{d_1} \geq 1/|\mathcal{A}|$ as well [47]. We are not only interested in minimizing the deception probabilities but also in minimizing the number of authentication rules, as this determines the amount of information to be exchanged between Alice and Bob in a secure fashion before they use the authentication code. Authentication codes (without secrecy) with minimum possible deception probabilities and the minimum number of encoding rules have been shown to be equivalent to orthogonal arrays by Stinson [48]. In the following result, the first part deals with codes in which $k \leq l + 1$, while the second part deals with codes in which $k \geq l + 1$.

Theorem 6.1 ([48]). *Suppose we have an authentication code without secrecy for k source states and having l authenticators, in which $P_{d_0} = P_{d_1} = 1/l$. Then*

1. $|\ \mathcal{E}\ | \geq l^2$, *and equality occurs if and only if the authentication matrix is an orthogonal array $OA(2, k, l)$ (with $\lambda = 1$) and the authentication rules are used with equal probability;*

2. $|\ \mathcal{E}\ | \geq k(l-1)+1$, *and equality occurs if and only if the authentication matrix is an orthogonal array $OA_\lambda(2, k, l)$ where*

$$\lambda = \frac{k(l-1)+1}{l^2},$$

and the authentication rules are used with equal probability.

7 Universal Hash Functions

Universal classes of hash functions were introduced by Carter and Wegman [11], and were studied further by Sarwate [38] and by Wegman and Carter [53]. Several applications of universal hash functions were discussed in [11] and [53]. The applications include complexity theory, randomized algorithms, associative memory, cryptography and authentication.

Let A and B be finite sets, and denote $a = |A|$ and $b = |B|$. We will assume that $a \geq b$. A function $h : A \rightarrow B$ will be termed a *hash function.* A finite set H of hash functions is $strongly - universal_2$ (or SU_2) if for every $x_1, x_2 \in A$ ($x_1 \neq x_2$) and for every $y_1, y_2 \in B$,

$$|\{h \in H : h(x_1) = y_1, h(x_2) = y_2\}| = |H|/|B|^2.$$

For a motivation of this definition see [53]. For practical applications, it is also important that $|H|$ is small. This is because $\log_2 |H|$ bits are needed to specify a hash function from the family. It is fairly straightforward to show that strongly universal hash functions are equivalent to orthogonal arrays. The following theorem can be found in [51].

Theorem 7.1. *If there exists an $OA_\lambda(2, k, n)$, then there exists an SU_2 class H of hash functions from A to B, where $|A| = k$, $|B| = n$ and $|H| = \lambda n^2$. Conversely, if there exists an SU_2 class H of hash functions from A to B, where $a = |A|$ and $b = |B|$, then there exists an $OA_\lambda(2, k, n)$, where $n = b$, $k = a$ and $\lambda = |H|/n^2$.*

This theorem helps in establishing lower bounds on the number of hash functions and in constructing classes of hash functions which meet these bounds. It is straightforward to extend the definition and the theorem to SU_t class of universal hash functions.

8 Threshold Schemes

In a bank, there is a vault which must be opened every day. The bank employs three senior tellers; but it is not desirable to entrust the combination to any one person. Hence, we want to design a system whereby any two of the three senior tellers can gain access to the vault, but no individual can do so. This problem can be solved by means of a threshold scheme.

Threshold schemes are actually a special case of secret sharing schemes. For an excellent survey of secret sharing schemes, the reader is referred to [49]. Threshold schemes were first described by Shamir [42] and Blakley [9] in 1979. Blakley's solution uses finite geometries, while Shamir's scheme is based on polynomial interpolation. Since then many constructions have been given for threshold schemes.

Informally a (t, w)-*threshold scheme* is a method of sharing a secret key K among a finite set $\mathcal{P}$ of w participants, in such a way that any t participants can compute the value of K, but no group of $t-1$ (or fewer) participants can do so. The value of K is chosen by a special participant called the *dealer*. The dealer is denoted by D and we assume $D \notin \mathcal{P}$. When D wants to share the key K among the participants in $\mathcal{P}$, he gives each participant some partial information called a *share*. The shares should be distributed secretly, so no participant knows the share given to another participant.

At a later time, a subset of participants $B \subseteq \mathcal{P}$ will pool their shares in an attempt to compute the secret key K. If $| B |\geq t$, then they should be able to compute the value of K as a function of the shares they collectively hold; if $| B |< t$, then they should not be able to compute K. In the example described above, we desire a $(2, 3)$-threshold scheme.

Often, we desire not only that an unauthorized subset of participants should be unable to compute the value of K by pooling their shares, but also they should be unable to determine anything about the value of K. Such a threshold scheme is called a *perfect threshold scheme*. Here, we will be concerned only about perfect threshold schemes.

We will use the following notation. Let $\mathcal{P} = \{P_i : 1 \leq i \leq w\}$ be the set of participants. $\mathcal{K}$ is the *key set* (i.e., the set of all possible keys); and $\mathcal{S}$ is the *share set* (i.e., the set of all possible shares). It is well known that in any perfect threshold scheme $| \mathcal{K} | \leq | \mathcal{S} |$. The schemes in which equality is attained are called *ideal threshold schemes*.

Orthogonal arrays come into picture once again by means of the following theorem due to E. Dawson, E.S. Mahmoodian and Alan Rahilly [16]. This theorem was first proved by Keith Martin in his Ph.D. thesis in 1992; However, only after several years it appeared in a paper by Jackson and Martin [26].

Theorem 8.1 ([16],[26]). *An ideal (t, w) threshold scheme with $| \mathcal{K} |= v$ exists if and only if an $OA(t, w+1, v)$ exists.*

The construction of the threshold scheme starting from the orthogonal array proceeds as follows. The first column of the OA corresponds to the dealer and the remaining w columns correspond to the w participants. To distribute a specific key K, the dealer selects a random row of the OA such that K appears in the first column and gives out the remaining w elements of the row as the shares. When t participants later pool their shares, the collective information will determine an unique row of the OA (as $\lambda = 1$) and hence they can compute K as the value of the first element in the row.

Can a group of $t - 1$ participants compute K ? Any possible value of the secret along with the actual shares of these $t - 1$ participants determine a unique row of the OA. Hence, no value of the secret can be ruled out. Moreover, it is clear that the $t - 1$ participants can obtain no information about the secret.

9 Perfect Local Randomizers

A (k, n) *sequence generator* stretches a binary random seed of length k to a binary pseudorandom sequence of length n. A (k, n) sequence generator is a (k, n, e) *perfect local randomizer* (PLR) if every subset of e of the n binary output random variables is a set of e independent and balanced random variables. The concept of perfect local randomizers was introduced and studied by Maurer and Massey [30]. They suggest the use of perfect local randomizers in key scheduling schemes where a small secret key must be stretched to a long key. It is obvious that a (k, n, e) perfect local randomizer is equivalent to an $OA_{2^{k-e}}(e, n, 2)$. This equivalence was observed by Maurer and Massey. Thus perfect local randomizers are really orthogonal arrays in disguise. Hence, we will not further elaborate on them.

10 Concluding Remarks

In this article, we surveyed several applications of orthogonal arrays to computer science. The diverse areas from where the applications are drawn attests to the ubiquitous nature of orthogonal arrays.

Although, we have surveyed several applications of orthogonal arrays, our survey is by no means complete due to lack of space. We briefly mention some other applications. Orthogonal arrays are used in the decorrelation theory of block ciphers (see [52]) and in the construction of almost k-wise independent sample spaces (see [28]). They are also used in ramp schemes, which are generalizations of threshold schemes (see [25]). Ordered orthogonal arrays are a generalization of orthogonal arrays and they are used in constructions of (t, m, s) nets (see [33]).

Orthogonal arrays are closely related to correlation-immune and resilient functions as well (see [20]). These functions are of fundamental importance in their own right and have a multitude of applications. Correlation-immune functions are used in stream ciphers as combining functions for running-key generators that are resistant to correlation attacks (see [37]). Resilient functions are used in the design of S-Boxes in block ciphers. Yet another application of resilient functions involves renewing a partially leaked cryptographic key, which is particularly relevant in the setting of quantum cryptography (see [6]). Finally resilient functions are also used in the generation of shared random strings in the presence of faulty processors. The last two applications were mentioned in [7] and [13].

References

[1] V.K. Agarwal and A.S.F. Fung, Multiple Fault Testing of Large Circuits by Single Fault Test Sets, *IEEE. Trans. Comput.*, **C-30(11)** (1981).

[2] Noga Alon, Laszlo Babai and Alon Itai, A Fast and Simple Randomized Parallel Algorithm for the Maximal Independent Set Problem, *Journal of Algorithms*, **7** (1986), pp. 567–583.

[3] Noga Alon and Joel H. Spencer, *The Probabilistic Method*, John Wiley & Sons, Inc., (1992).

[4] E. Bach and J. Shallit, *Algorithmic Number Theory-Vol. I Efficient Algorithms*, MIT Press, (1996).

[5] Paul H. Bardell, William H. McAnney, and Jacob Savir, *Built-in Test for VLSI–Pseudorandom Techniques*, John Wiley & Sons, (1987).

[6] C. H. Bennett, G. Brassard, and A. K. Ekert, Quantum cryptography, *Scientific American*, **267(4)** pp. 26–33. (1992).

[7] C. H. Bennett, G. Brassard, and J. M. Robert, Privacy amplification by public discussion, *SIAM J. Computing*, **17** (1988), pp. 210–229.

[8] A. Beutelspacher, Applications of finite geometry to cryptography, in *Geometries, Codes and Cryptography* pp. 161–186. Springer-Verlag, 1990.

[9] G. R. Blakely, Safeguarding cryptographic keys, in *Proc. N.C.C., AFIPS Conference Proceedings*, **48** (1979), pp. 313–317.

[10] F. Brglez, P. Powrall, and R. Hung, Applications of testability analysis from ATPG to critical delay path tracing, in *Proc. of International Test Conference*, pp. 705–712. 1984.

[11] L. Carter and M. Wegman, Universal Hash Functions, *J. Computer and System Sciences*, **18** (1979), pp. 143–154.

[12] Benny Chor and Oded Goldreich, On the Power of Two-Point Based Sampling, *Journal of Complexity*, **5** (1989), pp. 96–106.

[13] Benny Chor, Oded Goldreich, Johan Håstad, Joel Friedman, Steven Rudich, and Roman Smolensky, The Bit Extraction Problem or t-Resilient Functions, *IEEE Symposium on Foundations of Computer Science*, **26** (1985), pp. 396–407.

[14] C. J. Colbourn and P. C. Van Oorschot, Applications of combinatorial designs in computer science, *ACM Computing Surveys*, **21** (1989), pp. 223–250.

[15] C.J. Colbourn, J.H. Dinitz, and D.R. Stinson, Applications of Combinatorial Designs to Communications, Cryptography and Networking, *Surveys in Combinatorics*, pp 37–100, Cambridge University Press, (1999).

[16] E. Dawson, E. S. Mahmoodian, and Alan Rahilly, Orthogonal arrays and ordered threshold schemes, *Australasian J. of Combinatorics*, **8** (1993), pp. 27–44.

[17] P. Dukes and A. C. H. Ling, A combinatorial error bound for t-point based sampling, *Theoretical Computer Science*, **310** (2004), pp. 479–488.

[18] E. B. Eichelberger and T. W. Williams, A Logic Design Structure for LSI Testability, in *Proc. of the 14th Design Automation Conference*, June 1977.

[19] E. N. Gilbert, F. J. MacWilliams, and N. J. A. Sloane, Codes which detect deception, *Bell System Tech. Journal*, **53** (1974), pp. 405–424.

[20] K. Gopalakrishnan and D. R. Stinson, Three characterizations of non-binary correlation immune and resilient functions, *Designs, Codes and Cryptography*, **5** (1995), pp. 241–251.

[21] K. Gopalakrishnan and D. R. Stinson, A simple analysis of the error probability of two-point based sampling, *Information Processing Letters*, **60** (1996), pp. 91–96.

[22] Rajiv Gupta, Scott A. Smolka, and Shaji Bhaskar, On randomization in sequential and distributed algorithms, *ACM Computing Surveys*, **26(1)** (1994), pp. 7–86.

[23] John P. Hayees, *Computer Architecture*, McGraw Hill, (1978).

[24] A. S. Hedayat, N. J. A. Sloane, and J. Stuffken, *Orthogonal Arrays: Theory and Applications*, Springer-Verlag, New York (1999).

[25] W. A. Jackson and K. Martin, A combinatorial interpretation of ramp schemes, *Australasian Journal of Combinatorics*, **14** (1996), pp. 51–60.

[26] W.A. Jackson and K. Martin, Combinatorial models for perfect secret sharing schemes, *Journal of Combinatorial Mathematics and Combinatorial Computing*, **28** (1998), pp. 249–265.

[27] A. Joffe, On a set of almost deterministic k-independent random variables, *Ann. Probab.*, **2(1)** (1974), pp. 161–162.

[28] K. Kurosawa, T. Johannsson, and D. R. Stinson, Almost k-wise independent sample spaces and their cryptologic applications, *Journal of Cryptology*, **14** (2001), pp. 231–253.

[29] Michael Luby, A Simple Parallel Algorithm for the Maximal Independent Set Problem, *SIAM J. Comput.*, **15(4)** (1986), pp. 1036–1053.

[30] U. M. Maurer and J. L. Massey, Local randomness in pseudorandom sequences, *Journal of Cryptology*, **4(2)** (1991), pp. 135–149.

[31] R. Von Mises, *Probability, Statistics and Truth*, Macmillan, New York (1957).

[32] K. Mulmuley, U. V. Vazirani, and V. V. Vazirani, Matching is as easy as matrix inversion, *19th STOC*, pp. 345–354, (1987).

[33] Harald Niederreiter, Constructions of (t, m, s) nets and (t, s) sequences, *Finite Fields and Their Applications*, **11** (2005), pp. 578–600.

[34] K. R. Popper, *The Logic of Scientific Discovery*, London, (1959).

[35] D. K. Pradhan, *Fault Tolerant Computing: Theory and Techniques, volume 1*, Prentice Hall, Englewood Cliffs, NJ (1984).

[36] C. R. Rao, Factorial experiments derivable from combinatorial arrangements of arrays, *J. Royal Stat. Soc.*, **9** (1947), pp. 128–139.

[37] Rainer A. Rueppel, *Analysis and Design of Stream Ciphers*, Springer-Verlag, (1986).

[38] D. V. Sarwate, A note on universal classes of hash functions, *Information Processing Letters*, **10** (1980), pp. 41–45.

[39] J. Savir and P. H. Bardell, On Random Pattern Test Length, *IEEE. Trans. Comput.*, **C-33(6)** (1984), pp. 467–474.

[40] Jacob Savir, Gary S. Ditlow, and Paul H. Bardell, Random Pattern Testability, *IEEE Trans. Computers*, **C-33(1)** (1984), pp. 79–90.

[41] S. C. Seth, B. B. Bhattacharya, and V. D. Agrawal, An exact analysis for efficient computation of random pattern testability in combinational circuits, *Digest of papers, FTCS-16*, (1986), pp. 318–323.

[42] A. Shamir, How to share a secret, *Comm. ACM.*, **22** (1979), pp. 612–613.

[43] G. J. Simmons, Message authentication: a game on hypergraphs, *Congressus Numerantium*, **45** (1984), pp. 161–192.

[44] G. J. Simmons, A survey of information authentication, *Contemporary Cryptology, The Science of Information Integrity*, pp. 379–419, IEEE Press, (1992).

[45] N. J. A. Sloane, Error-correcting codes and cryptography, *The Mathematical Gardner*, pp. 346–382, Prindle, Weber and Schmidt, (1981).

[46] Thomas H. Spencer, Provably Good Pattern Generators for a Random Pattern Test, *Algorithmica*, **11(5)** (1994), pp. 429–442.

[47] D. R. Stinson, The combinatorics of authentication and secrecy codes, *Journal of Cryptology*, **2** (1990), pp. 23–49.

[48] D. R. Stinson, Combinatorial characterizations of authentication codes, *Designs, Codes and Cryptography*, **2** (1992), pp. 175–187.

[49] D. R. Stinson, An explication of secret sharing schemes, *Designs, Codes and Cryptography*, **2** (1992), pp. 357–390.

[50] D. R. Stinson, Combinatorial Designs and Cryptography, *K. Walker, editor,* Surveys in Combinatorics, pp. 257–287. Cambridge University Press, (1993).

[51] D. R. Stinson, Combinatorial techniques for universal hashing, *Journal of Comp. and Sys. Sci.*, **48(2)** (1994), pp. 337–346.

[52] Serge Vaudenay,Decorrelation: A theory for block cipher security,*Journal of Cryptology*, **16** (2003), pp. 249–286.

[53] M. N. Wegman and J. L. Carter, New hash functions and their use in authentication and set equality, *J. Computer and System Sciences*, **22** (1981), pp. 265–279.

Proc. of ICDM 2006, Pages 165–174

Forbidden Intersection Patterns in the Families of Subsets

Gyula O.H. Katona*
Rényi Institute, Budapest, Hungary
Email: ohkatona@renyi.hu

1 Introduction

Let $[n] = \{1, 2, \ldots, n\}$ be a finite set, $\mathcal{F} \subset 2^{[n]}$ a family of its subsets. In the present paper $\max |\mathcal{F}|$ will be investigated under certain conditions on the family $\mathcal{F}$. The well-known Sperner theorem ([10]) was the first such result.

Theorem 1.1. *If $\mathcal{F}$ is a family of subsets of $[n]$ without inclusion ($F, G \in \mathcal{F}$ implies $F \not\subset G$) then*

$$|\mathcal{F}| \leq \binom{n}{\lfloor \frac{n}{2} \rfloor}$$

holds, and this estimate is sharp as the family of all $\lfloor \frac{n}{2} \rfloor$-element subsets shows.

There is a very large number of generalizations and analogues of this theorem. Here we will consider only results when the condition on $\mathcal{F}$ excludes certain configurations what can be expressed by inclusion, only. That is, no intersections, unions, etc. are involved. The first such generalization was obtained by Erdős [4]. The family of k distinct sets with mutual inclusions, $F_1 \subset F_2 \subset \ldots F_k$ is called a *chain of lenght k*. It will be simply denoted by P_k. Let $\mathrm{La}(n, P_k)$ denote the largest family $\mathcal{F}$ without a chain of lenght k.

Theorem 1.2 ([4]). $\mathrm{La}(n, P_{k+1})$ *is equal to the sum of the k largest bimomial coefficients of order n.*

Let V_r denote the *r-fork*, that is the following family of distinct sets: $F \subset G_1$, $F \subset G_2, \ldots F \subset G_r$. The quantity $\mathrm{La}(n, V_r)$, that is, the largest family on n elements containing no V_r was first (asymptotically) determined for $r = 2$.

Theorem 1.3 ([7]).

$$\binom{n}{\lfloor \frac{n}{2} \rfloor}\left(1 + \frac{1}{n} + O\left(\frac{1}{n^2}\right)\right) \leq \mathrm{La}(n, V_2) \leq \binom{n}{\lfloor \frac{n}{2} \rfloor}\left(1 + \frac{2}{n}\right).$$

The first result for general r is contained in the following theorem.

Theorem 1.4 ([11]).

$$\binom{n}{\lfloor \frac{n}{2} \rfloor}\left(1 + \frac{r}{n} + O\left(\frac{1}{n^2}\right)\right) \leq \mathrm{La}(V_{r+1}) \leq \binom{n}{\lfloor \frac{n}{2} \rfloor}\left(1 + 2\frac{r^2}{n} + o\left(\frac{1}{n}\right)\right).$$

*The work was supported by the Hungarian National Foundation for Scientific Research grant numbers NK0621321, AT048826, the Bulgarian National Science Fund under Grant IO-03/2005 and the projects of the European Community: INTAS 04-77-7171, COMBSTRU–HPRN-CT-2002-000278, FIST–MTKD-CT-2004-003006.

The constant in the second term in the upper estimate was recently improved.

Theorem 1.5 ([1]).

$$\binom{n}{\lfloor \frac{n}{2} \rfloor}\left(1+\frac{r}{n}+O\left(\frac{1}{n^2}\right)\right) \le \mathrm{La}(V_{r+1}) \le \binom{n}{\lfloor \frac{n}{2} \rfloor}\left(1+2\frac{r}{n}+o\left(\frac{1}{n}\right)\right).$$

See some remarks in Section 5 explaining why this second term is difficult to improve any more.

The aim of the present paper is to introduce some recent results and show a method, proving good upper estimates, developed recently.

2 Notations, Definitions

A *partially ordered set*, shortly *poset* P is a pair $P = (X, \le)$ where X is a (in our case always finite) set and $\le$ is a relation on X which is *reflexive* ($x \le x$ holds for every $x \in X$), *antisymmetric* (if both $x \le y$ and $x \ge y$ hold for $x, y \in X$ then $x = y$) and transitive ($x \le y$ and $y \le z$ always implies $x \le z$). It is easy to see that if $X = 2^{[n]}$ and the $\le$ is defined as $\subseteq$, then these conditions are satisfied, that is the familiy of all subsets of an n-element set ordered by inclusion form a poset. We will call this poset the *Boolean lattice* and denote it by B_n.

The definition of a *subposet* is obvious: $R = (Y, \le_2)$ is a subposet of $P = (X, \le_1)$ iff there is an injection α of Y into X is such a way that $y_1, y_2 \in Y, y_1 \le_2 y_2$ implies $\alpha(y_1) \le_1 \alpha(y_2)$. On the other hand R is an *induced subposet* of P when $\alpha(y_1) \le_1 \alpha(y_2)$ holds iff when $y_1 \le_2 y_2$. If $P = (X, \le)$ is a poset and $Y \subset X$ then the poset *spanned by* Y *in* P is defined as $(Y, \le^*)$ where $\le^*$ is the same as $\le$, for all the pairs taken from Y. Given a "small" poset R, $\mathrm{La}(n, R)$ denotes the maximum number of elements of $Y \subset 2^{[n]}$ (that is, the maximum number of subsets of $[n]$) such that R is not a subposet of the poset spanned by Y in B_n.

Redefine our "small" configurations in terms of posets. The chain P_k contains k elements: $a_1, \dots, a_k$ where $a_1 < \dots < a_k$. The *r-fork* contains $r + 1$ elements: $a, b_1, \dots, b_r$ where $a < b_1, \dots a < b_r$. It is easy to see that the definitions of $\mathrm{La}(n, P_k)$, $\mathrm{La}(n, V_r)$, in Sections 1 and 2 agree. In the rest of the paper we will use the two different terminology alternately. In the definition of $\mathrm{La}(n, R)$ we mean non-induced subposets, that is, if $R = V_2$ then P_3 is also excluded as a subposet.

A poset is *connected* if for any pair (z_0, z_k) of its elements there is a sequence $z_1, \dots, z_{k-1}$ such that either $z_i < z_{i+1}$ or $z_i > z_{i+1}$ holds for $0 \le i < k$. If the poset is not connected, maximal connected subposets are called its *connected components.* Given a family $\mathcal{F}$ of subsets of $[n]$, it spans a poset in B_n. We will consider its connected components in two different ways. First as posets themself, secondly as they are represented in B_n. In the latter case the sizes of in the sets are also indicated. A *full chain* in B_n is a family of sets $A_0 \subset A_1 \subset \dots \subset A_n$ where $|A_i| = i$. We say that a (full) chain *goes through* a family (subposet) $\mathcal{P}$ if their intersection is non-empty, that is if it "goes through" at least one member of the family.

3 Lubell's Proof of the Sperner Theorem

The number of full chains in $[n]$ is $n!$ since the choice of a full chain is equivalent to the choice of a permutation of the elements of $[n]$. On the other hand, the number of full chains going through a given set F of f elements is $f!(n - f)!$ since the chain "must grow" within F until it "hits" F and outside after that. Suppose that the family $\mathcal{F}$ of subsets of $[n]$ is without inclusion ($F, G \in \mathcal{F}$ implies $F \not\subset G$). Then a full chain cannot go through two members of $\mathcal{F}$. Therefore the set of full chains going through distinct members of $\mathcal{F}$ must be disjoint. Hence we have

$$\sum_{F\in\mathcal{F}} |F|!(n-|F|)! \leq n!.$$

Dividing the inequality by $n!$

$$\sum_{F\in\mathcal{F}} \frac{1}{\binom{n}{|F|}} \leq 1 \tag{3.1}$$

is obtained. As $\binom{n}{|F|} \leq \binom{n}{\lfloor \frac{n}{2} \rfloor}$, then

$$\frac{|\mathcal{F}|}{\binom{n}{\lfloor \frac{n}{2} \rfloor}} = \sum_{F\in\mathcal{F}} \frac{1}{\binom{n}{\lfloor \frac{n}{2} \rfloor}} \leq 1$$

follows, the theorem is proved.

Let us remark that inequality (3.1) is important on its own right and is called the YBLM-inequality (earlier LYM, see [3, 8, 9, 12]). □

4 The Method, Illustrated with an Old Result

Lubell's proof easily applies for Theorem 1.2, however, surprisingly it was not exploited for proving theorems of the present type. The reason might be that not the "excluded" configurations should be considered when using the idea, but the "allowed induced posets". (See later.)

Following the definition of the r-fork, let us define the r-*brush* (in a poset) which contains $r+1$ elements: $a, b_1, \ldots, b_r$ where $a > b_1, \ldots a > b_r$ and is the "complement" of the r-fork. Theorem 1.3 gives the best expected asymptotic upper bound up to the second term for V_2 in the Boolean lattice. It is easy to see that it implies the same solution for Λ_2. However the result is very different when both of them are excluded. Our notation $\mathrm{La}(n, R)$ is extended in an obvious way for the case when two subposets R_1 and R_2 are excluded: $\mathrm{La}(n, R_1, R_2)$.

Theorem 4.1 ([7]).

$$\mathrm{La}(n, V_2, \Lambda_2) = 2\binom{n-1}{\lfloor \frac{n-1}{2} \rfloor}.$$

Proof. The construction giving the equality is the following:

$$\left\{F \subset [n] : 1 \notin F, |F| = \left\lfloor \frac{n-1}{2} \right\rfloor\right\} \cup \left\{F \subset [n] : 1 \in F, |F| = \left\lfloor \frac{n+1}{2} \right\rfloor\right\}.$$

The non-trivial part of the proof is the verification of the upper bound.

Let $\mathcal{F}$ be a family of subsets of $[n]$ which contains neither a V_2 nor a Λ_2 as subposet. Therefore it cannot contain a P_3 either. Consider the connected components of the poset spanned by $\mathcal{F}$. It is obvious that a connected component can be either a one element poset P_1 or a P_2. Let α_1 and α_2 be their respective numbers. Then

$$|\mathcal{F}| = \alpha_1 + 2\alpha_2. \tag{4.1}$$

We will now determine the minimum number of full chains going through a one or two-element component. Let $P(1;a)$ be a one-element component which is an a-element set. The number $c(P(1;a))$ of full chains going through $P(1;a)$ is $a!(n-a)!$. Therefore

$$\frac{c(P(1;a))}{n!} = \frac{1}{\binom{n}{a}}.$$

It takes on its minimum at the value $a = \lfloor \frac{n}{2} \rfloor$. Hence we obtained

$$\left\lfloor \frac{n}{2} \right\rfloor ! \left\lceil \frac{n}{2} \right\rceil ! \leq c(P(1;a)). \tag{4.2}$$

The two-element component consisting of an a-element A and a b-element B ($A \subset B$) ($a < b$) subset is denoted by $P(2;a,b)$. The number $c(P(2;a,b))$ of full chains going through (at least one element of) $P(2;a,b)$ is

$$c(P(2;a,b)) = a!(n-a)! + b!(n-b)! - a!(b-a)!(n-b)!. \tag{4.3}$$

Divide it by $n!$.

$$\frac{c(P(2;a,b))}{n!} = \frac{1}{\binom{n}{a}} + \frac{1}{\binom{n}{b}} - \frac{1}{\binom{n}{b}\binom{b}{a}} = \frac{1}{\binom{n}{a}} + \frac{1}{\binom{n}{b}}\left(1 - \frac{1}{\binom{n}{a}}\right). \tag{4.4}$$

Suppose first that a is fixed and is $\leq \lfloor \frac{n-1}{2} \rfloor$. Then (4.4) takes on its minimum for $b = \lfloor \frac{n+1}{2} \rfloor$. Fix b here and consider the following variant of (4.4):

$$\frac{c(P(2;a,b))}{n!} = \frac{1}{\binom{n}{a}} + \frac{1}{\binom{n}{b}} - \frac{1}{\binom{n}{a}\binom{n-a}{n-b}} = \frac{1}{\binom{n}{b}} + \frac{1}{\binom{n}{a}}\left(1 - \frac{1}{\binom{n-a}{n-b}}\right). \tag{4.5}$$

This is a monotone decreasing function of a in the interval $0 \leq a \leq \lceil \frac{n}{2} \rceil$. Therefore the pair giving the minimum in this case is $a = \lfloor \frac{n-1}{2} \rfloor, b = \lfloor \frac{n+1}{2} \rfloor$.

Suppose now that $a \geq \lfloor \frac{n}{2} \rfloor$. Then b can be chosen to be $a+1$ by (4.4), and (4.3) becomes $na!(n-a-1)!$ It achieves its minimum at $a = \lfloor \frac{n-1}{2} \rfloor$, again. We obtained

$$n \left\lfloor \frac{n-1}{2} \right\rfloor ! \left\lceil \frac{n-1}{2} \right\rceil ! \leq c(P(2;a,b)). \tag{4.6}$$

Observe that a full chain cannot go through two distinct components, therefore

$$\sum_{P_1 \text{is a component}} c(P_1) + \sum_{P_2 \text{is a component}} c(P_2) \leq n!$$

holds. The left hand side can be lowerestimated by (4.2) and (4.6):

$$\alpha_1 \left\lfloor \frac{n}{2} \right\rfloor ! \left\lceil \frac{n}{2} \right\rceil ! + \alpha_2 n \left\lfloor \frac{n-1}{2} \right\rfloor ! \left\lceil \frac{n-1}{2} \right\rceil ! \leq n!. \tag{4.7}$$

(4.1) has to be maximized with respect to (4.7). Rewrite (4.7) a little bit:

$$\alpha_1 \left\lfloor \frac{n}{2} \right\rfloor ! \left\lceil \frac{n}{2} \right\rceil ! + 2\alpha_2 \frac{n}{2} \left\lfloor \frac{n-1}{2} \right\rfloor ! \left\lceil \frac{n-1}{2} \right\rceil ! \leq n!. \tag{4.8}$$

Compare the coefficients of α_1 and $2\alpha_2$ in (4.8).

$$\left\lfloor \frac{n}{2} \right\rfloor ! \left\lceil \frac{n}{2} \right\rceil ! \geq \frac{n}{2} \left\lfloor \frac{n-1}{2} \right\rfloor ! \left\lceil \frac{n-1}{2} \right\rceil ! \tag{4.9}$$

holds (with equality for even n). Replacing the coefficient of α_1 in (4.8) using (4.9), the inequality

$$(\alpha_1 + 2\alpha_2)\frac{n}{2}\left\lfloor\frac{n-1}{2}\right\rfloor!\left\lceil\frac{n-1}{2}\right\rceil! \leq n!$$

is obtained, what results in

$$|\mathcal{F}| = \alpha_1 + 2\alpha_2 \leq \frac{n!}{\frac{n}{2}\lfloor\frac{n-1}{2}\rfloor!\lceil\frac{n-1}{2}\rceil!} = 2\binom{n-1}{\lfloor\frac{n-1}{2}\rfloor}.$$

□

5 A Construction = a Lower Estimate

Although we concentrate in this paper on the upper estimates, it seems to be important to show the construction serving as a lower estimate in Theorem 1.3

The consctruction for a family avoiding a V_2 is the following. Take all the sets of size $\lfloor\frac{n}{2}\rfloor$ and a family $A_1, \ldots A_m$ of $\lfloor\frac{n}{2}\rfloor + 1$-element sets satisfying the condition $|A_i \cap A_j| < \lfloor\frac{n}{2}\rfloor$ for every pair $i < j$. It is easy to see that this family contains no V_2. We only have to maximize m. Since the $\lfloor\frac{n}{2}\rfloor$-element subsets of the A_is are all distinct, we have

$$m\left(\left\lfloor\frac{n}{2}\right\rfloor + 1\right) \leq \binom{n}{\lfloor\frac{n}{2}\rfloor}.$$

This gives the upper estimate

$$m \leq \binom{n}{\lfloor\frac{n}{2}\rfloor}\frac{2}{n}. \tag{5.1}$$

There is a very nice construction (see [5]) of such sets A_i with

$$m = \binom{n}{\lfloor\frac{n}{2}\rfloor + 1}\frac{1}{n} = \binom{n}{\lfloor\frac{n}{2}\rfloor}\left(\frac{1}{n} + O(\frac{1}{n^2})\right).$$

It is a longstanding conjecture of coding theory what the right constant is here, 1 or 2. Or if the limit exists at all?

This is why there is a disturbing factor 2 between the second terms of the lower and upper estimates in Theorem 1.3. This gap cannot be bridged without solving the problem in coding theory mentioned above.

6 The Upper Estimate in Theorem 1.3

This theorem already has two different proofs in [7] and [2], however each of these proofs needed an *ad hoc* idea. Our new method also works here. It needs some tedious calculations, but the principal idea is as easy as in the previous case.

Suppose that $\mathcal{F}$ contains no V_2 as a subposet. Then it cannot contain a P_3 either. It is easy to deduce that the components of the poset spanned by $\mathcal{F}$ are all of type Λ_r where $0 \leq r$. This is a new phenomenon! The sizes of the components are unbounded. Yet, the method works.

Let $\gamma(r)$ $(0 \leq r)$ denote the number of components of form Λ_r. Then

$$|\mathcal{F}| = \sum_{r=0}^{\infty}(r+1)\gamma(r). \tag{6.1}$$

is obvious. This has to be maximized under a linear condition obtained from the fact that the chains going through distinct components must be distinct.

Let $\Lambda(r; u, u_1, \ldots, u_r)$ $(u > u_1, \ldots, u_r)$ denote the component Λ_r represented by sets of sizes $u, u_1, \ldots, u_r$, respectively. $c(\Lambda(r; u, u_1, \ldots, u_r))$ denotes the number of full chains going through this component. The following lemma gives a good lower estimate on this number. For the proof see [6].

Lemma 6.1. *Suppose* $6 \leq n, 1 \leq r$. *Then*

$$u^*!(n-u^*)! + ru^*!u^*(n-u^*-1)! \leq c(\Lambda(r; u, u_1, \ldots, u_r))$$
$$(u > u_1, \ldots, u_r)$$

holds where $u^* = u^*(n) = \frac{n}{2} - 1$ *if* n *is even,* $u^* = \frac{n-1}{2}$ *if* n *is odd and* $r - 1 \leq n$, *while* $u^* = \frac{n-3}{2}$ *if* n *is odd and* $n < r - 1$.

In the case $r = 0$ *the inequality* $\lfloor \frac{n}{2} \rfloor! \lceil \frac{n}{2} \rceil! \leq \Lambda(0; u)$ *holds.*

We will actually need a lower estimate on the number of full chains going through the component, divided by the number of elements of this component.

$$u^*!u^*(n-u^*-1)! \leq \frac{u^*!(n-u^*)! + ru^*!u^*(n-u^*-1)!}{r+1} \quad (0 \leq r) \tag{6.2}$$

is a consequence of the lemma and the remark on the case $r = 0$.

Let $\mathcal{C}_i$ $(i \leq K = \sum_{r=0}^{\infty} \gamma(r))$ be the components spanned by $\mathcal{F}$. Each $\mathcal{C}_i$ is equal to a $\Lambda(r; u, u_1, \ldots, u_r)$ for some parameters. $c(\mathcal{C}_i)$ denotes the number of chains going through the component $\mathcal{C}_i$. The full chains going through distinct components are distinct. This implies

$$n! \geq \sum_{i=1}^{K} c(\mathcal{C}_i). \tag{6.3}$$

Hence by Lemma 6.1 we obtain

$$n! \geq \sum_{r=1}^{K} (u^*!(n-u^*)! + ru^*!u^*(n-u^*-1)!)\gamma(r). \tag{6.4}$$

(6.1) has to be maximized with respect to (6.4). Slightly modify the right hand side and use (6.2):

$$\sum_{r=1}^{K} \frac{u^*!(n-u^*)! + ru^*!u^*(n-u^*-1)!}{r+1}(r+1)\gamma(r)$$
$$\geq \sum_{r=1}^{K} u^*!u^*(n-u^*-1)!(r+1)\gamma(r). \tag{6.5}$$

From (6.4), (6.5) and (6.1) we obtain

$$n! \geq u^*!u^*(n-u^*-1)! \sum_{r=1}^{K} (r+1)\gamma(r) = u^*!u^*(n-u^*-1)!|\mathcal{F}|,$$

that is,

$$|\mathcal{F}| \leq \frac{n!}{u^*!u^*(n-u^*-1)!}.$$

This is equal to

$$\binom{n}{\lfloor\frac{n}{2}\rfloor}\frac{\frac{n}{2}}{\frac{n}{2}-1},\qquad \binom{n}{\lfloor\frac{n}{2}\rfloor}\frac{\frac{n+1}{2}}{\frac{n-1}{2}},\qquad \binom{n}{\lfloor\frac{n}{2}\rfloor}\frac{\frac{n-1}{2}}{\frac{n-3}{2}}$$

in the cases $u^* = \frac{n}{2}-1, \frac{n-1}{2}$ and $\frac{n-3}{2}$, respectively. These are all equal to

$$= \binom{n}{\lfloor\frac{n}{2}\rfloor}\Big(1+\frac{2}{n}+O\Big(\frac{1}{n^2}\Big)\Big).$$

□

7 Excluding the *N*

The poset N contains 4 distinct elements a, b, c, d satisfying $a < c, b < c, b < d$. In the Boolean lattice a subposet N consists of four disticts subsets satisfying $A \subset C, B \subset C, B \subset D$. It is somewhat surprising that excluding N the result is basically the same as in the case of V_2. The goal of the present section is to sketch the proof of the following theorem.

Theorem 7.1 ([6]).

$$\binom{n}{\lfloor\frac{n}{2}\rfloor}\left(1+\frac{1}{n}+O\left(\frac{1}{n^2}\right)\right) \le La(n,N) \le \binom{n}{\lfloor\frac{n}{2}\rfloor}\left(1+\frac{2}{n}+O\left(\frac{1}{n^2}\right)\right)$$

holds.

The lower estimate is obtained from Theorem 1.3, since $\mathrm{La}(n, V_2) \le \mathrm{La}(n, N)$.

Let $\mathcal{F}$ be a family of subsets of $[n]$ contaning no four distinct members forming an N. Consider the poset $P(\mathcal{F})$ spanned by $\mathcal{F}$ in B_n. Its connected components are denoted by $\mathcal{C}_1, \dots, \mathcal{C}_K$. and $c(\mathcal{C}_i)$ denotes the number of full chains going through $\mathcal{C}_i$. Since a full chain cannot go through two distinct components. the following inequality holds.

$$\sum_{i=1}^{K} c(\mathcal{C}_i) \le n!. \tag{7.1}$$

What can these components be? A component might be a P_3, but no component can contain a P_3 as a proper subposet, since adding one more element to P_3 an N is created no matter which element of P_3 is in relation with the new element. Furthermore, if $a < b$ are two elements of a component then a and b cannot be both comparable in the component with some other distinct elements c, d, only in the way $c < a < b < b$ what is a P_3. But one of them can be comparable with many others. Therefore the following ones are the only possible components:

$$a < b < c, \tag{7.2}$$

$$a < b_i (1 \le i \le r) \text{ where } 0 \le r, \tag{7.3}$$

$$a > b_i (1 \le i \le r) \text{ where } 0 \le r. \tag{7.4}$$

These are denoted by $P(3), V(r), \Lambda(r)$ in this order. However the numbers of full chains going through these posets depend on the sizes of the elements of the poset, that is the sizes of the members of the family $\mathcal{C}_i$. This is why we introduce the notation $P(3; u, v, w)$ for the posets of type $P(3)$ where $|a| = u < |b| = v < |c| = w$. The analogous notations are $V(r; u, u_1, \ldots, u_r)$ $(u < u_1, \ldots, u_r)$ and $\Lambda(r; v, v_1, \ldots, v_r)$ $(v < v_1, \ldots, v_r)$. All the components $\mathcal{C}_i$ in (7.1) are of this form.

An upper bound is sought for

$$|\mathcal{F}| = \sum_{i=1}^{K} |\mathcal{C}_i| \tag{7.5}$$

where $|P(3)| = 3, |V(r)| = |\Lambda(r)| = r + 1$ are obvious. This upper bound will be determined entirely on the basis of (7.1). Denote the numbers of $\mathcal{C}_i$s of type $L(3), V(r), \Lambda(r)$ by $\alpha, \beta(r), \gamma(r)$ respectively. Then (7.5) can be written in the form

$$|\mathcal{F}| = 3\alpha + \sum_{r=0}^{\infty}(r+1)\beta(r) + \sum_{r=0}^{\infty}(r+1)\gamma(r). \tag{7.6}$$

If the minima (or good lower bounds)

$$\min_{u,v,w} c(P(3; u, v, w)), \quad \min_{u,u_1,\ldots,u_r} c(V(r; u, u_1, \ldots, u_r)), \quad \min_{v,v_1,\ldots,v_r} c(\Lambda(r; v, v_1, \ldots, v_r)) \tag{7.7}$$

are determined then (7.1) leads to a linear combination of $\alpha, \beta(r)$, and $\gamma(r)$. That is, one linear combination, namely (7.5) has to be maximized under the condition that another combination is bounded from above. Therefore our main problem is now to determine the minima in (7.7). The first one is solved in the following lemma (for the proof see [6]).

Lemma 7.2. $c(P(3; u, v, w))$ $(u < v < w)$ *takes its minimum for the values* $u = \lfloor \frac{n}{2} \rfloor - 1, v = \lfloor \frac{n}{2} \rfloor, w = \lfloor \frac{n}{2} \rfloor + 1$, *that is,*

$$\left(\left\lfloor \frac{n}{2} \right\rfloor - 1\right)! \left(\left\lceil \frac{n}{2} \right\rceil - 1\right)! \left(\left\lfloor \frac{n}{2} \right\rfloor^2 - n\left\lfloor \frac{n}{2} \right\rfloor + n^2 - 1\right) \leq c(L(3; u, v, w)).$$

We already have a good lower estimate for the last minimum in (7.7): Lemma 6.1. Moreover, the middle minimum in (7.7) is the same, since the excluded poset is the complement of the previous one. We obtain the following inequality from (7.1) and the lower estimates.

$$\begin{aligned} n! \geq & \left(\left\lfloor \frac{n}{2} \right\rfloor - 1\right)! \left(\left\lceil \frac{n}{2} \right\rceil - 1\right)! \left(\left\lfloor \frac{n}{2} \right\rfloor^2 - n\left\lfloor \frac{n}{2} \right\rfloor + n^2 - 1\right)\alpha \\ & + \sum_{r=1}^{K}(u^*!(n-u^*)! + ru^*!u^*(n-u^*-1)!)\beta(r) \\ & + \sum_{r=1}^{K}(u^*!(n-u^*)! + ru^*!u^*(n-u^*-1)!)\gamma(r). \end{aligned} \tag{7.8}$$

(7.6) will be maximized with respect to (7.8). For this aim, let us modify (7.8):

$$\begin{aligned} n! \geq & \frac{\left(\left\lfloor \frac{n}{2} \right\rfloor - 1\right)! \left(\left\lceil \frac{n}{2} \right\rceil - 1\right)! \left(\left\lfloor \frac{n}{2} \right\rfloor^2 - n\left\lfloor \frac{n}{2} \right\rfloor + n^2 - 1\right)}{3} 3\alpha \\ & + \sum_{r=1}^{K} \frac{u^*!(n-u^*)! + ru^*!u^*(n-u^*-1)!}{r+1}(r+1)\beta(r) \end{aligned}$$

$$+\sum_{r=1}^{K} \frac{u^*!(n-u^*)! + ru^*!u^*(n-u^*-1)!}{r+1}(r+1)\gamma(r). \tag{7.9}$$

(6.2) gives a good lower estimate on the ratios in all terms, except the first one. Surprisingly, the same estimate is valid for the first term:

$$u^*!u^*(n-u^*-1)! \le$$
$$\frac{1}{3}\left(\left\lfloor\frac{n}{2}\right\rfloor - 1\right)!\left(\left\lceil\frac{n}{2}\right\rceil - 1\right)!\left(\left\lfloor\frac{n}{2}\right\rfloor^2 - n\left\lfloor\frac{n}{2}\right\rfloor + n^2 - 1\right)$$

as it can be easily checked for each u^*. All ratios can be replaced by $u^*!u^*(n-u^*-1)!$ in (7.9):

$$n! \ge u^*!u^*(n-u^*-1)!(3\alpha + \sum_{r=0}^{\infty}(r+1)\beta(r) + \sum_{r=0}^{\infty}(r+1)\gamma(r)) = |\mathcal{F}|u^*!u^*(n-u^*-1)!.$$

The proof can be finished like in the previous section. □

8 Concluding Remarks

1. Lemma 7.2 is much easier than Lemma 6.1. That is, we obtained Theorem 7.1 almost free after having the proof Theorem 1.3 with our method. This probably will happen often. The solution for a given excluded configuration can be obtained by putting together estimates for "allowed" posets, which have been already solved for other excluded patterns.

2. We see that the method can be applied for many other problems, on the other hand it does not seem to be sufficient for all excluded posets. Our research was not extensive enough to see the limits: what kind of problems can be solved, what cannot be solved using this idea.

For instance Theorem 1.5 has a quite easy proof in [1], on the other hand we do not see its proof with the present method. The family of "allowed" components seems to be too rich to handle.

3. Let us mention one more recent result. Four disticts subsets satisfying $A \subset C, A \subset D, B \subset C, B \subset D$ are called a *butterfly* and are denoted by B.

Theorem 8.1 ([2]). *Let* $n \ge 3$. *Then* $\mathrm{La}(n, B) = \binom{n}{\lfloor n/2 \rfloor} + \binom{n}{\lfloor n/2 \rfloor + 1}$.

When V_2 was excluded, the largest family consisted of the largest level, plus a $\frac{1}{n}$ part of the next level. Excluding somewhat less, the N, the result is the same. However excluding B there is a considerable jump: the largest family consists of the two largest levels. We do not see the reasons. When and why does a jump occur?

References

[1] Annalisa De Bonis and Gyula O.H. Katona, Forbidden r-forks, submitted.

[2] Annalisa De Bonis, Gyula O.H. Katona and Konrad J. Swanepoel, Largest family without $A \cup B \subseteq C \cup D$, *J. Combin. Theory Ser. A*, **111** (2005), pp. 331–336.

[3] B. Bollobás, On generalized graphs, *Acta. Math. Acad. Sci. Hungar.*, **16** (1965), pp. 447–452.

[4] P. Erdős, On a lemma of Littlewood and Offord, *Bull. Amer. Math. Soc.*, **51** (1945), pp. 898–902.

[5] R.L. Graham and H.J.A. Sloane, Lower bounds for constant weight codes, *IEEE IT*, **26** pp. 37–43.

[6] Jerrold R. Griggs and Gyula O.H. Katona, No four sets forming an N, submitted.

[7] G.O.H. Katona and T.G. Tarján, Extremal problems with excluded subgraphs in the n-cube, *Lecture Notes in Math.*, **1018** pp. 84–93.

[8] D. Lubell, A short proof of Sperner's lemma, *J. Combin. Theory*, **1** (1966), pp. 299.

[9] L.D. Meshalkin, A generalization of Sperner's theorem on the number of subsets of a finite set, *Teor. Verojatnost. i Primen.*, **8** (1963), pp. 219–220. (in Russian with German summary).

[10] E. Sperner, Ein Satz über Untermegen einer endlichen Menge, *Math. Z.*, **27** (1928), pp. 544–548.

[11] Hai Tran Thanh, An extremal problem with excluded subposets in the Boolean lattice, *Order*, **15** (1998), pp. 51–57.

[12] K. Yamamoto, Logarithmic order of free distributive lattices, *J. Math. Soc. Japan*, **6** (1954), pp. 347–357.

Proc. of ICDM 2006, Pages 175–181

Directed Odd Cycle Decompositions of Some Regular Digraphs

R.S. MANIKANDAN
Department of Mathematics, Anna University Tiruchirappalli, Tiruchirappalli – 620 024, India.
P. PAULRAJA AND S. SIVASANKAR
Department of Mathematics, Annamalai University, Annamalainagar 608 002, India

Abstract. In this paper, it is proved that if $m \geq 3$, t odd, $3 \leq t \leq mn$, $gcd(m(m-1), t) \leq m$ and $t \mid m(m-1)n$, then $\vec{C}_t$, a directed cycle of length t, decomposes $(K_m \circ \overline{K}_n)^*$, where $\circ$ denotes the wreath product of graphs and $(K_m \circ \overline{K}_n)^*$ denotes the symmetric digraph of $(K_m \circ \overline{K}_n)$; in fact, $K_m \circ \overline{K}_n$ is the complete m partite graph in which each part has n vertices. As a consequence, it is shown that for any odd prime p, $p \leq mn$, if p divides the number of arcs in the complete m-partite symmetric digraph D^*, in which each part has n vertices, then $\vec{C}_p$ decomposes D^*. Further, it is shown that if both t and m are odd, $m \geq 3$, $3 \leq t \leq mn$, $gcd(\frac{m(m-1)}{2}, t) \leq m$ and $t \mid \frac{m(m-1)}{2}n$, then C_t decomposes $K_m \circ \overline{K}_n$.

Keywords: Wreath product, Latin square, directed k-cycle decomposition.

1 Introduction

Unless stated otherwise, by a graph, we mean a simple graph. A k-regular graph G is called *Hamilton cycle decomposable* if G is decomposable into $\frac{k}{2}$ Hamilton cycles when k is even and into $\frac{k-1}{2}$ Hamilton cycles together with a perfect matching when k is odd.

For a graph G, G^* denotes the symmetric digraph of G; that is, the digraph obtained from G by replacing each of its edge by a symmetric pair of arcs. For a graph G and a positive integer λ, $G(\lambda)$ is the graph obtained from G by replacing each of its edge by λ parallel edges. If $H_1, H_2, \ldots, H_k$ are edge-disjoint subgraphs of G such that $\bigcup_{i=1}^{k} H_i = G$, then we write it as $H_1 \oplus H_2 \oplus \ldots \oplus H_k = G$; if each $H_i \simeq H$, then we write $H \mid G$.

The complete graph on m vertices is denoted by K_m and its complement is denoted by $\overline{K}_m$. Let C_k (resp. $\vec{C}_k$) denote a cycle (resp. directed cycle) of length k and let P_k (resp. $\vec{P}_k$) denote a path (resp. directed path) on k vertices. For a digraph D, $A(D)$ denotes its arc set.

For two simple graphs (resp. digraphs) G and H their *wreath product*, denoted by $G \circ H$, has vertex set $V(G) \times V(H)$ in which $(g_1, h_1)(g_2, h_2)$ is an edge (resp. arc) whenever (g_1, g_2) is an edge (resp. arc) in G, or $g_1 = g_2$ and (h_1, h_2) is an edge (resp. arc) in H. Similarly, the *tensor product* of the graphs (resp. digraphs) G and H, denoted by $G \times H$, has vertex set $V(G) \times V(H)$ in which $(g_1, h_1)(g_2, h_2)$

is an edge (resp. arc) whenever (g_1, g_2) is an edge (resp. arc) in G and (h_1, h_2) is an edge (resp. arc) in H.

We shall use the following notation throughout this paper. Let G and H be simple graphs with vertex sets $V(G) = \{x_0, x_1, \ldots, x_{m-1}\}$ and $V(H) = \{y_0, y_1, \ldots, y_{n-1}\}$, respectively. Then $V(G \circ H) = V(G) \times V(H)$ and, for our convenience, we write $V(G \circ H) = \bigcup_{i=0}^{m-1} X_i$, where $X_i = \{x_i\} \times V(H)$; we can also write $V(G \circ H) = \bigcup_{j=0}^{n-1} Y_j$, where $Y_j = V(G) \times \{y_j\}$. Further, in the sequel, we shall denote the vertices of X_i by $\{x_j^i \,|\, 0 \leq j \leq n-1\}$ and the vertices of Y_j by $\{x_j^i \,|\, 0 \leq i \leq m-1\}$, where x_j^i stands for the vertex (x_i, y_j). We shall call X_i, $0 \leq i \leq m-1$, the *i th layer of* $G \circ H$ and Y_j, $0 \leq j \leq n-1$, the *j th column of* $G \circ H$. It is clear that $G \circ \overline{K}_m$ is an m-partite graph with parts $X_0, X_1, \ldots, X_{m-1}$.

If G is a bipartite graph with bipartition (X, Y), where $X = \{x_0, x_1, \ldots, x_{n-1}\}$, $Y = \{y_0, y_1, \ldots, y_{n-1}\}$, and if G contains the set of edges $F_i(X, Y) = \{x_j y_{i+j} \mid 0 \leq j \leq n-1\}$, where addition in the subscript is taken modulo n, $0 \leq i \leq n-1$, then we say that G has the *1-factor of jump i from X to Y*. Note that $F_i(X, Y) = F_{n-i}(Y, X)$, $0 \leq i \leq n-1$. Clearly, if $G = K_{n,n}$, then $E(G) = \bigcup_{i=0}^{n-1} F_i(X, Y)$. In a bipartite graph with bipartition (X, Y) with $|X| = |Y|$, if $x_i y_j$ is an edge, then $x_i y_j$ is called an edge of *jump* $j - i$ *from X to Y*, if $i \leq j$, or $n - (i - j)$ if $i > j$; the same edge is said to be of *jump* $i - j$ *from Y to X*, if $i \geq j$, or $n - (j - i)$ if $i < j$.

A *latin square* of order n is an $n \times n$ array, each cell of which contains exactly one of the symbols in $\{1, 2, \ldots, n\}$, such that each row and each column of the array contains each of the symbols in $\{1, 2, \ldots, n\}$. A latin square is said to be *idempotent* if the cell (i, i) contains the symbol i, $1 \leq i \leq n$. A latin square is said to be *commutative* if the cell (i, j) and (j, i) contain the same symbol, for all $1 \leq i, j \leq n$. Definitions which are not seen here can be found in [3] or [5].

We consider here the problem of decomposing $(K_m \circ \overline{K}_n)^*$ into directed cycles of odd length; clearly, $K_m \circ \overline{K}_n$ is the complete m-partite graph with n vertices in each of its parts. One may also consider a C_k-decomposition of any even regular complete m-partite graph G when k divides the number of edges of G and $k \leq |V(G)|$. This problem was completely settled by Sotteau [15] when $m = 2$. Further, Cavenagh [6] showed that the complete tripartite graph, namely, $K_3 \circ \overline{K}_m$ can be decomposed into k-cycles if and only if $k \leq 3m$ and k divides $3m^2$.

A C_5-decomposition of a complete tripartite graph was first considered by Mahmoodian and Mirzakhani [11] and additional results were obtained by Cavenagh [6], Cavenagh and Billington [8] and Cavenagh [7].

In [4], Billington et al. had solved the problem of decomposing $(K_m \circ \overline{K}_n)(\lambda)$ into 5-cycles. It is proved in [12–14] that the necessary conditions for the existence of C_p-decompositions of $K_m \times K_n$ and $K_m \circ \overline{K}_n$ are also sufficient, where $p \geq 5$ is a prime.

If $K_m \circ \overline{K}_n$ is an odd regular graph, then it can't be decomposed it into cycles of length k even if k divides the number of edges in $K_m \circ \overline{K}_n$ and $k \leq mn$. Hence we consider here the problem of decomposing $(K_m \circ \overline{K}_n)^*$, the symmetric digraph of $K_m \circ \overline{K}_n$, into cycles, whenever the divisibility condition is satisfied.

Here, we prove the following main

Theorem 1.1. *If* $m \geq 3$, *t is odd,* $3 \leq t \leq mn$, $gcd(m(m-1), t) \leq m$ *and* $t \,|\, m(m-1)n$, *then* $\vec{C}_t \,|\, (K_m \circ \overline{K}_n)^*$, *except possibly when* $(t, m) = (3, 6)$.

Similar techniques employed in the proof of the above theorem also yields the following

Theorem 1.2. *If both t and $m \geq 3$ are odd, $3 \leq t \leq mn$, $gcd(\frac{m(m-1)}{2}, t) \leq m$ and $t \mid \frac{m(m-1)}{2}n$, then $C_t \mid K_m \circ \overline{K}_n$.*

2 Proof of the Main Theorem

For disjoint subsets X and Y of vertices in a graph G, let $E(X, Y)$ denote the set of edges having one end in X and the other end in Y. For subsets S and T of vertices of a digraph D and vertices $x, y \in V(D)$, let (x, S) (resp. (T, y)) denote the set of arcs with tail at x (resp. head at y) and head at the vertices of S (resp. tail at the vertices of T).

We prove Theorem 1.1 after proving some lemmas.

Lemma 2.1. *If $t \geq 3$ and $n \geq 1$, then $\vec{C}_t \mid \vec{C}_t \circ \overline{K}_n$.*

Proof. Let $V(\vec{C}_t) = \{x_0, x_1, \ldots, x_{t-1}\}$ and $V(\overline{K}_n) = \{y_0, y_1, \ldots, y_{n-1}\}$; then $V(\vec{C}_t \circ \overline{K}_n) = \bigcup_{i=0}^{t-1} X_i$, where $X_i = \{x_0^i, x_1^i, \ldots, x_{n-1}^i\}$. Let $\vec{F}_\ell(X_i, X_{i+1})$ denote the n arcs of jump ℓ each from X_i to X_{i+1} in $\vec{C}_t \circ \overline{K}_n$. Then $A(\vec{C}_t \circ \overline{K}_n) = \bigcup_{i=0}^{n-1} A_i$, where $A_i = (x_i^0, X_1) \cup \vec{F}_i(X_1, X_2) \cup \vec{F}_i(X_2, X_3) \cup \ldots \cup \vec{F}_i(X_{t-2}, X_{t-1}) \cup (X_{t-1}, x_i^0)$, for $0 \leq i \leq n-1$. Clearly, each A_i consists of n arc-disjoint directed cycles of length t. Thus $\vec{C}_t \mid \vec{C}_t \circ \overline{K}_n$. □

In [9], it is shown that a $C_t \circ \overline{K}_n$ is Hamilton cycle decomposable and also Cavenagh has supplied an alternative proof in [6]; by simple modification of the Cavenagh's proof, that is, by assigning orientation to the Hamilton cycles obtained in the proof, in [6], we obtain the proof of the following lemma.

Lemma 2.2. *If both $n \geq 1$ and $t \geq 3$ are odd, then $\vec{C}_t \circ \overline{K}_n$ is directed Hamilton cycle decomposable.* □

We use the following theorem in the proof of the main theorem

Theorem 2.3 ([1]). *For positive integers m and n, with $2 \leq m \leq n$, $\vec{C}_m \mid K_n^*$ if and only if m divides the number of arcs in K_n^* and $(n, m) \neq (4, 4), (6, 3), (6, 6)$.* □

First we define a graph H_m which is used in the proof of Theorems 1.1 and 1.2; let $V(H_m) = X \cup Z$, $H_m[X] = K_m$, $H_m[Z] = \overline{K}_m$ and $E(X, Z) = E(K_{m,m} - F_0)$, where F_0 is the 1-factor of jump 0, $X = \{x_0, x_1, x_2, \ldots, x_{m-1}\}$ and $Z = \{z_0, z_1, z_2, \ldots, z_{m-1}\}$. Let T denote the transitive tournament of order 3.

Lemma 2.4. *For $m \geq 3$, $\vec{C}_3 \mid H_m^*$ and $T \mid H_m^*$, where H_m and T are as defined above.*

Proof. As defined earlier, let $V(H_m^*) = X \cup Z$, where $X = \{x_0, x_1, x_2, \ldots, x_{m-1}\}$, $Z = \{z_0, z_1, z_2, \ldots, z_{m-1}\}$ and $H_m^*[X]$, the subdigraph induced by X, is a K_m^* and the subdigraph induced by Z is a $\overline{K}_m$, that is, an independent set with m vertices. Clearly, $H_m^*[X] = K_m^* = \vec{C}_{m-1}^0 \oplus \vec{C}_{m-1}^1 \oplus \vec{C}_{m-1}^2 \oplus \ldots \oplus \vec{C}_{m-1}^{m-1}$, by Theorem 2.3, where $\vec{C}_{m-1}^i$ denotes a directed cycle of length $m-1$ and, we assume that $\vec{C}_{m-1}^i$, $0 \leq i \leq m-1$, misses the vertex x_i of K_m^*. To each directed cycle

$\vec{C}^i_{m-1}$, $0 \le i \le m-1$, of K^*_m ($\subset H^*_m[X]$) of the decomposition, we construct a subdigraph G_i as follows: $G_i = \vec{C}^i_{m-1} \oplus (z_i, V(\vec{C}^i_{m-1})) \oplus (V(\vec{C}^i_{m-1}), z_i)$ of H^*_m where z_i is the vertex of Z corresponding to $x_i \in X$, the vertex missed by $\vec{C}^i_{m-1}$.

Next we show that the subdigraph G_i of H^*_m is $\vec{C}_3$ decomposable as well as T decomposable, where T denotes the transitive tournament of order 3. First we shall show that $\vec{C}_3$ decomposes G_i. To each arc (x_k, x_ℓ) in $\vec{C}^i_{m-1}$ ($\subset H^*_m[X]$), consider the $\vec{C}_3 = (x_k, x_\ell, z_i, x_k)$ in G_i. (Note that $z_i \in Z$ is the corresponding vertex of $x_i \in X$ and x_i is missed by $\vec{C}^i_{m-1}$). Hence $\vec{C}_3 \,|\, G_i$. Next we show that $T \mid G_i$; to each arc (x_k, x_ℓ) in $\vec{C}^i_{m-1}$ ($\subset H^*_m[X]$), consider the subset of arcs $\{(x_k, x_\ell), (z_i, x_\ell), (x_k, z_i)\}$ of G_i which induces T. Thus $T \mid G_i$. □

Proof of Theorem 1.1. Let the $gcd(m(m-1), t) = k$. As t is odd, k is odd. Let us assume that $\frac{t}{k} = \ell$. As $\frac{t}{k}$ and $\frac{m(m-1)}{k}$ are relatively prime, $\frac{t}{k}$ divides n, by hypothesis. Let $n = (\frac{t}{k})b = \ell b$. We complete the proof in two cases.

Case 1. $k \ge 3$ is odd.

Let us obtain a new digraph $(K_m \circ \overline{K}_b)^*$, from $(K_m \circ \overline{K}_n)^*$, as follows: identify every ℓ vertices in each part of $(K_m \circ \overline{K}_n)^*$, into a single vertex and there is a symmetric pair of arcs between a pair of vertices in the new digraph if and only if the corresponding 2ℓ vertices of $(K_m \circ \overline{K}_n)^*$ induce a $K^*_{\ell,\ell}$, the complete symmetric bipartite digraph; the resulting digraph is $(K_m \circ \overline{K}_b)^*$. As $k \,|\, m(m-1)$ and $k \le m$, by hypothesis, $\vec{C}_k \,|\, K^*_m$, by Theorem 2.3, except when $(m, k) = (6, 3)$. Hence $(K_m \circ \overline{K}_b)^* = K^*_m \circ \overline{K}_b = (\vec{C}_k \circ \overline{K}_b) \oplus (\vec{C}_k \circ \overline{K}_b) \oplus \ldots \oplus (\vec{C}_k \circ \overline{K}_b)$. Now $\vec{C}_k \,|\, (\vec{C}_k \circ \overline{K}_b)$, by Lemma 2.1, and hence $\vec{C}_k \mid (K_m \circ \overline{K}_b)^*$. Clearly, a $\vec{C}_k$ in $(K_m \circ \overline{K}_b)^*$ corresponds to a subgraph of $(K_m \circ \overline{K}_n)^*$ isomorphic to $\vec{C}_k \circ \overline{K}_\ell$ (since each vertex of $\overline{K}_b$ arises out of identification of ℓ vertices of $\overline{K}_n$). If ℓ is odd, then $\vec{C}_k \circ \overline{K}_\ell$ can be decomposed into directed Hamilton cycles, that is, $t = k\ell$ cycles, by Lemma 2.2; if ℓ is even, then orienting the edges of the Hamilton cycles of a Hamilton cycle decomposition of $C_k \circ \overline{K}_\ell$ [9] into directed Hamilton cycles, that is, directed cycles of length $k\ell = t$, yields a directed t-cycle decomposition of $\vec{C}_k \circ \overline{K}_\ell$.

Case 2. $k = 1$.

As $k = 1$, $n = tb$. Let $V(K_m \circ \overline{K}_n)^* = \bigcup_{i=0}^{m-1} X_i$, where $X_i = \{x^i_0, x^i_1, \ldots, x^i_{n-1}\}$, for $0 \le i \le m-1$. As in the previous case, obtain the graph $(K_m \circ \overline{K}_t)^*$ from the graph $(K_m \circ \overline{K}_n)^*$ by identifying every b vertices in each part of $(K_m \circ \overline{K}_n)^*$, into a single vertex and join a pair of vertices by a symmetric pair of arcs if and only if the corresponding subsets of vertices induce a $K^*_{b,b}$ in $(K_m \circ \overline{K}_n)^*$. Let $V(K_t)$ be $\{z_0, z_1, \ldots, z_{t-1}\}$. In the sequel, for our convenience, we write the vertex set of $(K_m \circ \overline{K}_t)^* = \bigcup_{i=0}^{t-1} Z_i$, where $Z_i = V(K_m) \times \{z_i\} = \bigcup_{j=0}^{m-1} (x_j, z_i) = \{x^0_i, x^1_i, \ldots, x^{m-1}_i\}$; actually, Z_i is the ith column vertices of $(K_m \circ \overline{K}_t)^*$.

Clearly, K_t can be decomposed into paths $\mathcal{P}$ of length exactly $\frac{t-1}{2}$ such that each vertex of K_t is the origin of exactly one of these paths; this is achieved as follows: let $V(K_t) = \{z_0, z_1, \ldots, z_{t-1}\}$ and let $s_j = 1-2+3-4+\ldots+(-1)^{j+1}j$, $1 \le j \le \frac{t-1}{2}$. Let $P^i = z_i\, z_{i+s_1}\, z_{i+s_2} \cdots z_{i+s_{\frac{t-1}{2}}}$, $0 \le i \le t-1$, be a path of length $\frac{t-1}{2}$, where additions in the subscripts are taken modulo t. Thus $\mathcal{P} = \{P^i \,|\, 0 \le i \le t-1\}$ is a path decomposition of K_t such that each vertex of K_t is the origin of exactly one of the paths of $\mathcal{P}$. Corresponding to each path $P^i \in \mathcal{P}$, $0 \le i \le t-1$, of K_t, we define a subdigraph G_i, $0 \le i \le t-1$, of $(K_m \circ \overline{K}_t)^*$, as follows: let the vertices of G_i be the vertices of the columns of $(K_m \circ \overline{K}_t)^*$ corresponding to the vertices of P^i, that is, if $z_i \in P^i$ then consider the column Z_j the arc set of G_i consists of the arcs in the subdigraph of $(K_m \circ \overline{K}_t)^*$ induced by the column of vertices Z_i, (corresponding to the origin z_i of P^i), namely, the arcs of a K_m^*, together with the set of arcs of $(K_{m,m} - F_0)^*$ corresponding to each edge of P^i, where F_0 is the 1-factor of jump 0 in $K_{m,m}$; consequently, the arc set of G_i consists of the arcs of a K_m^* and the arcs of $\frac{t-1}{2}$ copies of $(K_{m,m} - F_0)^*$, see Figure 2.1. Since each vertex of K_t happens to be the origin of exactly one path P^i of $\mathcal{P}$, $\bigcup_{i=0}^{t-1} G_i = (K_m \circ \overline{K}_t)^*$.

Next we prove that the graph G_i, $0 \le i \le t-1$, has a $\vec{C}_t$ decomposition. We know that $\vec{C}_3 \mid H_m^*$ and $T \mid H_m^*$, where H_m^* and T are defined in Lemma 2.4. Using these decompositions, according as $\frac{t-1}{2}$ is even or odd, we shall obtain a $\vec{C}_t$ decomposition of G_i. Corresponding to each $\vec{C}_3$ or T of a decomposition of H_m^* into $\vec{C}_3$ or T, we describe a method of finding a $\vec{C}_t$ in G_i.

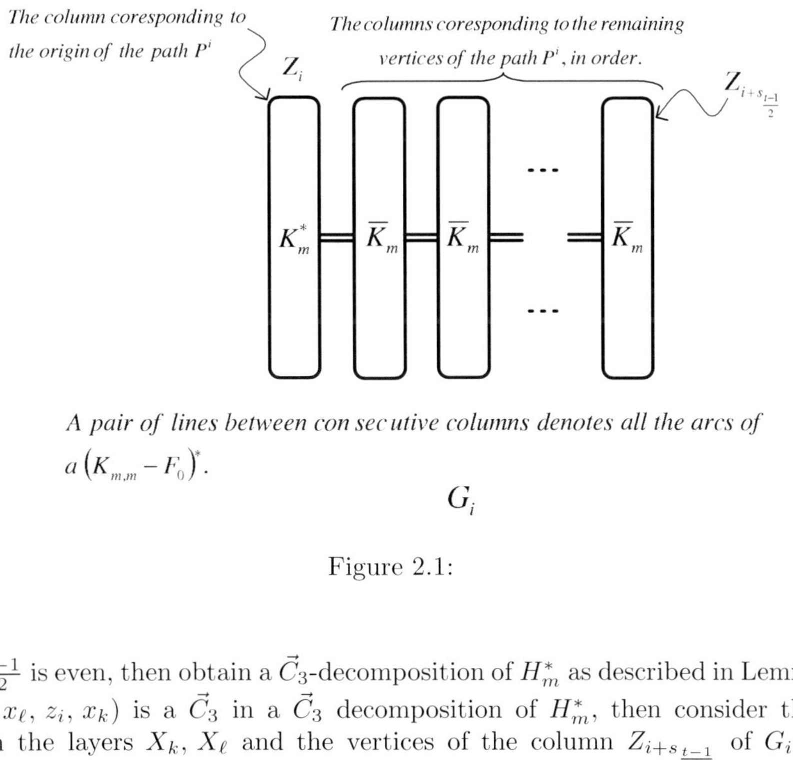

Figure 2.1:

If $\frac{t-1}{2}$ is even, then obtain a $\vec{C}_3$-decomposition of H_m^* as described in Lemma 2.4. If (x_k, x_ℓ, z_i, x_k) is a $\vec{C}_3$ in a $\vec{C}_3$ decomposition of H_m^*, then consider the vertices in the layers X_k, X_ℓ and the vertices of the column $Z_{i+s_{\frac{t-1}{2}}}$ of G_i. Then define a directed cycle as shown in the Figure 2.2 (a), (compare a $\vec{C}_3$ decomposition of H_m^* in Lemma 2.4). Hence a $\vec{C}_3$ decomposition of H_m^* yields a $\vec{C}_t$-decomposition of G_i. Similarly, if $\frac{t-1}{2}$ is odd, then obtain a T-decomposition of H_m^*, as described in Lemma 2.4. Corresponding to a T in the T-decomposition of H_m^* obtain a $\vec{C}_t$-decomposition of G_i as described in Figure 2.2 (b). As each $\vec{C}_t$ in $(K_m \circ \overline{K}_t)^*$ corresponds to a $\vec{C}_t \circ \overline{K}_b$ in $(K_m \circ \overline{K}_n)^*$, $(K_m \circ \overline{K}_n)^* = K_m^* \circ \overline{K}_n = (\vec{C}_t \circ \overline{K}_b) \oplus (\vec{C}_t \circ \overline{K}_b) \oplus \ldots \oplus (\vec{C}_t \circ \overline{K}_b)$. Now $\vec{C}_t \,|\, \vec{C}_t \circ \overline{K}_b$, by Lemma 2.1. This completes the proof of the theorem. □

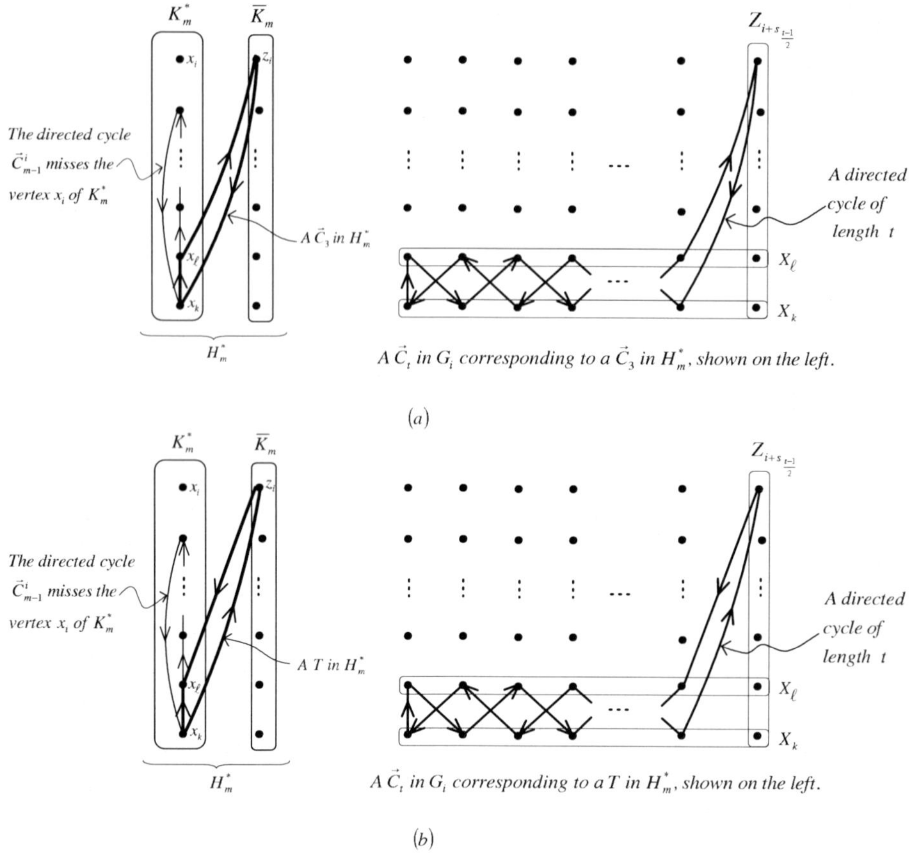

Figure 2.2: (a) Corresponding to a $\vec{C}_3$ in H_m^* a $\vec{C}_t$ in G_i is shown. (b) Corresponding to a T in H_m^* a $\vec{C}_t$ in G_i is shown.

Corollary 2.5. *For $m \geq 3$ and p, a prime, $3 \leq p \leq mn$, $\vec{C}_p \,|\, (K_m \circ \overline{K}_n)^*$ if and only if $p \,|\, m(m-1)n^2$, except possibly when $(p, m) = (3, 6)$.* □

The proof of Theorem 1.2 is similar to the proof of Theorem 1.1 and hence we omit the proof of it.

Acknowledgments: The last two authors would like to thank the Department of Science and Technology, Government of India, New Delhi, for partial financial assistance through the Project grant No: SR/S4/MS: 217/03.

References

[1] B. Alspach, H. Gavlas, M. Šajna and H. Verrall, Cycle decompositions IV: complete directed graphs and fixed length directed cycles, *J. Combinatorial Theory (Ser. A)*, **103** (2003), pp. 165–208.

[2] B. Alspach and H. Gavlas, Cycle decompositions of K_n and $K_n - I$, *Combinatorial Theory (Ser. B)*, **81** (2001), pp. 77–99.

[3] R. Balakrishnan and K. Ranganathan, *A Text Book of Graph Theory*, Springer-Verlag, New York (2000).

[4] E.J. Billington, D.G. Hoffman and B.M. Maenhaut, Group divisible pentagon systems, *Utilitas Math.*, **55** (1999), pp. 211–219.

[5] J.A. Bondy and U.S.R. Murty, *Graph Theory with Applications*, Elsevier, New York (1976).

[6] N.J. Cavenagh, Decompositions of complete tripartite graphs into k-cycles, *Australasian J. Combinatorics*, **18** (1998), pp. 193–200.

[7] N.J. Cavenagh, Further decompositions of complete tripartite graphs into 5-cycles, *Discrete Math.*, **256** (2002), pp. 55–81.

[8] N.J. Cavenagh and E.J. Billington, On decomposing complete tripartite graphs into 5-Cycles, *Australasian J. Combinatorics*, **22** (2000), pp. 41–62.

[9] R. Laskar, Decompositions of some complete graphs into Hamilton cycles, in *Proc. 5th Hungarian coll., Keszthely 1976, North Holland* (1978), pp. 705–716.

[10] C.C. Lindner and C.A. Rodger, *Design Theory*, CRC Press, New York (1997).

[11] E.S. Mahmoodian and M. Mirzakhani, Decomposition of complete tripartite graphs into 5-cycles, in *Combinatorics and Advances (Eds. C.J. Colbourn and E.S. Mahmoodian)*, Kluwer Academic Publishers, pp. 235–241. (1995).

[12] R.S. Manikandan and P. Paulraja, C_5-decompositions of some regular graphs, *Australasian J. Combinatorics*, **37** (2007), pp. 285–293.

[13] R.S. Manikandan and P. Paulraja, C_7-decompositions of some regular graphs, Submitted.

[14] R.S. Manikandan and P. Paulraja, C_p-decompositions of some regular graphs, *Discrete Math.*, **306** (2006), pp. 429–451.

[15] D. Sotteau, Decomposition of $K_{m,n}$ ($K^*_{m,n}$) into cycles (circuits) of length $2k$, *J. Combinatorial Theory (Ser. B)*, **30** (1981), pp. 75–81.

Proc. of ICDM 2006, Pages 183–194

Packing and Domination Invariants on Cartesian Products and Direct Products

DOUGLAS F. RALL
Furman University,Department of Mathematics,Greenville, SC 29613 USA
Email: doug.rall@furman.edu

Abstract. The dual notions of domination and packing in finite simple graphs were first extensively explored by Meir and Moon in [15]. Most of the lower bounds for the domination number of a nontrivial Cartesian product involve the 2-packing, or closed neighborhood packing, number of the factors. In addition, the domination number of any graph is at least as large as its 2-packing number, and the invariants have the same value for any tree. In this paper we survey what is known about the domination, total domination and paired-domination numbers of Cartesian products and direct products. In the process we highlight two other packing invariants that each play a role similar to that played by the 2-packing number in dominating Cartesian products.

Keywords: graph products, paired-domination, k-packing, Vizing's conjecture, open packing, total domination, direct product
AMS subject classification: 05C69, 05C70, 05C05

1 Introduction

One of the most well-known open problems in domination theory is the conjecture made by V. G. Vizing [19] almost forty years ago. The conjecture states that the domination number of the Cartesian product of any pair of graphs is no less than the product of their domination numbers. The closed neighborhood, or distance-two, packing number plays an important role in many of the attacks on this conjecture. In this paper we consider two other types of packing invariants and show how each plays a similar role in the study of a generalized domination number on a particular graph product.

1.1 Domination and Packing

For notation and graph theory terminology we generally follow [9]. Let G be a finite, simple, undirected graph with vertex set $V = V(G)$ and edge set $E = E(G)$. The *open neighborhood* of a vertex $v \in V$ is $N_G(v) = \{ u \in V \mid uv \in E \}$, while its *closed neighborhood* is $N_G[v] = N(v) \cup \{v\}$. For a set $S \subseteq V$, $N_G(S)$ denotes $\bigcup_{v \in S} N_G(v)$ and $N_G[S] = N_G(S) \cup S$. If the graph is clear from the context, then we omit the subscript on these neighborhood names. The neighborhood of an edge $e = uv$ is $N(\{u,v\}) = N[\{u,v\}]$. For a set $S \subseteq V$, the subgraph of G induced by S is denoted by $\langle S \rangle_G$ or $\langle S \rangle$ and the complement of G by $\overline{G}$. A family $\{S_k\}_{k \in I}$ of subsets of vertices in G is a *cover* of (or *covers*) G if $V(G) = \cup_{k \in I} S_k$. The family is a *packing* if the subsets are pairwise disjoint. A *matching* in G is a collection of edges that, as sets of vertices, is a packing. A matching M is a *perfect* matching of G if it covers G when M is considered as a collection of 2-element subsets of vertices.

For a positive integer k, Meir and Moon [15] defined a *k-packing* in G to be a set $A \subseteq V$ such that for every pair of distinct vertices u and v in A, the distance between u and v in G is more than k. For $k = 2$ this is equivalent to requiring that $\{N[x]\}_{x \in A}$ is a packing in the sense defined in the previous paragraph. A vertex subset B is an *open packing* in G if the collection of open neighborhoods, $\{N(x)\}_{x \in B}$, is a packing in G. The *k-packing number* is the order of a largest k-packing of G and is denoted $\rho_k(G)$, while the *open packing number*, $\rho^{\mathrm{o}}(G)$, is the order of a largest open packing in G. Of course, 1-packings (independent sets) have played a prominent role in graph theory almost since its formal beginnings. We will use the more common notation $\alpha(G)$, instead of $\rho_1(G)$, for the *vertex independence number* of G, the cardinality of a largest independent set of vertices in G.

A subset D of vertices is a *dominating set* of G if every vertex x in V either belongs to D or is adjacent to a vertex in D. Equivalently, the family $\{N[x]\}_{x \in D}$ of closed neighborhoods covers G. Such a dominating set is *minimal* if no proper subset of D is a dominating set. The *domination number*, $\gamma(G)$, is the minimum cardinality of a dominating set of G. A set S of vertices is a *total dominating* set if the family $\{N(x)\}_{x \in S}$ of open neighborhoods covers G, and the *total domination number*, $\gamma_t(G)$, equals the minimum cardinality of a total dominating set of G. To guarantee that a graph has a total dominating set we must assume it has no isolated vertices. Hence we assume this to be the case for all graphs in the remainder of the paper. A dominating set D whose induced subgraph $\langle D \rangle$ contains a perfect matching is called a *paired-dominating* set, and the *paired-domination number* $\gamma_{\mathrm{pr}}(G)$ is the cardinality of a smallest such set. The vertices saturated by the edges of any maximal matching in G constitute a paired dominating set. Thus, $\gamma_{\mathrm{pr}}(G)$ is defined for all graphs under consideration and is no larger than twice the order of a smallest maximal matching. It follows from their definitions that $\gamma(G) \le \gamma_t(G) \le \gamma_{\mathrm{pr}}(G)$. The path of order five shows that all of the inequalities can be strict; all three values are equal on a path of order four. A dominating set D of cardinality $\gamma(G)$ is called a $\gamma(G)$-set or a γ-set if G is understood from context. We define γ_t-set, γ_{pr}-set, ρ_k-set and ρ^{o}-set similarly.

The fact that $\rho_2(G) \le \gamma(G)$ for every G follows directly from the observations that a γ-set intersects every closed neighborhood and a ρ_2-set contains at most one vertex from each closed neighborhood. In an entirely similar way, and based only on the definitions, it is immediate that $\rho^{\mathrm{o}}(G) \le \gamma_t(G)$ for every G that has no isolated vertices. In [3] Brešar, et al demonstrated a connection, although not so direct, between paired-domination and 3-packings. We summarize these relationships in the following.

Proposition 1.1. *For any graph G without isolated vertices,*

(i) $\gamma(G) \ge \rho_2(G)$;

(ii) $\gamma_t(G) \ge \rho^{\mathrm{o}}(G)$; *and*

(iii) ([3]) $\gamma_{\mathrm{pr}}(G) \ge 2\rho_3(G)$.

Proof. The first two inequalities follow as above. Let D be a $\gamma_{\mathrm{pr}}(G)$-set and let C be a $\rho_3(G)$-set. Suppose $|D| = 2k$. For each vertex $x \in C$, let $p(x)$ be a vertex such that $p(x) \in D \cap N[x]$ and let $D' = \cup_{x \in C}\{p(x)\}$. Since C is a 3-packing,

$|C| = |D'|$ and D' is an independent set in $\langle D\rangle$. Now, $\langle D\rangle$ is covered by k edges and consequently

$$\gamma_{\rm pr}(G) = |D| \geq 2|D'| = 2\rho_3(G)\,. \qquad \square$$

1.2 Product Graphs

For product graphs we follow the book by Imrich and Klavžar [12]. Specifically, the *Cartesian product*, $G_1 \,\Box\, G_2$, of two graphs G_1 and G_2 has the ordinary (set) Cartesian product $V(G_1) \times V(G_2)$ as its vertex set. Two vertices $u = (u_1, u_2)$ and $v = (v_1, v_2)$ are adjacent in $G_1 \,\Box\, G_2$ if they are equal in one coordinate and adjacent in the other. (That is, for exactly one of $i = 1$ or $i = 2$, $u_i = v_i$ and u_{3-i} is adjacent to v_{3-i} in G_{3-i}.) The *direct product*, $G_1 \times G_2$, has the same vertex set as the Cartesian product, and u and v are adjacent in $G_1 \times G_2$ precisely when u_1v_1 is an edge in G_1 and u_2v_2 is an edge in G_2.*

For a vertex g in G, the subgraph of $G \,\Box\, H$ induced by the set $\{(g, h) \mid h \in V(H)\}$ is isomorphic to H and is denoted by ${}^gH$. For $h \in V(H)$, G^h is the subgraph, isomorphic to G and induced by $\{(g, h) \mid g \in V(G)\}$. Both of these subgraphs are called *fibers* of $G \,\Box\, H$. In a direct product $G \times H$ we also refer to these as fibers, but in this case they are independent sets in $G \times H$.

Depending on the particular graph product and the particular property of vertex subsets, it may be easy to establish that the (set) Cartesian product of two sets with the property retains that property in the product graph. For example, if I_j is an independent set in the graph G_j for $j = 1, 2$, then the set $I_1 \times I_2$ is independent in the graph $G_1 \,\Box\, G_2$. Because α measures the largest order of an independent set, it follows that $\alpha(G_1 \,\Box\, G_2) \geq \alpha(G_1)\alpha(G_2)$. On the other hand, if D_i is a dominating set of G_i for $i = 1, 2$, then $D_1 \times D_2$ dominates $G_1 \,\Box\, G_2$ only if $D_i = V(G_i)$ for at least one of the values of i.

For a given graphical invariant σ and given graph product $\otimes$ it is thus natural to investigate the behavior of σ on $\otimes$. It is sometimes the case that the value $\sigma(G \otimes H)$ depends directly on the two values $\sigma(G)$ and $\sigma(H)$ for all pairs of graphs G and H. We say that σ is *supermultiplicative* (respectively, *submultiplicative*) on $\otimes$ if $\sigma(G \otimes H) \geq \sigma(G)\sigma(H)$ (respectively, $\sigma(G \otimes H) \leq \sigma(G)\sigma(H)$) for all pairs G and H. A class of graphs $\mathcal{C}$ is called a *universal multiplicative class for σ on $\otimes$* if for every graph H, it follows that $\sigma(G \otimes H) = \sigma(G)\sigma(H)$ whenever G is from the class $\mathcal{C}$. (We only require the given inequality or equality to hold for those graphs for which the invariant is defined.)

Possibly the most well-known problem in this part of domination theory is the conjecture of V. G. Vizing [19] that dates from 1968.

Vizing's Conjecture. *The domination invariant γ is supermultiplicative on the Cartesian product.* $\square$

We will say that Vizing's conjecture holds for (or is satisfied by) the graph G if $\gamma(G \,\Box\, H) \geq \gamma(G)\gamma(H)$ for every H.

In the remainder of the paper we summarize what is known on questions related to whether various domination invariants are sub- or super- multiplicative on the graph products $\Box$ and $\times$. In the process we illustrate the importance of a packing invariant in three of these cases and show similarities between these packing-domination pairs.

*The direct product has gone by a number of different names in the literature including categorical, tensor, cardinal, etc.

2 Domination

2.1 Cartesian Product

For nearly a dozen years no positive results were published on Vizing's conjecture. In 1979 Barcalkin and German [2] showed that the conjecture holds for a large class of graphs that includes some well-known families of graphs.

Theorem 2.1 ([2]). *If G is the spanning subgraph of a graph G' such that $\gamma(G) = \gamma(G')$ and G' has $\gamma(G')$ complete subgraphs that cover G', then Vizing's conjecture holds for G.*

Let us say that a graph G is a BG-graph[†] if it satisfies the hypothesis of Theorem 2.1. Suppose $A = \{v_1, \ldots, v_k\}$ is a $\rho_2(G)$-set and that $k = \gamma(G)$. The set A may not dominate G; in any case, let $B = V(G) - N[A]$. Add edges to G so that each closed neighborhood $N_G[v_i]$ becomes a complete subgraph. If B is not empty, add edges to make $N[v_1] \cup B$ a complete subgraph. Call this graph with added edges G'. It is clear that $\rho_2(G') = k$ and thus G is shown to be a BG-graph. The following result of Meir and Moon [15] then shows that every tree is a BG-graph.

Theorem 2.2 ([15]). *If T is any tree, then $\gamma(T) = \rho_2(T)$.*

The proof given in [15] is by induction on the order of the tree. One can also prove this theorem by an argument similar to that given in [17] for Theorem 3.5 by considering the intersection graph on the closed neighborhoods of T. In [2] it is also shown that graphs with domination number two are BG-graphs. One can readily prove that every cycle is a BG-graph. It is not always an easy matter to determine if a given graph is a BG-graph since there are potentially many ways to add edges to form the required number of complete subgraphs that cover the graph while at the same time not reducing the domination number. An example of a small graph that is not a BG-graph and yet satisfies Vizing's conjecture is the cubic graph F of order eight that is formed from $K_{3,3}$ by "expanding" one vertex to a K_3. For every pair u, v of non-adjacent vertices in F, $\gamma(F + uv) = 2 < 3 = \gamma(F)$, and clearly F is not covered by three complete subgraphs proving that F is not a BG-graph. Part (i) of the next result shows that F satisfies the conjecture.

Theorem 2.3. *Vizing's conjecture holds for any graph G satisfying one of the following conditions:*

(i) ([18]) $\gamma(G) = 3$;

(ii) ([6]) $\gamma(G) = \rho_2(G) + 1$;

(iii) ([1]) *G is chordal.*

Another approach—instead of showing that the conjecture holds for a new class of graphs—was taken by Clark and Suen. They were the first to show that the domination number of the Cartesian product of two graphs is always bounded below by a positive constant multiple of the product of their domination numbers. Their result has also proven to be useful in studying other domination parameters on Cartesian products.

Theorem 2.4 ([5]). *For any graphs G and H, $\gamma(G \,\Box\, H) \geq \frac{1}{2}\gamma(G)\gamma(H)$.*

[†]It should be noted that Barcalkin and German called this collection of graphs the A-class. BG-graphs is our term to give them credit for this significant result.

Attempts to increase the constant in the above have so far proven to be futile. A proof of Vizing's conjecture would, of course, solve the following problem. But, short of that, finding a solution would represent significant progress towards settling the conjecture.

Problem 2.5. *Find a constant $c > \frac{1}{2}$ such that, for all graphs G and H,*

$$c \cdot \gamma(G)\gamma(H) \leq \gamma(G \,\square\, H).$$

In contrast to what we will show in the case of total domination of direct products, the following result of Hartnell and Rall implies there is no nontrivial universal multiplicative class for γ on $\square$.

Theorem 2.6 ([8]). *If T is a tree that has a vertex adjacent to at least two leaves, then $\gamma(G \,\square\, T) > \gamma(G)\gamma(T)$ for every connected graph G of order at least two.*

Jacobson and Kinch demonstrated the importance of 2-packings in the study of dominating sets in Cartesian products. They used the fact that the closed neighborhood of a vertex (u, v) in $G \,\square\, H$ is contained in $N_G[u] \times V(H)$ and that any dominating set of $G \,\square\, H$ must contain at least $\gamma(H)$ vertices from $N_G[u] \times V(H)$ to prove the following result. Note that the bound might provide little information about the domination number of the Cartesian product (e.g., if both graphs have a small 2-packing number).

Theorem 2.7 ([13]). *For any pair of graphs G and H,*

$$\gamma(G \,\square\, H) \geq \max\{\gamma(G)\rho_2(H), \gamma(H)\rho_2(G)\}.$$

We summarize three important connections between packings and domination, at least as they relate to dominating Cartesian products.

Theorem 2.8. *Let G and H be any graphs and let T be any tree. Then*

(i) $\gamma(G) \geq \rho_2(G)$;

(ii) $\gamma(G \,\square\, H) \geq \max\{\gamma(G)\rho_2(H), \gamma(H)\rho_2(G)\}$;

(iii) $\gamma(T) = \rho_2(T)$.

2.2 Direct Product

The domination invariant γ is not supermultiplicitive on direct products. For example, as shown in [16], if G is a complete graph of even order at least six with a perfect matching removed, then $\gamma(G \times G) = 3 < \gamma(G)\gamma(G)$. In [14] Klavžar and Zmazek exhibited an infinite collection of graphs, $\{G_n\}$, such that $\gamma(G_n \times G_n) \leq \frac{7}{9}\gamma(G_n)\gamma(G_n)$. Also, γ is not submultiplicative on $\times$ as illustrated by $\gamma(K_3 \times K_3) = 3 > \gamma(K_3)\gamma(K_3)$. However, using Theorem 3.3 and the easily established fact that $\gamma_t(G) \leq 2\gamma(G)$ it follows immediately that

$$\gamma(G \times H) \leq 4\gamma(G)\gamma(H). \tag{1}$$

It thus seems reasonable to ask for the largest value of the constant c and the smallest value of C such that for every pair of graphs G and H,

$$c \cdot \gamma(G)\gamma(H) \leq \gamma(G \times H) \leq C \cdot \gamma(G)\gamma(H). \tag{2}$$

The above examples show that c can be at most $3/4$ while C must be at least 3. At the present time we do not know of a constant $0 < c < 1$ such that the first inequality in (2) holds for all graphs G and H. A recent paper by Brešar, Klavžar and Rall established $C = 3$ as the correct value in the second inequality.

Theorem 2.9 ([4]). *For any graphs G and H, $\gamma(G \times H) \leq 3\gamma(G)\gamma(H)$.*

The authors of [4] show that if G is a graph such that $\gamma(G) > \rho_2(G)$, then for every H, $\gamma(G \times H) < 3\gamma(G)\gamma(H)$. However, choose each of G and H to be a graph having a ρ_2-set that is also a dominating set and then add two pendant vertices to each vertex in each of these γ-sets. Denote the resulting graphs G^+ and H^+, respectively. They prove that $\gamma(G^+ \times H^+) = 3\gamma(G^+)\gamma(H^+)$, thus proving that the bound in Theorem 2.9 is sharp.

Nowakowski and Rall [16] proved that if G and H have no isolated vertices, then

$$\gamma(G \times H) \geq \max\{\gamma_t(H)\rho_2(G), \gamma_t(G)\rho_2(H)\}. \tag{3}$$

For any tree T, it then follows from Theorem 2.1 that

$$\gamma(T \times H) \geq \rho_2(T)\gamma_t(H) = \gamma(T)\gamma_t(H) \geq \gamma(T)\gamma(H).$$

Hence, with appropriate restrictions on one of the factors, c can be chosen to be 1 in (2). In [4] it is shown that if $\gamma(G) = \gamma_t(G)$ and $\gamma(H) = \gamma_t(H)$, then $\gamma(G \times H) \leq \gamma(G)\gamma(H)$. If, in addition, either $\rho_2(G) = \gamma(G)$ or $\rho_2(H) = \gamma(H)$, then by appealing to (3) we see that $\gamma(G \times H) = \gamma(G)\gamma(H)$. That is, if additional conditions are imposed on the factors, then we are able to say more about the constants in (2). For the general case we pose the following question.

Question 2.1. *What is the largest constant c such that, for all graphs G and H,*

$$c \cdot \gamma(G)\gamma(H) \leq \gamma(G \times H)\,?$$

3 Total Domination

3.1 Cartesian Product

The total domination invariant, γ_t, is neither submultiplicative nor supermultiplicative on Cartesian products as $\gamma_t(K_n \,\Box\, K_n)$ shows for $n \geq 5$ and $n = 3$, respectively. In fact, by letting n be an arbitrary integer larger than five in the above, we see that the ratio of $\gamma_t(G \,\Box\, H)$ to $\gamma_t(G)\gamma_t(H)$ can be made arbitrarily large. That is, there does not exist a constant C such that $\gamma_t(G \,\Box\, H) \leq C \cdot \gamma_t(G)\gamma_t(H)$. In this section we show—in the other direction—that for any pair of graphs the above ratio is at least $1/6$ and that if one of the graphs is, for example a tree, then it is at least $1/2$.

If D is any $\gamma(G)$-set, then let D' be the set obtained by enlarging D to include a vertex v' from the open neighborhood of each vertex v that is isolated in the subgraph $\langle D \rangle$. It is clear that D' is a total dominating set of G and that $|D'| \leq 2|D|$. Thus, $\gamma_t(G) \leq 2\gamma(G)$, and by combining this with the bound of Clark and Suen from Section 2.1 we get the following chain of inequalities

$$\gamma_t(G)\gamma_t(H) \leq 4\gamma(G)\gamma(H) \leq 8\gamma(G \,\Box\, H) \leq 8\gamma_t(G \,\Box\, H),$$

proving the ratio is always at least $1/8$. This was first observed by Henning and Rall in [11] in the only study to date of the multiplicative nature of γ_t on $\Box$.

By employing a technical counting argument based around a partition of $V(G)$ that is formed via a minimum paired-dominating set of G, the authors were able to establish the following result that provided the first nontrivial lower bound for $\gamma_t(G \,\square\, H)$ as a multiple of $\gamma_t(G)\gamma_t(H)$.

Theorem 3.1 ([11]). *For any graphs G and H without isolated vertices,*

$$\gamma_t(G)\gamma_t(H) \le 6\gamma_t(G \,\square\, H).$$

We do not know of any pair of graphs for which the six in the established bound cannot be replaced by two. As is true in many of these situations, if one factor is appropriately restricted then the bound can be improved. We give the proof of this result to illustrate the interplay between several of the domination and packing invariants.

Theorem 3.2 ([11]). *Let G and H be graphs without isolated vertices and assume that $\rho_2(G) = \gamma(G)$. Then*

$$\gamma_t(G)\gamma_t(H) \le 2\gamma_t(G \,\square\, H).$$

Proof. Let D be a minimum total dominating set of the product $G \,\square\, H$, and let $S = \{g_1, \ldots, g_{\gamma(G)}\}$ be a $\rho_2(G)$-set. Choose a partition $\{V_1, \ldots, V_{\gamma(G)}\}$ of $V(G)$ such that for $i = 1, \ldots, \gamma(G)$, $N_G[g_i] \subseteq V_i$. For $i = 1, \ldots, \gamma(G)$, let $D_i = D \cap (V_i \times V(H))$. Further, let S_i be a minimum set of vertices in $G \,\square\, H$ that every vertex of the fiber ${}^{g_i}H$ has a neighbor in S_i and such that S_i contains as many vertices in ${}^{g_i}H$ as possible. Then, $S_i \subseteq V_i \times V(H)$. If S_i contains a vertex x not in ${}^{g_i}H$, then x has a unique neighbor $x' \in {}^{g_i}H$. Replacing x in the set S_i with any neighbor of x' in ${}^{g_i}H$ (notice that such a vertex exists since H has no isolated vertices) produces a new minimum set S_i' of vertices in $G \,\square\, H$ such that ${}^{g_i}H \subseteq N(S_i')$ and S_i' contains more vertices in ${}^{g_i}H$ than does S_i, a contradiction. Hence, $S_i \subseteq H_i$, and so S_i is a total dominating set of the fiber ${}^{g_i}H$. Since ${}^{g_i}H$ is isomorphic to H, it follows that $|S_i| \ge \gamma_t(H)$. But ${}^{g_i}H \subseteq N(D_i)$ as well, and hence $|D_i| \ge |S_i|$. Thus,

$$\gamma_t(G \,\square\, H) = |D| = \sum_{i=1}^{\gamma(G)} |D_i| \ge \sum_{i=1}^{\gamma(G)} |S_i| \ge \sum_{i=1}^{\gamma(G)} \gamma_t(H) = \gamma(G)\gamma_t(H) \ge \frac{1}{2}\gamma_t(G)\gamma_t(H). \qquad \square$$

In particular, by Theorem 2.1 it follows that if T is any tree of order at least two, then

$$\gamma_t(T)\gamma_t(H) \le 2\gamma_t(T \,\square\, H). \tag{4}$$

In [11] Henning and Rall showed the bound of Theorem 3.2 is sharp and, in fact, showed that for connected graphs, equality holds in (4) if and only if $\gamma_t(T) = 2\gamma(T)$ and $H = K_2$. This is further evidence for an affirmative answer to the following.

Question 3.1. *Is $\frac{1}{2} \cdot \gamma_t(G)\gamma_t(H) \le \gamma_t(G \,\square\, H)$ for all graphs G and H without isolated vertices?*

3.2 Direct Product

In this section we will show that open packings assume a role for total domination in direct products that is strikingly similar to that played by 2-packings relative to ordinary domination in Cartesian products. In particular, we prove a result that corresponds to Theorem 2.8.

If R is a total dominating set of G and S is a total dominating set of H, it follows immediately from the definition of the direct product that every vertex (g, h) of $G \times H$ is adjacent to a vertex in $R \times S$. An immediate consequence of this is that γ_t is submultiplicative on the direct product.

Theorem 3.3 ([16]). *If G and H have no isolated vertices, then* $\gamma_t(G \times H) \leq \gamma_t(G)\gamma_t(H)$.

A lower bound for the total domination number of a direct product can be expressed in terms of the open packing numbers and total domination numbers of the factor graphs in a manner analogous to Theorem 2.1 in the case of Cartesian products.

Theorem 3.4 ([17]). *For any graphs G and H with no isolated vertices,*

$$\gamma_t(G \times H) \geq \max\{\gamma_t(G)\rho^o(H), \gamma_t(H)\rho^o(G)\}.$$

Proof. It suffices to show that $\gamma_t(G \times H) \geq \gamma_t(G)\rho^o(H)$. Fix a vertex h in H, let B be a $\rho^o(H)$-set and let S be a $\gamma_t(G \times H)$-set. For $g \in V(G)$ the open neighborhood of (g, h) is given by $N((g, h)) = N_G(g) \times N_H(h)$. Then, S contains a vertex (x, y) adjacent to (g, h) such that $(x, y) \in \cup_{w \in N(h)} G^w$. Let $S_h = \{\, x \mid (x, y) \in S \cap (\cup_{w \in N(h)} G^w) \,\}$. It follows that S_h is a total dominating set of G. But now

$$S \supseteq \bigcup_{h \in B} [S \cap (\bigcup_{w \in N(h)} G^w)],$$

and thus

$$\gamma_t(G \times H) = |S| \geq \sum_{h \in B} |S \cap (\bigcup_{w \in N(h)} G^w)| \geq \sum_{h \in B} |S_h| \geq \rho^o(H)\gamma_t(G). \qquad \square$$

Putting Theorem 3.4 together with the fact that γ_t is submultiplicative on $\times$ we can now show that there is a nontrivial universal multiplicative class $\mathcal{U}$, for γ_t on $\times$. Any graph G such that $\gamma_t(G) = \rho^o(G)$ belongs to $\mathcal{U}$, for then

$$\gamma_t(G)\gamma_t(H) = \rho^o(G)\gamma_t(H) \leq \gamma_t(G \times H) \leq \gamma_t(G)\gamma_t(H).$$

As noted in [17], the complete bipartite graphs, cycles of order divisible by four as well as any graph G constructed by adding at least one endvertex adjacent to each vertex of any connected graph F belong to $\mathcal{U}$. Again, in a result reminiscent of Theorem 2.1 for 2-packings and domination in trees, Rall proved the following result for open packings and total domination in trees. As a consequence, every nontrivial tree belongs to $\mathcal{U}$.

Theorem 3.5 ([17]). *If T is any tree of order at least two, then $\gamma_t(T) = \rho^o(T)$. Thus, for any graph G with no isolated vertices, $\gamma_t(T \times G) = \gamma_t(T)\gamma_t(G)$.*

Combining Proposition 1.1(ii), Theorem 3.4 and Theorem 3.5 we observe that there is an analogous relationship between open packings and total domination in direct products as there is between 2-packings and ordinary domination in Cartesian products. See Theorem 2.8.

Theorem 3.6. *Let G and H be any graphs without isolated vertices,and let T be any tree of order at least two. Then*

(i) $\gamma_t(G) \geq \rho^o(G)$;

(ii) $\gamma_t(G \times H) \geq \max\{\gamma_t(G)\rho^o(H), \gamma_t(H)\rho^o(G)\}$;

(iii) $\gamma_t(T) = \rho^o(T)$.

We close this section with a question about which essentially nothing is known, except that if such a constant exists then it can be at most 3/4.

Question 3.2. *Is there a positive constant c (and if so, what is the largest) such that*

$$c \cdot \gamma_t(G)\gamma_t(H) \leq \gamma_t(G \times H)$$

hold for all graphs G and H without isolated vertices?

4 Paired-Domination

4.1 Cartesian Product

The invariant $\gamma_{\rm pr}$ is neither supermultiplicative nor submultiplicative on $\square$. For example,

$$\gamma_{\rm pr}(P_2 \square P_2) = 2 < \gamma_{\rm pr}(P_2)\gamma_{\rm pr}(P_2) \quad \text{and}$$
$$\gamma_{\rm pr}(K_{2n} \square K_{2n}) = 2n > 4 = \gamma_{\rm pr}(K_{2n})\gamma_{\rm pr}(K_{2n}).$$

The latter example shows that the ratio of $\gamma_{\rm pr}(G)\gamma_{\rm pr}(H)$ to $\gamma_{\rm pr}(G \square H)$ can be arbitrarily close to zero. On the other hand, this ratio cannot be larger than 8 as is shown by the following inequality chain,

$$\gamma_{\rm pr}(G)\gamma_{\rm pr}(H) \leq 4\gamma(G)\gamma(H) \leq 8\gamma(G \square H) \leq 8\gamma_{\rm pr}(G \square H). \tag{5}$$

The second inequality above follows from Theorem 2.4 while the first is a consequence of the easily established fact that the paired-domination number of a graph is at most twice its domination number. It is not known whether a smaller constant can replace the 8 in the above inequality. However, Brešar, Henning and Rall showed that 2 will suffice with an appropriate restriction placed on one of the factor graphs. An infinite family of graphs was described to show this bound is sharp.

Theorem 4.1 ([3]). *If $\gamma_{\rm pr}(G) = 2\rho_3(G)$, then for every H with no isolated vertices,*

$$\gamma_{\rm pr}(G)\gamma_{\rm pr}(H) \leq 2\gamma_{\rm pr}(G \square H).$$

In particular, they showed that the hypothesis holds if G is any nontrivial tree, thus establishing a 3-packing, paired-domination result for trees that is similar to Theorem 2.1 and Theorem 3.5. We omit the proof.

Theorem 4.2 ([3]). *For every nontrivial tree T, $\gamma_{\rm pr}(T) = 2\rho_3(T)$.*

From Inequality (5) and Proposition 1.1 it follows that $\gamma_{\rm pr}(G \square H) \geq \frac{1}{4}\rho_3(G)\gamma_{\rm pr}(H)$. A stronger lower bound for the paired-domination number of the Cartesian product in terms of the 3-packing and paired-domination numbers of the factors was established in [3]. Note the similarity of this result to those given earlier in Theorem 2.1 and Theorem 3.4.

Theorem 4.3 ([3]). *For any graphs G and H without isolated vertices,*

$$\gamma_{\rm pr}(G \,\square\, H) \geq \max\{\gamma_{\rm pr}(G)\rho_3(H), \gamma_{\rm pr}(H)\rho_3(G)\}.$$

Proof. Let $S = \{v_1, \ldots, v_{\rho_3(G)}\}$ be a $\rho_3(G)$-set. For $i = 1, \ldots, \rho_3(G)$, let $V_i = N_G[v_i]$, and so the sets V_i are pairwise disjoint. Let D be a $\gamma_{\rm pr}(G \,\square\, H)$-set. Fix a perfect matching M of $\langle D \rangle_{G \,\square\, H}$. For $u \in D$ the unique vertex $v \in D$ such that $uv \in M$ is called the partner of u with respect to M. For $i = 1, \ldots, \rho_3(G)$, let D_i consist of all vertices of D in $V_i \times V(H)$ together with their partners (some of which may possibly not belong to $V_i \times V(H)$), and let $H_i = V({}^{v_i}H) = \{v_i\} \times V(H)$. Since S is a 3-packing in G, and since every vertex of a paired-dominating set has a unique partner, we note that the sets D_i are pairwise disjoint. Hence

$$|D| \geq \sum_{i=1}^{\rho_3(G)} |D_i|\,.$$

For $i = 1, \ldots, \rho_3(G)$, let S_i be a minimum set of vertices in $G \,\square\, H$ such that (i) S_i dominates H_i and (ii) the subgraph induced by S_i in $G \,\square\, H$ contains a perfect matching M_i. Among all such sets S_i, we choose S_i to contain as many vertices in H_i as possible. Since the set D_i satisfies conditions (i) and (ii), we have that

$$|D_i| \geq |S_i|\,.$$

If S_i contains a vertex x not in $V_i \times V(H)$, then x is the partner of a vertex x' (with respect to the perfect matching M_i) not in H_i that uniquely dominates a vertex $y \in H_i$. Replacing x in the set S_i with the vertex y produces a new set of vertices in $G \,\square\, H$ that satisfies conditions (i) and (ii) but that contains more vertices in H_i than does S_i, a contradiction. Hence, $S_i \subseteq V_i \times V(H)$.

We show next that $S_i \subseteq H_i$. Suppose S_i contains a vertex u not in H_i. Let u' be the partner of u (with respect to M_i). If $u' \notin H_i$, then let w (respectively, w') be the neighbor of u (respectively, u') in $G \,\square\, H$ that belongs to H_i. We distinguish two cases with respect to w and w' being distinct.

Case 1. $w \neq w'$. Then neither w nor w' is in S_i (or else neither u nor u' would have been needed in S_i) and at least one of w and w' is not dominated by $S_i - \{u, u'\}$. So we can replace the pair u and u' in the set S_i with the pair w and w' to produce a new set of vertices in $G \,\square\, H$ that satisfies conditions (i) and (ii) but that contains more vertices in H_i than does S_i, contradicting our choice of S_i.

Case 2. $w = w'$. Then no vertex from $N[w] \cap H_i$ belongs to S_i (or again S_i would not have contained either u or u'). So by replacing u' by w in S_i we produce a new set that satisfies conditions (i) and (ii) but that contains more vertices in H_i than does S_i, a contradiction.

Both cases imply that $u' \in H_i$. If S_i contains every neighbor of u' in H_i, then we can simply delete u and u' from the set S_i to contradict our choice of S_i. Hence at least one neighbor u'' of u' in H_i does not belong to S_i. Replacing u in the set S_i with u'' produces a new set of vertices in $G \,\square\, H$ that satisfies conditions (i) and (ii) (with u' paired with u'' in the new set) but that contains more vertices in H_i than does S_i, a contradiction. Hence, $S_i \subseteq H_i$, and so S_i is a paired-dominating set of the subgraph ${}^{v_i}H$. Since ${}^{v_i}H$ is isomorphic to H it follows that

$$|S_i| \geq \gamma_{\rm pr}(H)\,.$$

$$\text{Thus, } \gamma_{\mathrm{pr}}(G \,\Box\, H) = |D| \geq \sum_{i=1}^{\rho_3(G)} |D_i| \geq \sum_{i=1}^{\rho_3(G)} |S_i| \geq \sum_{i=1}^{\rho_3(G)} \gamma_{\mathrm{pr}}(H) = \rho_3(G)\gamma_{\mathrm{pr}}(H)\,,$$

which establishes the desired bound. □

Again, as in the case of domination in Cartesian products (Theorem 2.8) and in total domination of direct products (Theorem 3.6) we summarize the relationships involving paired-domination and 3-packings. This result follows from Proposition 1.1(iii), Theorem 4.1 and Theorem 4.2.

Theorem 4.4. *Let G and H be any graphs and let T be any tree. Then*

(i) $\gamma_{\mathrm{pr}}(G) \geq 2\rho_3(G)$;

(ii) $\gamma_{\mathrm{pr}}(G \,\Box\, H) \geq \max\{\gamma_{\mathrm{pr}}(G)\rho_3(H), \gamma_{\mathrm{pr}}(H)\rho_3(G)\}$;

(iii) $\gamma_{\mathrm{pr}}(T) = 2\rho_3(T)$.

Rewriting inequality (5) we see that for every pair of graphs

$$\frac{1}{8} \cdot \gamma_{\mathrm{pr}}(G)\gamma_{\mathrm{pr}}(H) \leq \gamma_{\mathrm{pr}}(G \,\Box\, H)\,.$$

We close with the following question related to improving this bound.

Question 4.1. *What is the largest constant $c \geq \frac{1}{8}$ such that, for all graphs G and H,*

$$c \cdot \gamma_{\mathrm{pr}}(G)\gamma_{\mathrm{pr}}(H) \leq \gamma_{\mathrm{pr}}(G \,\Box\, H)\,?$$

4.2 Direct Product

If A and B are paired-dominating sets of G and H respectively, then clearly $A \times B$ dominates $G \times H$ and one easily shows that $\langle A \times B \rangle$ contains a perfect matching. Thus, γ_{pr} is submultiplicative on $\times$. That is, for every pair of graphs G and H with no isolates, $\gamma_{\mathrm{pr}}(G \times H) \leq \gamma_{\mathrm{pr}}(G)\gamma_{\mathrm{pr}}(H)$. In [4] it is shown that there are infinite families for which equality is actually achieved, although no universal multiplicative class is shown to exist. On the other hand, if S_n denotes the subdivided star with n pendant vertices, then it is shown that

$$\gamma_{\mathrm{pr}}(S_n \times S_n) \leq 2n^2 + 2n < 4n^2 = \gamma_{\mathrm{pr}}(S_n)\gamma_{\mathrm{pr}}(S_n)\,.$$

Question 4.2. *Is there a positive constant c (and if so, what is the largest) such that*

$$c \cdot \gamma_{\mathrm{pr}}(G)\gamma_{\mathrm{pr}}(H) \leq \gamma_{\mathrm{pr}}(G \times H)$$

hold for all graphs G and H without isolated vertices?

References

[1] R. Aharoni and T. Szabó, personal communication (2006).

[2] A. M. Barcalkin and L. F. German, The external stability number of the Cartesian product of graphs, *Bul. Akad. Shtiinl Moldoven*, **1** (1979), 5–8.

[3] B. Brešar, M. A. Henning and D. F. Rall, Paired-domination of Cartesian products of graphs, *Util. Math.*, **73** (2007), pp. 255–265.

[4] B. Brešar, S. Klavžar and D.F. Rall, Dominating direct products of graphs, *Discrete Math.*, **307** (2007), pp. 1636–1642.

[5] W.E. Clark and S. Suen, An inequality related to Vizing's conjecture, *Electron. J. Combin.* **7** (2000), #N4.

[6] B. Hartnell and D.F. Rall, Vizing's conjecture and the one-half argument, *Discuss. Math. Graph Theory* **15** (1995), 205–216.

[7] B. Hartnell and D.F. Rall, Domination in Cartesian products: Vizing's Conjecture, In [10], 163–189.

[8] B. Hartnell and D.F. Rall, Lower bounds for dominating Cartesian products, *J. Combin. Math. Combin. Comput.* **31** (1999), 219–226.

[9] T.W. Haynes, S.T. Hedetniemi and P.J. Slater, *Fundamentals of Domination in Graphs*, Marcel Dekker, New York (1998).

[10] T.W. Haynes, S.T. Hedetniemi and P.J. Slater (eds), *Domination in Graphs: Advanced Topics*, Marcel Dekker, New York (1998).

[11] M.A. Henning and D.F. Rall, On the total domination number of Cartesian products of graph, *Graphs Combin.* **21** (2005), 63–69.

[12] W. Imrich and S. Klavžar, *Product Graphs: Structure and Recognition*, John Wiley & Sons, Inc., New York (2000).

[13] M.S. Jacobson and L.F. Kinch, On the domination of the products of graphs II: trees, *J. Graph Theory* **10** (1986), 97–106.

[14] S. Klavžar and B. Zmazek, On a Vizing-like conjecture for direct product graphs, *Discrete Math.* **156** (1996), 243–246.

[15] A. Meir and J.W. Moon. Relations between packing and covering numbers of a tree, *Pacific J. Math.* **61** (1975), 225–233.

[16] R.J. Nowakowski and D.F. Rall, Associative graph products and their independence, domination and coloring numbers, *Discuss. Math. Graph Theory* **16** (1996), 53–79.

[17] D.F. Rall, Total domination in categorical products of graphs, *Discuss. Math. Graph Theory* **25** (2005), 35–44.

[18] L. Sun, A result on Vizing's conjecture, *Discrete Math.* **275** (2004), 363–366.

[19] V.G. Vizing, Some unsolved problems in graph theory, *Uspehi Mat. Nauk* **23** (1968) no. 6 (144), 117–134.

Proc. of ICDM 2006, Pages 195–203

Independent Domination and Graph Colorings

S. Arumugam, I. Sahul Hamid and A. Muthukamatchi
Core Group Research Facility (CGRF),
National Centre for Advanced Research in Discrete Mathematics (n-CARDMATH),
Department of Mathematics, Kalasalingam University, Anand Nagar, Krishnankoil-626190, India.
Email: s_arumugam_akce@yahoo.com

Abstract. In this paper we introduce three fundamental parameters which involve independent domination and chromatic number. Let G be a graph with chromatic number $\chi(G) = k$. The *dom-color number* $d_\chi(G)$ is the maximum number of color classes which are dominating sets of G, where the maximum is taken over all k-colorings of G. The *chromatic edge stability number* $es_\chi(G)$ is the minimum number of edges whose removal results in a graph G_1 with $\chi(G_1) = \chi(G) - 1$. The *chromatic bondage number* $\rho(G)$ is the minimum number of edges between two color classes in a k-coloring of G, where the minimum is taken over all k-colorings of G. We present several interesting results and unsolved problems on these parameters.

Keywords. Chromatic edge stability number, Chromatic bondage number, dom-color number, independent domination.
2000 Mathematics Subject Classification: 05C

1 Introduction

By a graph $G = (V, E)$, we mean a connected, finite, undirected graph with neither loops nor multiple edges. The order and size of G are denoted by n and m respectively. For graph theoretic terminology we refer to Harary [5].

Graph coloring theory and domination in graphs are two major areas within graph theory which have been extensively studied. Graph coloring deals with the fundamental problem of partitioning a set of objects into classes, according to certain rules. Time tabling, sequencing and scheduling problems in their many terms are basically of this nature. The fundamental parameter in the theory of graph coloring is the chromatic number $\chi(G)$ of a graph G which is defined to be the minimum number of colors required to color the vertices of G in such a way that no two adjacent vertices of G receive the same color. If $\chi(G) = k$, we say that G is k-chromatic. The chromatic number is a very well studied parameter whose history dates back to the famous four color problem and the early work of Kempe [8] in 1879 and Heawood [7] in 1890.

Another fastest growing area within graph theory is the study of domination and related subset problems such as independence, covering and matching. A comprehensive treatment of the fundamentals of domination is given in Haynes et al. [6].

A set $S \subseteq V$ is said to be a dominating set of G if every vertex v in $V - S$ is adjacent to a vertex in S. If further, S is independent, then S is called an independent dominating set of G. The minimum cardinality of an independent dominating set is called the *independent domination number* of G and is denoted by $i(G)$. An independent dominating set of G of cardinality $i(G)$ is called an $i(G)$-set of G. A partition of $V(G)$ into independent dominating sets of G is called an independent domatic partition of G and G is called indominable if it admits such a partition [1]. For any indominable graph G, the minimum order of an independent domatic partition of G is called the in-domatic number of G and is denoted by $d_i(G)$.

For any graph theoretic parameter the effect of removal of an edge on the parameter is of practical importance. Another related problem is the determination of the minimum number of edges whose removal alters the value of the parameter. For example, the minimum number of edges of G whose removal increases the domination number of G was studied by Bauer et al. [2] who called it as the edge stability number of G. Fink et al. [4] studied the same concept and used the term bondage number. We observe that the chromatic number of a graph can be reduced by removing a set of edges from the graph. Hence, analogous to the concept of bondage number for domination, it is natural to study the minimum number of edges whose removal reduces the chromatic number of the graph, which we call the chromatic edge stability number. We also consider another related parameter, namely, the minimum number of edges between two color classes in a k-coloring of G, where the minimum is taken over all k-colorings of G.

Walikar et al. [9] have proved the following theorem.

Theorem 1.1. *In a k-chromatic graph G, any k-coloring of G yields another k-coloring of G containing a color class which is a dominating set of G.*

Hence it is natural to consider the maximum number of color classes which are dominating sets of G, where the maximum is taken over all k-colorings of G, which we call the dom-color number of G.

In this paper we initiate a study of these three parameters.

We need the following definitions and theorem.

Definition 1.2. *The corona of two graphs G_1 and G_2 is the graph $G = G_1 \circ G_2$ formed from one copy of G_1 and $|V(G_1)|$ copies of G_2 where the ith vertex of G_1 is adjacent to every vertex in the ith copy of G_2.*

Definition 1.3. *A k-chromatic graph G is said to be uniquely colorable if every k-coloring of G induces the same partition of $V(G)$.*

Theorem 1.4 ([5], Page 138). *In the k-coloring of a uniquely k-colorable graph, the subgraph induced by the union of any two color classes is connected.*

2 Main Results

Definition 2.1. *The chromatic edge stability number of a graph G is defined to be the minimum number of edges of G whose removal results in a graph G_1 with $\chi(G_1) = \chi(G) - 1$ and is denoted by $es_\chi(G)$.*

Example 2.2.

(i) $es_\chi(G) = m$ if and only if G is bipartite.

(ii) $es_\chi(K_n) = 1$ for all $n \geq 2$.

(iii) $es_\chi(C_n) = 1$ if n is odd.

If $(V_1, V_2, \ldots, V_k)$ is a k-coloring of a k-chromatic graph G and if G_1 is the graph obtained from G by removing all the edges between two color classes V_i and V_j for some $i \neq j$, then $\chi(G_1) = k-1$. This observation motivates the following definition.

Definition 2.3. *The minimum number of edges between two color classes in a k-coloring of a k-chromatic graph G, where the minimum is taken over all k-colorings of G, is called the chromatic bondage number of G and is denoted by $\rho(G)$.*

Obviously, for any graph G, $es_\chi(G) \leq \rho(G)$. Further, for all the graphs given in Example 2.2, $es_\chi = \rho$. In the following lemma we give another family of graphs with this property.

Lemma 2.4. *For the complete tripartite graph $G = K_{\lambda,\lambda,\lambda}$, $es_\chi(G) = \lambda^2$.*

Proof. Since $\rho(G) = \lambda^2$, we have $es_\chi(G) \leq \lambda^2$. Further G is decomposable into λ^2 triangles and hence $es_\chi(G) \geq \lambda^2$. Thus $es_\chi(G) = \lambda^2$. □

For the complete k-partite graph $G = K_{\lambda_1,\lambda_2,\ldots,\lambda_k}$, we have $\rho(G) = min\{\lambda_i\lambda_j : 1 \leq i < j \leq k\}$. In this connection we pose the following conjecture.

Conjecture 2.5. For the complete k-partite graph $G = K_{\lambda_1,\lambda_2,\ldots,\lambda_k}$, $es_\chi(G) = \rho(G)$.

More generally, we have the following problem.

Problem 2.6. *Characterize graphs G for which $es_\chi(G) = \rho(G)$.*

The following theorem shows that the difference between ρ and es_χ can be made arbitrarily large.

Theorem 2.7. *Given two positive integers k and r with $k \geq 3$, there exists a graph G with $\chi = k$ and $\rho - es_\chi > r$.*

Proof. Let G_1 be a copy of K_k. For each edge $e = xy$ of G_1, we attach $r+1$ copies of K_k, say $G_{i_1}, G_{i_2}, \ldots, G_{i_{(r+1)}}$ such that $V(G_{i_j}) \cap V(G_1) = \{x, y\}$, where $1 \leq j \leq r+1$. Let G be the resultant graph. Clearly $\chi(G) = k$. If H is the subgraph of G obtained by removing the $\binom{k}{2}$ edges of G_1, we have $\chi(H) = k-1$. Further, if H_1 is any subgraph of G obtained by removing a set of m edges, where $m < \binom{k}{2}$, then H_1 contains a copy of K_k and hence $\chi(H_1) = k$. Thus $es_\chi(G) = \binom{k}{2}$. Now, let V_1 and V_2 be two color classes in a k-coloring of G. Since G_1 is a complete graph, there is a unique edge between any two color classes. Let the unique edge of G_1 between the color classes V_1 and V_2 be e_s. Now, each G_{i_j}, where $1 \leq i \leq \binom{k}{2}$, $i \neq s$ and $1 \leq j \leq r+1$, contains exactly one edge with ends in V_1 and V_2. Hence $\rho = \left[\binom{k}{2} - 1\right](r+1) + 1$. Hence $\rho - es_\chi = r\left[\binom{k}{2} - 1\right] > r$. □

Remark 2.8. Let G be a graph with $\chi(G) = k$ and let $\{V_1, V_2, \ldots, V_k\}$ be a k-coloring of G. Since the m edges of G are distributed between the $\binom{k}{2}$ pairs of color classes, it follows that there exist two color classes V_i and V_j such that the number of edges between V_i and V_j is at most $\frac{m}{\binom{k}{2}}$. Hence $es_\chi(G) \leq \rho(G) \leq \frac{m}{\binom{k}{2}}$. Further if the number of edges between a pair of color classes is greater than $\frac{m}{\binom{k}{2}}$, then there exist another pair of color classes such that the number of edges between them is less than $\frac{m}{\binom{k}{2}}$. Hence, if $es_\chi(G) = \frac{m}{\binom{k}{2}}$, then $\rho(G) = \frac{m}{\binom{k}{2}}$ and the number of edges between any two color classes in any k-coloring of G is $\frac{m}{\binom{k}{2}}$. However, the converse is not true. For example, for the uniquely 3-colorable graph G given in Figure 2.1, the number of edges between any two color classes in the 3-coloring of G is $3 = \frac{m}{\binom{k}{2}}$. However, $es_\chi(G) = 2$.

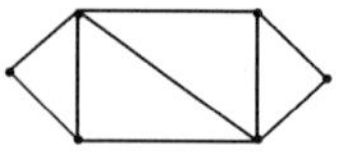

Figure 2.1:

In the following theorem we characterize k-chromatic graphs with $\Delta = n - 1$ for which $\rho = \frac{m}{\binom{k}{2}}$.

Theorem 2.9. *Let G be a k-chromatic graph with $\Delta = n - 1$. Then $\rho = \frac{m}{\binom{k}{2}}$ if and only if either G is complete or G has exactly one cut vertex and has ρ blocks each isomorphic to K_k.*

Proof. Let u be a vertex of G with deg $u = n - 1$.

Suppose $\rho = \frac{m}{\binom{k}{2}}$. Let $(V_1, V_2, \ldots, V_k)$ be any k-coloring of G with $V_1 = \{u\}$. Now, the number of edges between any pair of color classes (V_i, V_j) is ρ and hence $|V_i| = \rho$, for all i, where $2 \leq i \leq k$ and the subgraph induced by the ρ edges between V_i and V_j, where $2 \leq i < j \leq k$, is isomorphic to ρK_2. Hence if $\rho = 1$ it follows that G is complete. Now, suppose $\rho \geq 2$. Let $V_i = \{v_{i_1}, v_{i_2}, \ldots, v_{i_\rho}\}$ and $E(V_i \cup V_j) = \{v_{i_1}v_{j_1}, v_{i_2}v_{j_2}, \ldots, v_{i_\rho}v_{j_\rho}\}$, $2 \leq i < j \leq k$.

Now, for each $i = 1, 2, \ldots, \rho$, the subgraph of G induced by $\{u, v_{2_i}, v_{3_i}, \ldots, v_{k_i}\}$ is isomorphic to K_k. Hence G is the graph with exactly one cut-vertex and having ρ blocks each isomorphic to K_k. The converse is obvious. □

Theorem 2.10. *For any r-regular graph G on n vertices, $\rho(G) \leq (n-r)^2$. Further equality holds if and only if $n - r$ divides n and G is isomorphic to the complete k-partite graph $K_{\lambda,\lambda,\ldots,\lambda}$ where $\lambda = n - r$ and $k = \frac{n}{n-r}$.*

Proof. Since G is r-regular, the number of vertices in a color class in any k-coloring of G is at most $n - r$ and hence $\rho(G) \leq (n - r)^2$.

Suppose $\rho(G) = (n - r)^2$. Let $(V_1, V_2, \ldots, V_k)$ be a k-coloring of G. Since G is r-regular, $|V_i| \leq n - r$, for all $i = 1, 2, \ldots, k$. Suppose $|V_i| < n - r$ for some i. Then the number of edges between V_i and V_j for all $j \neq i$ is less than $(n - r)^2$, which is a contradiction. Hence $|V_i| = n - r$, for all $i = 1, 2, \ldots, k$. Since the number of edges between V_i and V_j, for all $i \neq j$, is $(n-r)^2$, we have G is a complete k-partite graph in which each part contains exactly $n - r$ vertices. Hence $k(n - r) = n$, so that $n - r$ divides n. The converse is obvious. □

Theorem 2.11. *Let G be a k-chromatic graph having a vertex u with* deg $u = n-1$. *Then $\rho = r$, where r is the minimum number of vertices in a color class other than $\{u\}$ in a k-coloring of G and the minimum is taken over all k-colorings of G.*

Proof. Since $\{u\}$ is a color class in any k-coloring of G, we have $\rho \leq r$. Suppose $(V_1, V_2, \ldots, V_{k-1}, \{u\})$ is a k-coloring of G such that the number of edges between V_1 and V_2 is less than r. Let $S = \{v \in V_1 : v$ is not adjacent to any vertex in $V_2\}$. Since $|V_1| \geq r$ and $|V_2| \geq r$, it follows that $|S| \geq |V_1| - r + 1$ and $V_2 \cup S$ is independent. Hence $(V_1 \backslash S, V_2 \cup S, V_3, \ldots, V_{k-1}, \{u\})$ is a k-coloring of G with $|V_1 \backslash S| < r$, which is a contradiction to the minimality of r. Thus $\rho = r$. □

Corollary 2.12. *For any k-chromatic graph G with* $\Delta = n - 1$, *we have* $\rho(G) \leq \beta_0(G)$, *where* $\beta_0(G)$ *is the independence number of G.*

Proof. Since $\rho = r$ and $r \leq \beta_0$, the result follows. □

Remark 2.13. The bound given in the above corollary is sharp. For example, for the graph G given in Figure 2.2, $\rho(G) = \beta_0(G) = 2$.

Figure 2.2:

Problem 2.14. *Characterize graphs G with* $\Delta = n - 1$ *for which* $\rho(G) = \beta_0(G)$.

We now proceed to investigate properties of graphs G for which $es_\chi(G) = 1$.

Remark 2.15. Let G be a k-chromatic graph. Then $es_\chi(G) = 1$ if and only if there exists a k-coloring $(V_1, V_2, \ldots, V_k)$ of G such that $|V_1| = 1$ and there is exactly one edge between V_1 and V_2.

Theorem 2.16. *Let G be a k-chromatic graph with* $es_\chi(G) = 1$. *Then there exist two adjacent vertices u and v such that* $\deg\ u + \deg\ v \geq 2(k-1)$.

Proof. It follows from the above theorem that there exists a k-coloring $(\{u\}, V_1, V_2, \ldots, V_{k-1})$ of G such that there is exactly one edge $e = uv$ between $\{u\}$ and V_1. Since u is adjacent to at least one vertex of V_i, for all $i \geq 1$, we have $\deg\ u \geq k-1$. We now claim that $\deg\ v \geq k-1$. Suppose $\deg\ v < k-1$. Then there exists a color class, say V_2, such that v is not adjacent to any vertex of V_2. Now, $(V_1 - \{v\}) \cup \{u\}, V_2 \cup \{v\}, V_3, \ldots, V_{k-1})$ is a $(k-1)$-coloring of G, which is a contradiction. Hence $\deg\ v \geq k-1$ so that $\deg\ u + \deg\ v \geq 2(k-1)$. □

Theorem 2.17. *Let G be a uniquely k-colorable graph. Then* $es_\chi(G) = 1$ *if and only if the k-coloring of G contains two singleton color classes.*

Proof. Let $(V_1, V_2, \ldots, V_k)$ be the k-coloring of G. Let $es_\chi(G) = 1$. Then by Theorem 2.15 there exist color classes say V_1 and V_2 such that $|V_1| = 1$ and there is exactly one edge between V_1 and V_2. By Theorem 1.3, we have $\langle V_1 \cup V_2 \rangle$ is connected and hence $|V_2| = 1$. The converse is obvious. □

Problem 2.18. *Characterize graphs G for which* $es_\chi(G) = 1$.

Remark 2.19. Since for any graph G, $1 \leq es_\chi(G) \leq m$, we have

(i) $2 \leq es_\chi(G) + es_\chi(\overline{G}) \leq \binom{n}{2}$ and

(ii) $1 \leq es_\chi(G) \cdot es_\chi(\overline{G}) \leq m\left[\binom{n}{2} - m\right]$.

Further, the following are equivalent.

(a) $es_\chi(G) + es_\chi(\overline{G}) = \binom{n}{2}$

(b) $es_\chi(G) \cdot es_\chi(\overline{G}) = m\left[\binom{n}{2} - m\right]$

(c) G is one of the graphs $P_3, \overline{P_3}, P_4, C_4$ and $\overline{C_4}$.

In fact, $es_\chi(G) + es_\chi(\overline{G}) = \binom{n}{2}$ if and only if $es_\chi(G) = |E(G)| = m$ and $es_\chi(\overline{G}) = |E(\overline{G})| = \binom{n}{2} - m$. Hence G and $\overline{G}$ are both bipartite, so that (a), (b) and (c) are equivalent.

It follows from Theorem 1.1. that any k-chromatic graph G admits a k-coloring in which at least one color class is a dominating set of G. This leads to the following definition.

Definition 2.20. *Let G be a graph with $\chi(G) = k$. Let $\mathcal{C} = (V_1, V_2, \ldots, V_k)$ be a k-coloring of G. Let $d_\mathcal{C}$ denote the number of color classes in $\mathcal{C}$ which are dominating sets of G. Then $d_\chi = \max\limits_{\mathcal{C}} d_\mathcal{C}$, where the maximum is taken over all k-colorings of G, is called the dom-color number of G.*

Example 2.21.

(i) $d_\chi = \chi(C_n) = 2$, if n is even. If n is odd, then $\chi(C_n) = 3$ and $d_\chi(C_n) = 2$.

(ii) $d_\chi(K_n) = \chi(K_n) = n$

(iii) For any connected bipartite graph G, $d_\chi(G) = \chi(G) = 2$.

Remark 2.22. Clearly $1 \le d_\chi(G) \le \chi(G)$. Further given integers a and b with $1 \le a \le b$ there exists a graph G such that $d_\chi(G) = a$ and $\chi(G) = b$. If $a = 1$, let $G = K_b \circ K_1$. If $a \ge 2$, let G be the graph obtained from K_b by attaching one pendant edge at $a - 1$ vertices of K_b.

Remark 2.23. Dunbar et al. [3] have introduced the concept of Fall coloring of a graph. A coloring $\mathcal{C} = \{V_1, V_2, \ldots, V_k\}$ of a graph G is called a *Fall coloring* if $N[v] \cap V_i \ne \emptyset$ for all $v \in V$ and for all i, $1 \le i \le k$. Thus if a graph G admits a Fall coloring, then every color class is a dominating set and hence G is indominable. The minimum integer k for which a graph G has a Fall coloring is called the Fall chromatic number of G and is denoted by $\chi_f(G)$. Clearly if $d_\chi(G) = \chi(G)$, then G admits a Fall coloring of G. However, if a graph G admits a Fall coloring, it is not necessary that $d_\chi(G) = \chi(G)$. For the cartesian product $G = C_4 \Box C_5$, it has been proved in [3] that $\chi_f(G) = 4$ and $\chi(G) = 3$. It is easily seen that G admits a 3-coloring in which two color classes are dominating sets and since $\chi_f(G) = 4$, it follows that $d_\chi(G) = 2 < \chi(G)$. However if G admits a Fall coloring and $\chi_f(G) = \chi(G)$, then $d_\chi(G) = \chi(G)$.

We now proceed to study graphs with $d_\chi(G) = \chi(G)$.

Theorem 2.24. *For any uniquely colorable graph G, $d_\chi(G) = \chi(G)$.*

Proof. Let $\mathcal{C} = (V_1, V_2, \ldots, V_k)$ be the k-coloring of G, where $\chi(G) = k$. Suppose there exists a color class in $\mathcal{C}$, say V_1 which is not a dominating set of G. Then there exists a vertex $v \in V_i$ where $i > 1$, such that v is not adjacent any of the vertices of V_1. Then $\mathcal{C}' = (V_1 \cup \{v\}, \ldots, V_i - \{v\}, V_{i+1} \ldots V_k)$ is a k-coloring of G, which is a contradiction. Hence each V_i is a dominating set of G. Hence $d_\chi(G) = \chi(G)$. □

Corollary 2.25. *If G is uniquely colorable graph, then $d_i(G) = \chi(G)$.*

Remark 2.26. The converse of Theorem 2.23 is not true. For the graph G given in Figure 2.3, $d_\chi(G) = \chi(G) = 3$ and G is not uniquely colorable.

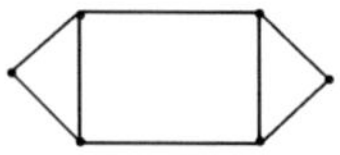

Figure 2.3:

Remark 2.27. If $d_\chi(G) = \chi(G)$, then G is indominable and $\chi \leq d_i$. This inequality can be strict. For the graph G given in Figure 2.4, $\chi(G) = 3$ and $d_i(G) = 4$.

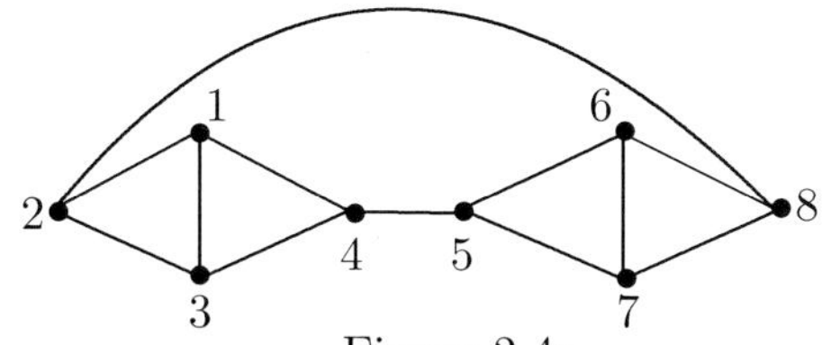

Figure 2.4:

Remark 2.28. For any indominable graph G, we have $d_\chi(G) \leq \chi(G) \leq d_i(G)$.

Theorem 2.29. *Let G be a k-chromatic graph with $\delta \geq 1$. If $\rho = \frac{m}{\binom{k}{2}}$, then every k-coloring of G is an independent domatic partition of G.*

Proof. Suppose $\rho = \frac{m}{\binom{k}{2}}$. Then the number of edges between any two color classes in any k-coloring of G is $\frac{m}{\binom{k}{2}}$. Let $(V_1, V_2, \ldots, V_k)$ be any k-coloring of G. We claim that each V_i is a dominating set of G. Suppose on the contrary that V_1 is not a dominating set of G. Then there exists a vertex v in a color class other than V_1, say V_2, such that v is not adjacent to any of the vertices of V_1. Hence $(V_1 \cup \{v\}, V_2 - \{v\}, V_3, \ldots, V_k)$ is a k-coloring of G and the number of edges between the color class $V_2 - \{v\}$ and some color class is less than $\frac{m}{\binom{k}{2}}$, which is a contradiction. Thus each V_i is a dominating set of G. □

Corollary 2.30. *Let G be a k-chromatic graph with $\delta \geq 1$ and $\rho = \frac{m}{\binom{k}{2}}$. Then $d_\chi(G) = \chi(G)$.*

Corollary 2.31. *For any k-chromatic graph G with $\delta \geq 1$ and $\rho = \frac{m}{\binom{k}{2}}$, $i(G) \leq \frac{n}{k}$ and this bound is sharp.*

Proof. Let $(V_1, V_2, \ldots, V_k)$ be a k-coloring of G. Since each V_i is a dominating set of G, $|V_i| \geq i(G)$ for all $i = 1, 2, \ldots, k$ and hence $i(G) \leq \frac{n}{k}$. This bound is sharp, since for the complete k-partite graph $G = K_{\lambda,\lambda,\ldots,\lambda}$, we have $i(G) = \frac{n}{k}$. □

Remark 2.32. The converse of Theorem 2.29 is not true. For the 3-chromatic graph G given in Figure 2.5,

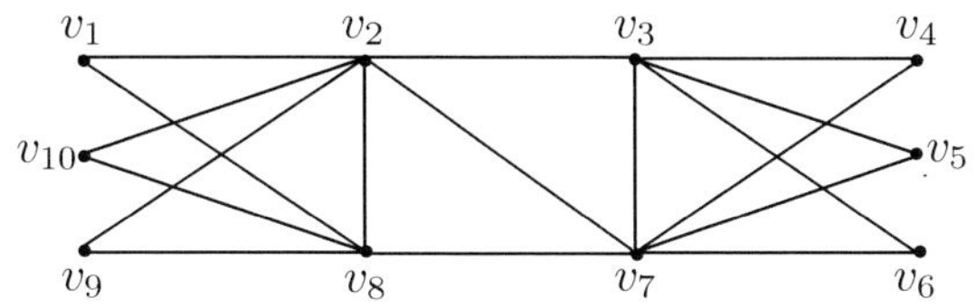

Figure 2.5:

$(\{v_2, v_4, v_5, v_6\}, \{v_3, v_8\}, \{v_1, v_7, v_9, v_{10})$ is the unique 3-coloring of G in which each color class is a dominating set of G. However $\rho = 5 \neq \frac{m}{\binom{k}{2}}$.

Problem 2.33. *Characterize k-chromatic graphs G with $\rho(G) = \frac{m}{\binom{k}{2}}$ for which $i(G) = \frac{n}{k}$.*

We observe that a k-chromatic graph G with $\rho(G) = \frac{m}{\binom{k}{2}}$ for which $i(G) = \frac{n}{k}$ must be a k-partite graph with a partition $(V_1, V_2, \ldots, V_k)$ such that each V_i is an $i(G)$-set of G and the number of edges in $\langle V_i \cup V_j \rangle$ is same for all $i \neq j$.

Theorem 2.29 leads to the following problem.

Problem 2.34. *Characterize k-chromatic graphs G in which every k-coloring of G is an independent domatic partition of G.*

Theorem 2.35. *For any uniquely colorable graph G, $\rho(G) \geq 2i(G) - 1$. Further, equality holds if and only if the k-coloring of G contains two color classes V_1 and V_2 such that $|V_1| = |V_2| = i(G)$ and $\langle V_1 \cup V_2 \rangle$ is a tree.*

Proof. Let $(V_1, V_2, \ldots, V_k)$ be the k-coloring of G. Since G is uniquely colorable, it follows from Theorem 2.24 that each V_i is a dominating set of G and hence $|V_i| \geq i(G)$. Now, it follows from Theorem 1.3 that the induced subgraph $\langle V_i \cup V_j \rangle$ is connected for all $i \neq j$ so that the number of edges in $\langle V_i \cup V_j \rangle$ is at least $2i(G)-1$. Thus $\rho(G) \geq 2i(G)-1$. Obviously, equality holds if and only if the k-coloring of G contains two color classes V_1 and V_2 such that $|V_1| = |V_2| = i(G)$ and $\langle V_1 \cup V_2 \rangle$ is a tree. □

Remark 2.36. The above bound is sharp. For the complete graph K_n, we have $\rho(K_n) = 1 = 2i(K_n) - 1$.

Problem 2.37. *Characterize uniquely colorable graphs G for which $\rho(G) = 2i(G) - 1$.*

The following are some interesting problems for further investigation.

Problem 2.38.

(i) *Characterize graphs G for which*

  (a) $es_\chi(G) = \frac{m}{\binom{k}{2}}$.

  (b) $\rho = \frac{m}{\binom{k}{2}}$.

  (c) $es_\chi(G) + es_\chi(\overline{G}) = 2$.

(ii) *Characterize graphs for which $d_\chi = 1$.*

(iii) *Characterize graphs for which $d_i(G) = \chi_f(G)$.*

(iv) *Determine the graph G for which $d_i(G) = \chi_f(G)$.*

References

[1] B.D. Acharya and H.B. Walikar, Indominable graphs cannot be characterized by a family of forbidden subgraphs, *Nat. Acad. Sci–Letters*, **4(1)** (1981), pp. 23–25.

[2] D. Bauer, F. Harary, J. Nieminen and C.L. Suffel, Domination alteration sets in graphs, *Discrete Math.*, **47** (1983), pp. 153–161.

[3] J.E. Dunbar, S.M. Hedetniemi, S.T. Hedetniemi, D.P. Jacobs, J. Kinsely, R.C. Laskar and D.F. Rall, Fall colorings of graphs, *J. Combin. Math. Combin. Comput.*,, **33** (2000), pp. 257–273.

[4] J. F. Fink, M.S. Jacobson, L.F. Kinch and J. Roberts, The bondage number of a graph, *Discrete Math.*, **86** (1990), pp. 47–57.

[5] F. Harary, Graph Theory, Addison-Wesley, Reading, Mass (1972).

[6] T.W. Haynes, S.T. Hedetniemi and P.J. Slater, *Fundamentals of Domination in Graphs*, Marcel Dekker, Inc., New York (1998).

[7] P.J. Heawood, Map colour theorems, *Quart. J. Math.*, **24** (1890), pp. 332–338.

[8] A.B. Kempe, On the geographical problem of four colors, *Amer. J. Math.*, **2** (1879), pp. 193–204.

[9] H.B. Walikar, B.D. Acharya, Kishori Narayankar and H.G. Shekharappa, Embedding Index of Nonindominable Graphs, in *Proceedings of the National Conference on Graphs, Combinatorics, Algorithms and Applications, Arulmigu Kalasalingam College of Engineering, Krishnankoil, Eds. S. Arumugam, B.D. Acharya and S.B. Rao* (2004), Narosa Publishing House, New Delhi pp. 173–179.

Proc. of ICDM 2006, Pages 205–216

Graph Equivalences and Decompositions Definable in Monadic Second-Order Logic. The Case of Circle Graphs

BRUNO COURCELLE
Bordeaux 1 University, LaBRI and Institut Universitaire de France
Email: courcell@labri.fr

1 Introduction

Many graph properties and graph transformations can be formalized in *Monadic Second-Order logic.* This language is the extension of *First-Order logic* allowing variables denoting sets of elements. In the case of graphs, these elements can be vertices, and in some cases edges. Monadic second-order graph properties can be checked in linear time on the class of graphs of *tree-width* at most k for any fixed k. These properties are *Fixed Parameter Linear*, for tree-width as a parameter.

Monadic second-order logic as a language for specifying graph properties is interesting from several different points of view: we already mentioned complexity, but another point of view is that of *Graph Grammars.* For logicians, monadic second-order logic is attractive because relatively many classes of structures have a *decidable theory* for this language.

In this communication we will discuss the point of view of *Graph Theory.* Many graph properties concerning colorings, forbidden configurations, connectivity are expressible in Monadic Second-Order logic, but also many graph theoretical constructions like the canonical decompositions of a graph in *2- and 3-connected components*, its *modular* and its *split decompositions.*

We will review a number of cases where a set of graphs or of combinatorial objects is characterized by a common hierarchical decomposition. In the cases we will consider, the decomposition can be formalized in monadic second-order logic and from it, all graphs or objects of the corresponding set can be defined by monadic second-order formulas with the help of auxiliary data like a k-tuple of sets of vertices or a linear order on the vertices.

This general description applies to the following sets of graphs or combinatorial objects :

1. the connected graphs having the same cycle matroid as a given graph,
2. the different planar embeddings of a planar connected graph,
3. the transitive orientations of a comparability graph,
4. the interval models of an interval graph,
5. the pairs of linear orders representing a partial order of dimension 2,
6. the chord representations of a circle graph,
7. the systems of intersecting closed curves in the plane having a same Gauss multiword.

In the first three sections, we will present the main concepts underlying these descriptions. In the fourth one we will apply them to *Circle Graphs.*

For fixed parameter tractability the references are the books [21] and [23]. For the use of monadic second-order logic in the theory of graph grammars the reference is the book chapter [4]. For the links between graph decompositions and the decidability of monadic second-order logic the references are the articles [5, 9, 17]. The results for the seven above cases are explicit or implicit in the articles [6–12].

2 Graph Properties Expressible in Monadic Second-Order Logic

Let $R = \{A, B, C, ...\}$ be a finite set of *relation symbols*; each of them, say A, is given with a nonnegative integer $\rho(A)$ called its *arity*. We denote by $\mathcal{STR}(R)$ the set of *finite* R-structures $S = \langle D_S, (A_S)_{A\in R}\rangle$ where $A_S \subseteq D_S^{\rho(A)}$ for each A. A simple graph G can be defined as the $\{edg\}$-structure $S(G) = \langle V_G, edg_G\rangle$ where V_G is the set of vertices and $edg_G \subseteq V_G \times V_G$ is a binary relation representing the edges. For undirected graphs, the relation edg_G is symmetric. If in addition we need vertex labels, we will represent them by unary relations. Graphs, either simple or not, can also be represented by the richer incidence structure $I(G) = \langle V_G \cup E_G, inc_G\rangle$, where E_G, the set of edges, is now part of the domain and $inc_G(e, u, v)$ holds iff e is an edge from u to v (or between u and v if G is undirected).

Monadic Second-Order logic is the extension of *First-Order logic* by variables denoting subsets of the domains of the considered structures, and new atomic formulas of the form $x \in X$ expressing the membership of x in a set X. (Uppercase letters will denote set variables, lowercase letters will denote first-order variables). In the sequel *MS* will abbreviate Monadic Second-order.

We denote by $MS(R, W)$ the set of *MS formulas* written with the set R of relation symbols and having their free variables in a set W consisting of *first-order as well as of set variables.* As a typical and useful example, we give an MS formula τ with free variables x and y expressing that (x, y) belongs to the reflexive and transitive closure of a binary relation A :

$$\forall X(x \in X \wedge \forall u, v[(u \in X \wedge A(u, v)) \Longrightarrow v \in X] \Longrightarrow y \in X).$$

If the relation A is not given in the structure but is defined by an MS formula, then one replaces $A(u, v)$ by this formula with appropriate substitutions of variables.

An *MS property* of the structures S of a class $\mathcal{C} \subseteq \mathcal{STR}(R)$ is a property $\mathcal{P}$ such that for all $S \in \mathcal{C}$:

$$\mathcal{P}(S) \text{ holds if and only if } S \models \varphi,$$

for some fixed formula φ in $MS(R, \varnothing)$. We say also that $\mathcal{P}$ is *MS expressible.* (The notation $S \models \varphi$ means that the logical formula φ is true in the structure S.)

Example. For each k one can construct a formula φ_k in $MS(\{edg\}, \varnothing)$ such that for every graph G:

$$S(G) \models \varphi_k \text{ if and only if } G \text{ is } k\text{-colorable}.$$

Here is the formula φ_3:

$$\exists X_1, X_2, X_3[\forall x(x \in X_1 \vee x \in X_2 \vee x \in X_3) \wedge$$
$$\forall x(\neg(x \in X_1 \wedge x \in X_2) \wedge \neg(x \in X_2 \wedge x \in X_3) \wedge \neg(x \in X_1 \wedge x \in X_3))$$

$\wedge\forall u, v(edg(u, v) \wedge u \neq v \Longrightarrow \neg(u \in X_1 \wedge v \in X_1) \wedge \neg(u \in X_2 \wedge v \in X_2)$
$\wedge\neg(u \in X_3 \wedge v \in X_3))]$.

Many properties based on the existence of paths like connectivity, strong connectivity, biconnectivity can be expressed in MS logic with the help of the above written formula τ that defines the reflexive and transitive closure of a binary relation.

If H is a simple loop-free undirected graph, then the property that a graph G contains H as a minor is MS expressible. Let H have vertices $1, ..., n$. Then G contains H as a minor iff it has pairwise disjoint sets of vertices $X_1, ..., X_n$ that induce connected subgraphs of G and are such that for every edge $i - j$ in H, there exists an edge of G linking one vertex of X_i and one of X_j. These conditions are easily expressible by an MS formula μ_H. Every minor-closed class of undirected graphs is characterized by finitely many excluded minors, and is thus definable by an MS formula. This is the case of planar graphs which are characterized by the formula $\neg\mu_{K_5} \wedge \neg\mu_{K_{3,3}}$.

All properties of a graph G considered above are expressed via the relational structure $S(G)$. More properties are MS expressible via the relational structure $I(G)$. In particular the existence of a Hamiltonian cycle in a graph G is MS expressible by using $I(G)$ and not by using $S(G)$. See [4, 5].

Every property that is MS expressible via the relational structure $I(G)$ is *fixed parameter linear*, where the parameter is tree-width. Every property that is MS expressible via the relational structure $S(G)$ is *fixed parameter cubic*, where the parameter is clique-width (this follows from [15] and [27], the latter article showing that cubic time is enough for finding an appropriate decomposition of the given graph). However, the paradigm *MS logic + bounded tree-width or bounded clique-width* extends to other algorithmic problems than just the verification of properties. One can compute efficiently optimization functions [16], one can enumerate queries efficiently ([1, 13, 22]), one can label graphs in order to facilitate answers to queries [18].

3 Monadic Second-Order Transductions

The construction of MS formulas for expressing graph properties is not always an easy task. Our articles contribute to building a toolbox, the big hammer of which is the notion of *Monadic Second-Order transduction* (*MS transduction* in short). We first present a very simple case.

The *edge-complement* of a simple loop-free undirected graph G, that we denote by $\overline{G}$, can be defined in logical terms. The edge relation of $\overline{G}$ is defined from that of G by:

$edg_{\overline{G}}(x, y) \Longleftrightarrow x \neq y \wedge \neg edg(x, y)$.

Hence we define $\overline{G}$ from G by defining its edge relation by a logical (here first-order) formula to be evaluated in G. We say that the edge-complement transformation is a first-order transduction.

The notion of an *MS transduction* generalizes this example on several respects, and in particular by the use of MS formulas instead of first-order ones. As in *Language Theory*, a binary relation $\mathcal{R} \subseteq \mathcal{A} \times \mathcal{B}$ where $\mathcal{A}$ and $\mathcal{B}$ are sets of relational structures is called a *transduction*: $\mathcal{A} \rightarrow \mathcal{B}$. An *MS transduction* transforms a structure S, given with an n-tuple of subsets of its domain called the *parameters*, into a structure T, the domain of which is a subset of $D_S \times [k]$. ($[k] = \{1, ..., k\}$). Furthermore, each such transduction, has an associated *backwards translation*, a mapping that transforms effectively every MS formula φ relative to T, possibly with free variables, into one, say $\varphi^{\#}$, relative to S having free variables corresponding to those of φ (k times as many actually) together with those denoting the parameters. This new formula expresses in S the property of T defined by φ. We now give some details. (The main reference for this section is [4].)

We let R and Q be two finite sets of relation symbols. Let W be a finite set of set variables, called the *parameters*. A (Q,R)-*definition scheme* is a tuple of formulas of the form:

$$\Delta = (\varphi, \psi_1, \cdots, \psi_k, (\theta_w)_{w \in Q^*k})$$
$$\text{where } k > 0, Q^*k := \{(q, \vec{j}) \mid q \in Q, \vec{j} \in [k]^{\rho(q)}\},$$
$$\varphi \in MS(R, W),\ \psi_i \in MS(R, W \cup \{x_1\}) \text{ for } i = 1, \cdots, k,$$
$$\text{and } \theta_w \in MS(R, W \cup \{x_1, \cdots, x_{\rho(q)}\}), \text{ for } w = (q, \vec{j}) \in Q^*k.$$

These formulas are intended to define a structure T in $\mathcal{STR}(Q)$ from a structure S in $\mathcal{STR}(R)$. Let $S \in \mathcal{STR}(R)$, let γ be a W-assignment in S. A Q-structure T with domain $D_T \subseteq D_S \times [k]$ is *defined in* (S, γ) by Δ if:

(i) $(S, \gamma) \models \varphi$

(ii) $D_T = \{(d, i) \mid d \in D_S, i \in [k], (S, \gamma, d) \models \psi_i\}$

(iii) for each q in Q : $q_T = \{((d_1, i_1), \cdots, (d_t, i_t)) \in D_T^t \mid (S, \gamma, d_1, \cdots, d_t) \models \theta_{(q,\vec{j})}\}$, where $\vec{j} = (i_1, \cdots, i_t)$ and $t = \rho(q)$.

By $(S, \gamma, d_1, \cdots, d_t) \models \theta_{(q,\vec{j})}$, we mean $(S, \gamma') \models \theta_{(q,\vec{j})}$, where γ' is the assignment extending γ, such that $\gamma'(x_i) = d_i$ for all $i = 1, \cdots, t$; a similar convention is used for $(S, \gamma, d) \models \psi_i$.

Since T is associated in a unique way with S, γ and Δ whenever it is defined, i.e., whenever $(S, \gamma) \models \varphi$, we can use the functional notation $def_\Delta(S, \gamma)$ for T. The *transduction defined by* Δ is the binary relation:

$$\mathcal{D}_\Delta := \{(S, T) \mid T = def_\Delta(S, \gamma) \text{ for some } W\text{-assignment } \gamma \text{ in } S\}.$$

Hence $\mathcal{D}_\Delta \subseteq \mathcal{STR}(R) \times \mathcal{STR}(Q)$. A transduction $f \subseteq \mathcal{STR}(R) \times \mathcal{STR}(Q)$ is an *MS transduction* if it is equal, up to isomorphism of structures, to $\mathcal{D}_\Delta$ for some (Q, R)-definition scheme Δ.

An MS-transduction is defined as a binary relation. Hence it can be seen as a "nondeterministic" partial function associating with an R-structure one or more Q-structures. However, it is not really nondeterministic because the different outputs come from different choices of parameters. In the case where $W = \emptyset$ it defines a partial function. It may also happen that different choices of parameters yield isomorphic output structures. This is the case in the example of edge contraction detailed below. We will refer to the integer k by saying that Δ and $\mathcal{D}_\Delta$ are *k-copying*; if $k = 1$ we will say that they are *noncopying*. A noncopying definition scheme can be written more simply: $\Delta = (\varphi, \psi, (\theta_q)_{q \in Q})$.

Example. *Edge contraction.*

We consider a graph G with two types of edges, the ordinary edges and the ε−edges. It is represented by a structure $\langle V_G, edg_G, \varepsilon - edg_G \rangle$ where the binary relation $\varepsilon - edg_G$ represents the ε−edges. We want to define the graph H obtained from G by the contraction of all ε−edges.

It is formally defined as $\langle V_H, edg_H \rangle$ where $V_H = V_G / \sim$, $\sim$ is the equivalence relation such that $x \sim y$ if and only if x and y are linked by an undirected path made of $\varepsilon-$edges, and $edg_H([u],[v])$ holds if and only if $x \in [u]$, $y \in [v]$ for some (x,y) in edg_G ($[u]$ denotes the equivalence class of u). The MS formula $\xi(x,y)$ defined as:

$$\forall X[(x \in X \wedge \forall\, u,v\{u \in X \wedge (\varepsilon - edg(u,v) \vee \varepsilon - edg(v,u)) \Longrightarrow v \in X\}) \Longrightarrow y \in X]$$

expresses $x \sim y$. For defining V_H we must select a set containing one and only one vertex of each equivalence class. This can be done with a set variable Y that will be a parameter of the MS transduction, satisfying the formula $\varphi(Y)$ saying that for every x there is a unique y such that $y \in Y \wedge \xi(x,y)$ holds. Edge contraction can thus be defined by the transduction with noncopying definition scheme $\Delta = (\varphi, \psi, \theta_{edg})$ where $\psi(Y,x)$ is $x \in Y$ and:

$\theta_{edg}(Y,x,y)$ is the formula $x \in Y \wedge y \in Y \wedge \exists u,v.[edg(u,v) \wedge \xi(x,u) \wedge \xi(y,v)]$.

Remark that the structures associated with all values of the parameter Y satisfying $\varphi(Y)$ are isomorphic. They only differ regarding the concrete subsets Y of V_G used as sets of vertices of H. □

The Fundamental Property of MS Transductions

The following proposition says that if $T = def_\Delta(S,\gamma)$, then the monadic second-order properties of T can be expressed as monadic second-order properties of (S,γ). The usefulness of definable transductions is based on this proposition.

Let $\Delta = (\varphi, \psi_1, \cdots, \psi_k, (\theta_w)_{w \in Q^*k})$ be a (Q,R)-definition scheme, written with a set of parameters W. Let V be a set of set variables disjoint from W. For every variable X in V, for every $i = 1, \cdots, k$, we let X_i be a new variable. We let $V' := \{X_i / X \in V,\ i = 1, \cdots, k\}$. Let S be a structure in $\mathcal{STR}(R)$ with domain D. For every mapping $\eta : V' \longrightarrow \mathcal{P}(D)$, we let $\eta^k : V \longrightarrow \mathcal{P}(D \times [k])$ be defined by $\eta^k(X) = \eta(X_1) \times \{1\} \cup \cdots \cup \eta(X_k) \times \{k\}$. With this notation we can state:

Proposition 1. *For every formula β in $MS(Q,V)$ one can construct a formula $\beta^\#$ in $MS(R, V' \cup W)$ such that, for every S in $\mathcal{STR}(R)$, for every assignment $\gamma : W \longrightarrow \mathcal{P}(D)$, for every assignment $\eta : V' \longrightarrow \mathcal{P}(D)$ we have :*

$(S, \eta \cup \gamma) \models \beta^\#$ if and only if :
$def_\Delta(S,\gamma)$ is defined, η^k is a V-assignment in $def_\Delta(S,\gamma)$,
and $(def_\Delta(S,\gamma), \eta^k) \models \beta$.

We call $\beta^\#$ the *backwards translation* of β relative to the transduction def_Δ. The composition of two transductions is defined as the composition of the corresponding binary relations (equivalently, multivalued functions). From Proposition 1 we get:

Proposition 2. 1) *The composition of two MS transductions is an MS transduction.*

2) *The inverse image of an MS-definable class of structures under an MS transduction is MS-definable.*

The mapping from $I(G)$ to the set of spanning forests of a graph G or to the set of its subgraphs, or of its minors (represented by their incidence structures) are MS transductions. The mapping from $S(G)$ to the set of its *induced* subgraphs for a simple graph G is an MS transduction (a subgraph H of G is represented by $S(H)$).

4 Graph Decompositions Defined by MS Transductions

Of particular interest are *graph decompositions*. They are useful for the construction of efficient algorithms, and also because they give structural descriptions of the considered graphs. The general results of [19, 20] define canonical decompositions which include as particular instances the *modular decomposition* and the *Tutte decomposition of a graph in 3-connected components.* (Tree-decompositions, rank decompositions [27], optimal decompositions for clique-width are not canonical.)

We show that these canonical decompositions can be defined by monadic second-order formulas "inside" the considered graphs. In our language the mapping from a (linearly ordered) graph to its decomposition of this type is an MS transduction. We first review informally the *modular decomposition.* (See [25, 26] for thorough studies of this notion).

Modular decomposition

The modular decomposition of a graph is a canonical expression of this graph in terms of (nested) substitutions. Let G and H be two simple loop-free undirected graphs with disjoint sets of vertices. The *substitution* of H for a vertex u of G is the graph $G[H/u]$ defined as the union of G and H, *minus* the vertex u and the incident edges, *plus* edges between w and all vertices of H for every edge $w - u$ in G. Substitutions can be done *in parallel* (disjoint graphs for distinct vertices) giving $G[H_1/u_1, ..., H_n/u_n]$ and *nested* in expressions like $G[H[L/u]/v]$.

Let K be a graph and M a set of vertices. The graph K can be expressed as $G[H/u]$ with $M = V_H$ iff M is a *module* of K, that is, a set of vertices such that every vertex not in M is linked either to no vertex of M or to all of them.

A module is *strong* if does not overlap any module. (Two sets *overlap* if they have a nonempty intersection and none is a subset of the other.) The strong modules of a graph G form a tree for inclusion, called its *modular decomposition.* This tree is the syntactic tree of a canonical expression of G in terms of substitutions. (The modular decomposition is called *substitution decomposition* in some works). This tree can be enrich by labels and additional edges, and this gives the *graph representation of the modular decomposition.* From it the graph G can be reconstructed.

The mapping from a graph G given with a linear order $\preccurlyeq$ of its vertices is an MS transduction. We sketch the construction which is given in [6, 10, 14]. There exists an MS formula $\varphi(X)$ that is valid in the given graph iff X is a strong modules. To build the tree of strong modules, we specify in each strong module that is not a singleton a *representing* vertex x that will be the corresponding non-leaf node of the modular decomposition. This specification is done by an MS formula $\psi(x, X)$. Furthermore we need that distinct strong modules are represented by distinct vertices. The order $\preccurlyeq$ on vertices is here useful. For a set $Y \subseteq V_G$, we let $fl(Y)$ be the $\preccurlyeq$-smallest vertex of Y. Each strong module M contains a unique maximal proper strong module N such that $fl(N)$ is minimal. (Maximal is understood for set inclusion). We take $fl(M - N)$ as the representative vertex of M. Two non-singleton strong modules are represented by different vertices. (This would not be the case if we would represent M by $fl(M)$). We can thus construct the tree of strong modules by an MS transduction as a tree with set of nodes $V_G \times \{1\} \cup R_G \times \{2\}$, where R_G is the set of vertices which represent non-singleton strong module. The remaining parts of the construction are routine. Note that this construction uses a linear order on vertices, but for any two linear orders, the corresponding outputs are isomorphic relational structures.

Split decomposition:
This decomposition due to Cunningham ([20]) generalizes the modular decomposition. We review it briefly. We only consider simple loop-free directed graphs. An undirected edge can be considered as a pair of opposite directed edges. Hence, this definition also applies to undirected graphs.

A *split* of a strongly connected graph G is a bipartition $\{A, B\}$ of V_G such that A and B have at least 2 elements and $E_G = E_{G[A]} \cup E_{G[B]} \cup (A_1 \times B_1) \cup (B_2 \times A_2)$ for some $A_i \subseteq A$, $B_i \subseteq B$. If $\{A, B\}$ is a split, then G can be expressed as the union of $G[A]$ and $G[B]$ linked by two directed, complete bipartite graphs. (Since G is strongly connected the set $(A_1 \times B_1) \cup (B_2 \times A_2)$ is not empty). If G is undirected, then strong connectivity is just connectivity.

The inverse of splitting is the *join operation*, defined as follows.

Let H and K be two disjoint graphs with distinguished vertices h in H and k in K. We define $H \boxtimes_{(h,k)} K$ as the graph with set of vertices $V_H \cup V_K - \{h, k\}$ and edges $x \longrightarrow y$ such that, either $x \longrightarrow y$ is an edge of H, or an edge of K, or we have $x \longrightarrow h$ in H and $k \longrightarrow y$ in K, or we have $h \longrightarrow y$ in H and $x \longrightarrow k$ in K.

If $\{A, B\}$ is a split, then $G = H \boxtimes_{(h,k)} K$ where H is $G[A]$ augmented with a new *marker* vertex h and edges $x \longrightarrow h$ whenever there are in G edges from x to some u in B, and edges $h \longrightarrow x$ whenever there are edges from some u in B to x. The graph K is defined similarly from $G[B]$, with a new vertex k. The graphs H and K have at least 3 vertices and strictly less vertices than G.

A *decomposition* of a strongly connected graph G is defined as follows: $\{G\}$ is the only decomposition of size 1; if $\{G_1, ..., G_n\}$ is a decomposition of size n, and $G_n = H \boxtimes_{(h,k)} K$, then $\{G_1, ..., G_{n-1}, H, K\}$ is a decomposition of G of size $n + 1$. The graphs G_i are called the *components* of the decomposition. The graph G can be reconstructed without ambiguity provided the marker vertices and their matchings are specified. The components of a decomposition form an unrooted tree for the "matching" relation. By decomposing iteratively a strongly connected graph and by using only *good splits* defined as those that do not overlap any other, one gets *the Unique (Canonical) Split Decomposition.* The restriction to good splits generalizes the restriction to strong modules. These restrictions are important for having canonical decompositions. Figure 4.1 shows an undirected graph G and Figure 4.2 its split decomposition.

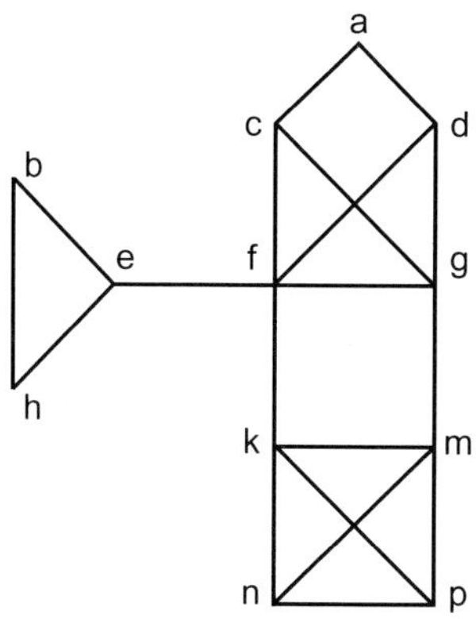

Figure 4.1: Graph G

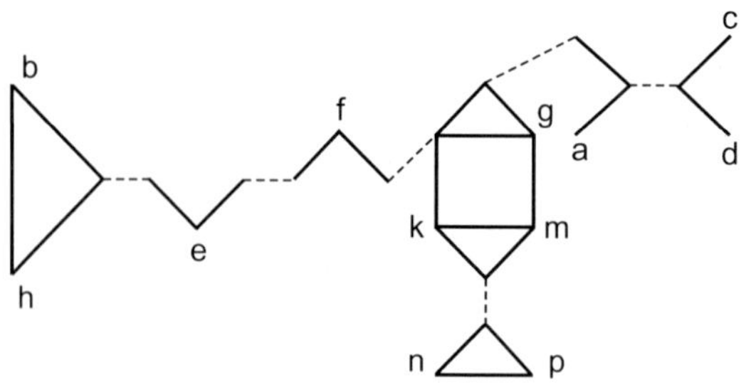

Figure 4.2: The split decomposition of G.

Theorem 1 ([6, 10]). *There are MS transductions that associate with a linearly ordered graph the graph representations of its modular and split decompositions. Up to isomorphism, the constructed relational structures do not depend on the chosen ordering of the vertices.*

5 Circle Graphs

A *circle graph* is the intersection graph of a set of chords of a circle. An equivalent combinatorial characterization can be given in terms of words *where each letter has two occurrences.* If a letter represents a chord, a set of chords is a word corresponding to the sequences of extremities of chords read around the circle, and the chords represented by a and b intersect iff the word can be written $aubvawbx$ for some words u, v, w, x.

Circle graphs which are *prime*, i.e., indecomposable for the *split decomposition* have *unique representations* as words (or as sets of chords), where *unique* is meant up to some obvious transformations. The main results are the following ones:

1. If two words define the same connected circle graph and have the same *subword of first occurrences of letters*, then they are equal.

2. If a circle graph is prime, the unique word representing it can be constructed by formulas of MS logic.

3. If a circle graph is not prime, then all its representations by double occurrence words can be defined from it and the linear orders of its set of vertices, by a fixed MS formula.

Result 2 uses MS formulas written with set predicates of the form $Even(X)$ expressing that a set X has even cardinality. These formulas will be called C_2MS formulas. They are needed because the proof uses a result by Courcelle and Oum [17] showing that the characterization of circle graphs by the three *excluded vertex-minors* given by Bouchet [3] is expressible by C_2MS formulas. *Even* helps to express computations in GF(2), and these computations occur because the notion of a vertex-minor is handled through that an *isotropic system*, which is a vector space over GF(2).

Result 3 makes a crucial use of Theorem 1 for *split decompositions.* Every circle graph has a canonical split decomposition the components of which are prime circle graphs, cliques and stars. For a circle graph, the prime components have constructible word representations by Result 2. The other components have word representations constructible by means of the linear order. By varying the linear order one gets all word representations (the orders corresponding to the first occurrence words of the word representations of the given graph actually suffice). The results of this section are proved in [11].

Concerning the unique representability of prime circle graphs, a quite similar result holds for *comparability graphs* (see [25]). Those which are undecomposable for the *modular decomposition* have *unique transitive orientations* (actually they have two, one and its reversal) and it is possible to construct these orientations by MS formulas by a proof similar to the one used for circle graphs. This proof uses a characterization of comparability graphs by forbidden induced subgraphs. This characterization is expressible in MS logic although there are infinitely many forbidden subgraphs. See [9].

Circle graphs.
Let A be a countable set called the *set of letters*. We let W be the set *double occurrence words i.e.,* of finite nonempty words over A having two occurrences or no occurrence of each letter. We let $V(w)$ be the set of letters occuring in w and $G(w)$ be the graph with set of vertices $V(w)$ and an undirected edge between a and b iff $w = u_1au_2bu_3au_4bu_5$ or $w = u_1bu_2au_3bu_4au_5$ for some $u_1, ..., u_5$ in A^*. The graphs $G(w)$ are also called *circle graphs*. It is clear that $G(w) = G(w')$ if $w' = \widetilde{w}$ (the *mirror image* of w) or if w and w' are *conjugate*, denoted by $w \sim w'$, which means $w = uv$ and $w' = vu$ for some u, v in A^*; w and w' are *equivalent*, denoted by $w \equiv w'$, iff either $w \sim w'$ or $\widetilde{w} \sim w'$. Two equivalent words represent the same circle graph. Figure 5.1 shows a circle graph G and its chord representation defined by the word : $axbcuyvbycauxv$.

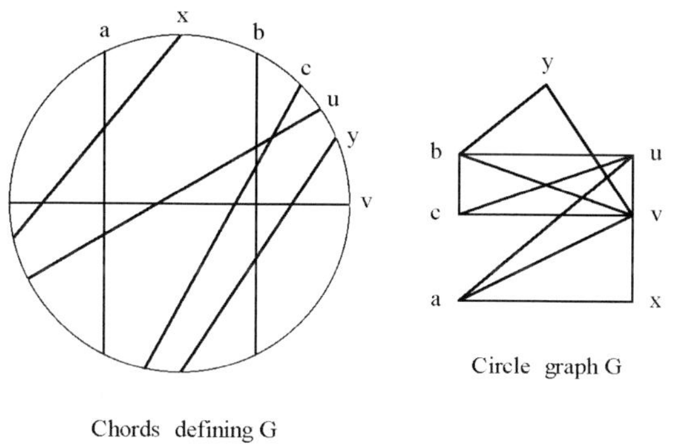

Figure 5.1: A circle graph and a chord representation of it.

A circle graph G is *uniquely representable* if $G = G(w) = G(w')$ implies $w \equiv w'$. The circle graphs with at most 3 vertices are uniquely representable, so are C_4, P_4, the graph K_4 minus one edge. The graphs K_4, S_3 are not. To take an example the star S_3 with center a is represented by the two inequivalent words $abcdadcb$ and $acbdadbc$.

Theorem 2 ([11]). *A circle graph with at least 5 vertices is uniquely representable iff it is prime for the split decomposition.*

Proof. "if" is proved in [2, 24]. "Only if" is claimed in [24] but not actually proved. □

Theorem 3 ([11]). *There exist C_2MS formulas that associate with every prime circle graph G a word $w \in W$ such that $G(w) = G$.*

A double occurrence word can be defined for logical treatment as a directed circuit with a binary relation representing pairs of positions bearing the same letter. Whether this letter is a or b does not matter.

Eulerian trails of 4-regular graphs.

We need a lemma concerning the Eulerian trails of 4-regular graphs without loops and multiple edges. We prove that these trails can be encoded by tuples of sets, hence defined and used by MS formulas.

Lemma. *There exist MS formulas that associate with every connected 4-regular undirected graph H the structures* $< V_H, edg_H, edg_{G(E)} >$ *for all Eulerian trails E of H, where* $edg_{G(E)}$ *is the edge relation of the corresponding circle graph G(E).*

Proof sketch of Theorem 3. Let $w \in W$, $a, b \in V(w)$, $a \neq b$. We say that a and b are *neighbours in* w if $w \equiv abw'$ for some w' in A^*. This means that on the representation of $G(w)$ by intersecting chords, chords a and b have two consecutive ends on the circle. If G is prime with at least 5 vertices, this notion depends only on G, and not on the word w representing it. Each letter occurring in w has *four different neighbours.* We let $N(G)$ be the *graph of neighbourhood*, with set of vertices $V(w)$ and an edge $a - b$ iff a and b are neighbours in G. This graph is 4-regular. We can prove that its adjacency relation is definable by a C_2MS formula over the given prime circle graph G, and that w can be constructed from $N(G)$. Figure 5.2 shows with solid lines the graph $N(G)$ for the graph G associated with the word : $axbcuyvbycauxv$. The dotted lines around the vertices show the Eulerian trail which corresponds to the chord representation of G, i.e., to the circular sequence of the ends of the chords corresponding to letters.

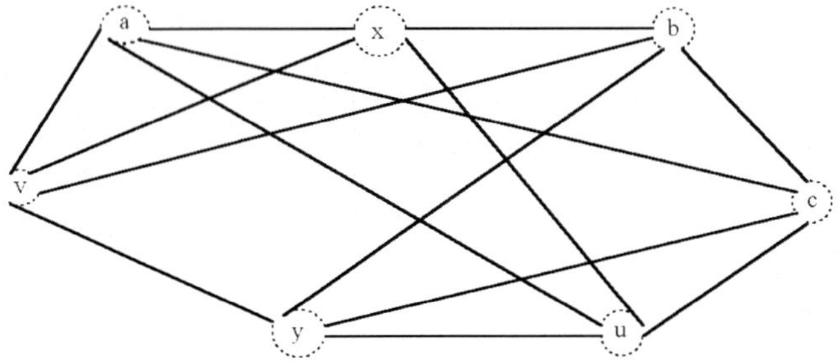

Figure 5.2: The neighbourhood graph *N(G)*.

For $a, b \in V_G(\subset A), a \neq b$, we let $G(a, b; u, v)$ be the graph G augmented with the path $a - u - v - b$ where $u, v \in A - V_G$.

Claim 1. $G(a, b; u, v)$ is a circle graph iff a, b are neighbours in G.

Claim 2. That a and b are neighbours in G is expressible by a C_2MS formula.

Hence the mapping associating $N(G)$ with a prime circle graph G is a C_2MS transduction. This claim uses the C_2MS characterization of circle graphs given in [17]. That a given graph G is a prime circle graph with at least 5 vertices is a C_2MS property. Assuming this satisfied, one can build from G and by C_2MS formulas (Claim 2) the 4-regular graph $N(G)$. This graph is connected and has an Eulerian trail F such that $G(F) = G$. The Eulerian trails of $N(G)$ are defined by tuples of sets of vertices (by the Lemma) and an MS formula can select the one defining F. □

First occurrence words:

For every word w in A^* we denote by $F(w)$ the subword of w consisting of the first occurrence of each letter. For an example, $F(abbdacdcefef) = abdcef$.

Theorem 4 ([11]). *If w, w' are double occurrence words such that $G(w) = G(w')$ is connected and $F(w) = F(w')$, then $w = w'$.*

Proof. The proof is by induction on the length of w. □

Theorem 5 ([11]). *There exist MS formulas that associate with $(G, \preccurlyeq)$ where G is a circle graph and $\preccurlyeq$ a linear order on V_G, the unique double occurrence word w representing it such that $F(w) = (V_G, \prec)$, provided such a word does exist.*

Proof sketch. On a structure given with a linear order, the set predicate $Even(X)$ can be expressed by an MS formula (see [4]). Hence on these structures, every C_2MS formula can be translated into an equivalent MS formula. In particular, an MS formula can check that the given graph is a circle graph. By Theorem 1 (Theorem 4.21 of [10]), one can construct from $(G, \preccurlyeq)$ its split decomposition $Split(G)$ by an MS transduction. For those components of $Split(G)$ which are prime, an MS formula can build double occurrence words representing them by Theorem 3. For the other components, which are isomorphic to stars and to cliques, the linear order $\preccurlyeq$ makes it possible to define by MS formulas representing double occurrence words. One obtains a tree of relational structures $S_1, ..., S_k$ representing double occurrence words for the k components of $Split(G)$. From this tree one can produce a double occurrence word for G. It remains to check that it matches the first occurrence word defined by $F(w) = (V_G, \prec)$. □

References

[1] G. Bagan, MSO queries on tree decomposable structures are computable with linear delay, *Computer Science Logic 2006, Lec. Notes Comput. Sci.*, **4207** (2006). pp. 167–181.

[2] A. Bouchet, Reducing prime graphs and recognizing circle graphs, *Combinatorica*, **7** (1987), pp. 243–254.

[3] A. Bouchet, Circle graph obstructions, *J. Combinatorial Theory B*, **60** (1994), pp. 107–144.

[4] B. Courcelle, The expression of graph properties and graph transformations in monadic second-order logic. In G. Rozenberg, editor, *Handbook of graph grammars and computing by graph transformations, Vol. 1: Foundations*, pages 313–400. World Scientific (1997).

[5] B. Courcelle, The Monadic Second-Order Logic of Graphs VIII: Orientations, *Ann. Pure Appl. Logic*, **72** (1995), pp. 103–143.

[6] B. Courcelle, The Monadic Second-Order Logic of Graphs X: Linear Orderings, *Theor. Comput. Sci.*, **160** (1996), pp. 87–143.

[7] B. Courcelle, The Monadic Second-Order Logic of Graphs XI: Hierarchical Decompositions of Connected Graphs, *Theor. Comput. Sci.*, **224** (1999), pp. 35–58.

[8] B. Courcelle, The monadic second-order logic of graphs XII: Planar Graphs and Planar Maps, *Theor. Comput. Sci.*, **237** (2000), pp. 1–32.

[9] B. Courcelle, The monadic second-order logic of graphs XV : On a Conjecture by D. Seese, *J. Applied Logic*, **4** (2006), pp. 79–114.

[10] B. Courcelle, The monadic second-order logic of graphs XVI : Canonical graph decompositions, *Logical Methods in Computer Science*, **2** (2006), pp. 1–46.

[11] B. Courcelle, Circle graphs and Monadic Second-order logic, *J. of Applied Logic*, in press.

[12] B. Courcelle, Diagonal walks in plane graphs and local duality, March 2005, http://www.labri.fr/perso/courcell/ActSci.html

[13] B. Courcelle, Linear delay enumeration and monadic second-order logic, March 2006, http://www.labri.fr/perso/courcell/ActSci.html, to appear in Discrete Applied Mathematics

[14] B. Courcelle and C. Delhommé, The modular decomposition of countable graphs: Constructions in Monadic Second-Order Logic, *Theoretical Computer Science*, **394** (2008), pp. 1–38.

[15] B. Courcelle, J. Makowsky and U. Rotics, On the fixed parameter complexity of graph enumeration problems definable in monadic second-order logic, *Discrete Applied Mathematics*, **108** (2001), pp. 23–52.

[16] B. Courcelle and M. Mosbah, Monadic Second-Order Evaluations on Tree-Decomposable Graphs, *Theor. Comput. Sci.*, **109** (1993), pp. 49–82.

[17] B. Courcelle and S. Oum, Vertex-minors, monadic second-order logic and a conjecture by Seese, *J. Comb. Theor. B*, **97** (2007), pp. 91–126.

[18] B. Courcelle and R. Vanicat, Query efficient implementation of graphs of bounded clique-width, *Discrete Applied Mathematics*, **131** (2003), pp. 129–150.

[19] W. Cunnigham and J. Edmonds, A combinatorial decomposition theory, *Can. J. of Math.*, **XXXII** (1980), pp. 734–765.

[20] W. Cunnigham, Decomposition of directed graphs, *SIAM J. Algebraic Discrete Methods*, **3** (1982), pp. 214–228.

[21] R. Downey and M. Fellows, Parameterized Complexity, Springer-Verlag (1999).

[22] J. Flum, M. Frick and M. Grohe, Query evaluation via tree-decompositions. *J. ACM*, **49** (2002), pp. 716–752.

[23] J. Flum and M. Grohe, Parameterized Complexity Theory, Springer (2006).

[24] C. Gabor, W. Hsu and K. Supowit, Recognizing circle graphs in polynomial time, *J. of ACM*, **36** (1989), pp. 435–473.

[25] D. Kelly, Comparability graphs, in *Graphs and order*, I. Rival ed., D. Reidel Pub. Co., 1985, pp. 3–40.

[26] R. Möhring and F. Radermacher, Substitution decomposition for discrete structures and connections with combinatorial optimization, *Annals of Discrete Maths*, **19** (1984), pp. 257–356.

[27] S. Oum, Approximating Rank-Width and Clique-Width Quickly, *Proc. WG 2005, Lec. Notes Comput. Sci.*, **3787** (2005), pp. 49–58.

Proc. of ICDM 2006, Pages 217–226

Restrained Domination in Graphs

JOHANNES H. HATTINGH
Department of Mathematics and Statistics,
Georgia State University, Atlanta, GA 30303 U.S.A.

Abstract. In this paper, we survey results on a variation of standard domination, namely restrained domination.

1 Introduction

Graph theory terminology not presented here can be found in [3]. Let $G = (V, E)$ be a graph with n vertices. A set $S \subseteq V$ is a *dominating set* if every vertex not in S is adjacent to a vertex in S. The *domination number* of G, denoted by $\gamma(G)$, is the smallest cardinality of a dominating set.

Domination may be viewed in various metatheoretical ways, one of which we now describe. Consider an arbitrary set S and its complement $V - S$ in a graph $G = (V, E)$. Consider also the subgraphs $\langle S\rangle$ and $\langle V - S\rangle$ and the bipartite subgraph $\langle S, V - S\rangle$ generated by all edges between S and $V - S$. Note that $G = \langle S\rangle \cup \langle V - S\rangle \cup \langle S, V - S\rangle$. Consider four values:

- the minimum degree of vertices $v \in \langle S\rangle$, denoted by δ_S,
- the minimum degree of vertices $v \in S$ in $\langle S, V - S\rangle$, denoted by $\delta_{S,V-S}$,
- the minimum degree of vertices $v \in V-S$ in $\langle V-S, S\rangle$, denoted by $\delta_{V-S,S}$, and
- the minimum degree of vertices $v \in \langle V - S\rangle$, denoted by δ_{V-S}.

Many dominating concepts can be defined in terms of various combinations of these four values. In what follows, X will denote that the particular value does not matter.

δ_S	$\delta_{S,V-S}$	$\delta_{V-S,S}$	δ_{V-S}	Type of domination
X	X	≥ 1	X	S is a dominating set
$= 0$	X	≥ 1	X	S is an independent dominating set
≥ 1	X	≥ 1	X	S is a total dominating set
$= 1$	X	≥ 1	X	S is a paired dominating set
X	$= 1$	$= 1$	X	there is a perfect matching between S and $V - S$
X	X	$= 1$	X	S is a perfect dominating set
X	X	≥ 1	≥ 1	S is a restrained dominating set
≥ 1	X	≥ 1	≥ 1	S is a total restrained dominating set
X	X	≤ 1	X	S is a nearly perfect dominating set
X	≥ 1	≥ 1	X	both S and $V - S$ are dominating sets
X	X	$\geq n$	X	S is an n-dominating set
X	$\leq k$	≥ 1	X	S is a capacity k dominating set

Here, we continue to study the case when $\delta_{V-S,S} \geq 1$ and $\delta_{V-S} \geq 1$. A set satisfying $\delta_{V-S,S} \geq 1$ and $\delta_{V-S} \geq 1$ is called a *restrained dominating set*. The discussion above is based on Telle and Proskurowski [20].

Equivalently, a set $S \subseteq V$ is a restrained dominating set if every vertex in $V - S$ is adjacent to a vertex in S and another vertex in $V - S$. Note that every graph has a restrained dominating set, since $S = V$ is such a set. Let $\gamma_r(G)$ denote the size of a smallest restrained dominating set. We will call a set S a *γ_r-set* if S is a restrained dominating set of cardinality $\gamma_r(G)$, and use similar notation for other parameters.

It is clear from their definition that $\gamma(G) \leq \gamma_r(G)$ for any graph G.

2 Bounds

In this section, we survey results that determine exact values for γ_r for certain classes of graphs, mention best possible upper and lower bounds for γ_r, and review results that characterize graphs achieving these bounds.

Let K_n, C_n and P_n denote, respectively, the complete graph, the cycle and the path of order n. Also, let $K_{n_1,\ldots,n_t}$ denote the complete multipartite graph with vertex set $S_1 \cup \cdots \cup S_t$ where $|S_i| = n_i$ for $1 \leq i \leq t$. We call $K_{1,n-1}$ a *star*. A *subdivision* of an edge uv is obtained by introducing a new vertex w and replacing the edge uv with the edges uw and wv. A *spider* is a tree obtained from $K_{1,r}$, $r \geq 1$, by subdividing all of its edges. A *wounded spider* is a tree obtained from $K_{1,r}$, $r \geq 1$, by subdividing at most $r - 1$ of its edges. Thus, the star, $K_{1,r}$, is also a wounded spider. In a tree, a *stem* is a vertex adjacent to a leaf (a vertex of degree one).

The following results were obtained by Domke, Hattingh, Hedetniemi, Laskar and Markus [6].

Proposition 2.1. *If $n \neq 2$ is a positive integer, then $\gamma_r(K_n) = 1$.*

Proposition 2.2. *If $n \geq 2$ is an integer, then $\gamma_r(K_{1,n-1}) = n$.*

Proposition 2.3. *If n_1 and n_2 are integers such that $\min\{n_1, n_2\} \geq 2$, then $\gamma_r(K_{n_1,n_2}) = 2$.* .

Proposition 2.4. *If $t \geq 3$ is an integer, then*

$$\gamma_r(K_{n_1,\ldots,n_t}) = \begin{cases} 1 & \text{if } \min\{n_1,\ldots,n_t\} = 1 \\ 2 & \text{otherwise} \end{cases} .$$

Proposition 2.5. *If $n \geq 1$ is an integer, then $\gamma_r(P_n) = n - 2\lfloor \frac{n-1}{3} \rfloor$.*

Proposition 2.6. *If $n \geq 3$, then $\gamma_r(C_n) = n - 2\lfloor \frac{n}{3} \rfloor$.*

It is clear that $\gamma_r(G) \leq n$ for any graph G of order n. The following result shows that the star $K_{1,n-1}$ is the only connected graph G of order n for which $\gamma_r(G) = n$.

Proposition 2.7. *Let G be a connected graph of order n. Then $\gamma_r(G) = n$ if and only if G is a star.*

Theorem 2.8. *If T is a tree of order $n \geq 3$, then $\gamma_r(T) = n - 2$ if and only if T is obtained from P_4, P_5 or P_6 by adding zero or more leaves to the stems of the path.*

Theorem 2.9. *Let G be a connected graph of order n containing a cycle. Then $\gamma_r(G) = n - 2$ if and only if G is C_4 or C_5 or G can be obtained from C_3 by attaching zero or more leaves to at most two of the vertices of the cycle.*

Corollary 2.10. *If G is a graph of order n, then $\gamma_r(G) = n$ if and only if G is a disjoint union of stars and isolated vertices. Furthermore, $\gamma_r(G) = n-2$ if and only if exactly one of the components of G is isomorphic to a graph given in Theorem 2.8 or 2.9 and every other component is a star or P_1.*

Theorem 2.11. *If T is a tree of order $n \geq 3$, then $\gamma_r(T) \geq \Delta(T)$. Furthermore, $\gamma_r(T) = \Delta(T)$ if and only if T is a wounded spider which is not a star.*

Proposition 2.12. *If G is a graph, then $\gamma_r(G) = 1$ if and only if $G \cong K_1 + H$ where H is a graph with no isolated vertices.*

McCuaig and Shepherd [17] have shown that if a connected graph G of order n has minimum degree at least 2 and is not one of seven exceptional graphs, then $\gamma(G) \leq \frac{2}{5}n$.

McCuaig and Shepherd also characterized those connected graphs of order n with minimum degree at least two satisfying $\gamma(G) \geq 2n/5$ which are edge-minimal.

Let $\mathcal{B}$ be the collection of graphs shown in Figure 2.1.

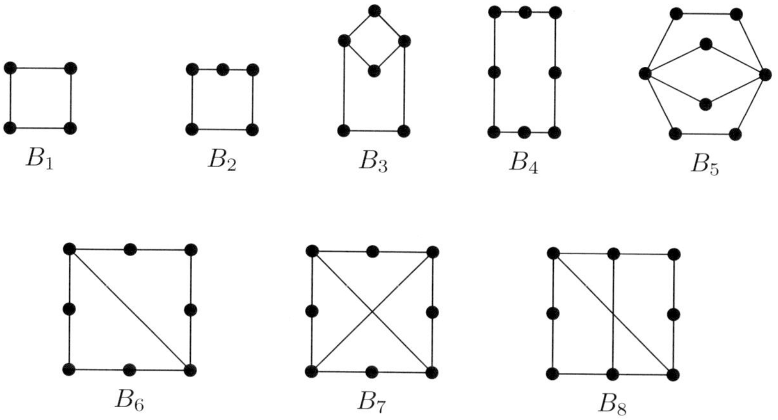

Figure 2.1: The collection $\mathcal{B}$ of graphs.

The following result was established in [8].

Theorem 2.13 (Domke, Hattingh, Henning and Markus). *Let G be a connected graph of order $n \geq 3$ with $\delta(G) \geq 2$. If $G \notin \mathcal{B}$, then*

$$\gamma_r(G) \leq (n-1)/2.$$

Henning [14] characterized those connected graphs of order n with minimum degree at least two satisfying $\gamma(G) \geq (n-1)/2$ which are edge-minimal.

The following results were established by Dankelmann, Day, Hattingh, Henning, Markus and Swart in [4].

Theorem 2.14. *If G is a graph of order n with $\delta(G) \geq 2$, then $\gamma_r(G) \leq n - \Delta(G)$.*

Theorem 2.15. *Let G be a connected graph of order n with $\delta(G) \geq 2$ that is both triangle-free and C_5-free. Then, $\gamma_r(G) = n - \Delta(G)$ if and only if G is obtained from a disjoint union of stars, each of order at least three, by adding a new vertex and joining it to all the leaves.*

In order to characterize the connected triangle-free graphs that achieve equality in Theorem 2.14, a family $\mathcal{F}$ of graphs is introduced as follows. Let G be obtained from a disjoint union of $s \geq 1$ stars, each of order at least three, and $t \geq 0$ copies of K_2 by adding a new vertex v and joining it to all the leaves of the stars of order at least three and to one leaf of each of the K_2s. Let $B = V(G)\setminus N[v]$ and let $D \subset B$ be the set of leaves of G. Let F be the graph obtained from G by adding any number of edges among vertices of B in such a way that in the graph F we have that (i) D is an independent set, (ii) $\deg(u, B) = 1$ for every $u \in D$, and (iii) $\deg(u, B) \leq \deg(u, N(v))$ for every $u \in B \setminus D$. Let $\mathcal{F}$ be the family of all such graphs F.

Theorem 2.16. *Let G be a connected triangle-free graph of order n with $\delta(G) \geq 2$. Then, $\gamma_r(G) = n - \Delta(G)$ if and only if $G = C_5$ or $G \in \mathcal{F}$.*

3 Trees for with Equal Domination and Restrained Domination Numbers

A graph G is called γ-excellent if every vertex of G belongs to some $\gamma(G)$-set. Results on γ-excellent trees can be found in [2, 9, 18] and elsewhere.

A subset $S \subseteq V$ is a *packing* in G if the vertices of S are pairwise at distance at least three apart in G. The *packing number* $\rho(G)$ is the maximum cardinality of a packing in G.

We define a *labeling* of a tree T as a function $S\colon V(T) \to \{A, B\}$. The label of a vertex v is also called its *status*, denoted $\mathrm{sta}(v)$.

By a *labeled* K_1 we shall mean a K_1 whose vertex is labeled with status B. Let $\mathcal{T}$ be the family of trees that can be labeled so that the resulting family of labeled trees contain a labeled K_1 and is closed under the two operations **O1** and **O2** listed below, which extend the tree T by attaching a tree to the vertex $v \in V(T)$, called the *attacher*.

- **O1.** Join a vertex v of status A to a leaf of the path xy; set $\mathrm{sta}(x) = A$ and $\mathrm{sta}(y) = B$.

O1:

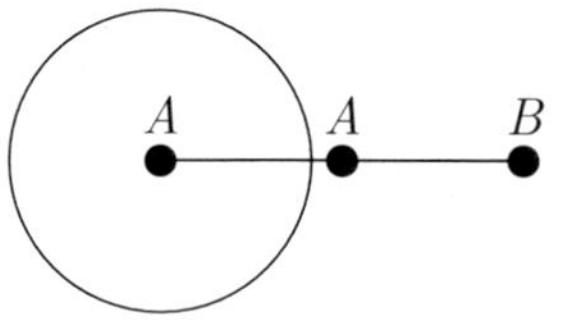

- **O2.** Join a vertex v of status B to a leaf of the path xyz; set $\mathrm{sta}(x) = \mathrm{sta}(y) = A$ and $\mathrm{sta}(z) = B$.

O2:

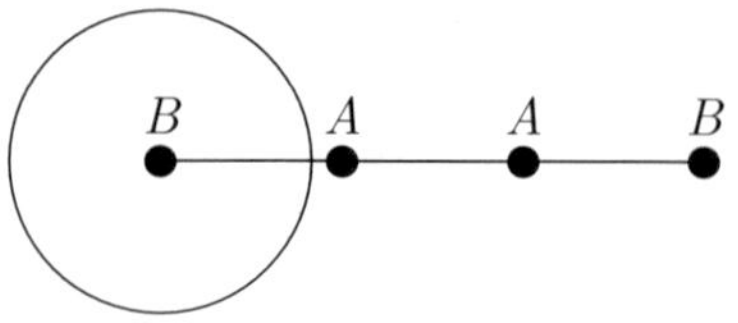

The following result is due to Dankelmann, Hattingh, Henning and Swart [5].

Theorem 3.1. *Let T be a tree. Then the following statements are equivalent:*
(i) $T \in \mathcal{T}$;
(ii) *T has a unique $\rho(T)$-set and this set is a dominating set of T;*
(iii) $\gamma(T) = \gamma_r(T)$;
(iv) *T is γ-excellent and $T \neq K_2$.*

4 Bounds on Trees

In this section, we survey further results which involve bounds on $\gamma_r(T)$ for a tree T.

The following result is due to Dankelmann, Hattingh, Henning and Swart [5].

Theorem 4.1. *If T is a tree of order n with ℓ leaves, then $\gamma_r(T) \leq (n + \ell + 1)/2$ with equality if and only if T is a nontrivial star.*

The following result is due to Domke, Hattingh, Henning and Markus [7].

Theorem 4.2. *If T is a tree of order $n \geq 1$, then $\gamma_r(T) \geq \lceil (n+2)/3 \rceil$.*

We denote the set of leaves of a tree T by $L(T)$. For $v \in V(T)$ and $\ell \in L(T)$, the path $vx_1 \ldots x_k\ell$ is called a *$v - L$ endpath* if $\deg x_i = 2$ for each i. If the vertex v need not be specified, a $v - L$ path is also called an *endpath.*

Let $\mathcal{T}$ be the class of all trees T of order n such that $\gamma_r(T) = \lceil \frac{n+2}{3} \rceil$. The trees in $\mathcal{T}$ were constructively characterized by Hattingh and Plummer in [13]. In order to state the characterization, define three simple operations on a tree T.

O1. Join a leaf or a stem, or a vertex v or x of T on an endpath $vxyz$ to a vertex of K_1, where $n(T) \equiv 1 \bmod 3$.

O2. Join a stem, or a vertex v of T which lies on an endpath vxz to a leaf of P_2, where $n(T) \equiv 0 \bmod 3$ or $n(T) \equiv 1 \bmod 3$.

O3. Join a leaf of T to ℓ disjoint copies of P_3 for some $\ell \geq 1$.

Let $\mathcal{C}$ be the class of all trees obtained from P_2 or P_4 by a finite sequence of Operations **O1**–**O3**.

Theorem 4.3 (Hattingh and Plummer). *$T \in \mathcal{T}$ if and only if $T \in \mathcal{C}$.*

5 A Nordhaus-Gaddum Type Result

Nordhaus and Gaddum provided best possible bounds on the sum of the chromatic numbers of a graph and its complement in [19]. A corresponding result for the domination number was presented by Jaeger and Payan in [16]: if G is a graph of order $n \geq 2$, then $3 \leq \gamma(G) + \gamma(\overline{G}) \leq n + 1$.

In this section we survey results on the sum of the restrained domination numbers of a graph and its complement.

The following result is due to Domke, Hattingh, Hedetniemi, Laskar and Markus [6].

Theorem 5.1. *If G is a graph of order $n \geq 2$, then $4 \leq \gamma_r(G) + \gamma_r(\overline{G})$. If G is a graph of order $n \geq 2$ such that $G \not\cong P_3$ and $\overline{G} \not\cong P_3$, then $\gamma_r(G) + \gamma_r(\overline{G}) \leq n + 2$.*

The following results are due to Hattingh, Jonck, Joubert and Plummer [12].

In order to characterize the extremal graphs achieving the lower bound of four, define $\mathcal{H}$ as the family of graphs G of order n where G or $\overline{G}$ is one of the following four types:

Type 1. $V(G) = \{x, y, z\} \cup X$. Moreover:
P1.1: x is adjacent to each vertex of $\{y, z\} \cup X$,
P1.2: each vertex of $\{y, z\} \cup X$ is adjacent to some vertex of $\{y, z\} \cup X$,
P1.3: each vertex of X is non-adjacent to some vertex of $\{y, z\}$ and non-adjacent to some vertex in X.

Type 2. $V(G) = \{x, y\} \cup X$. Moreover:
P2.1: each vertex of X is adjacent to exactly one vertex of $\{x, y\}$ and also non-adjacent to exactly one vertex of $\{x, y\}$,
P2.2: each vertex of X is non-adjacent to some vertex of X,
P2.3: each vertex of X is adjacent to some vertex of X.

Type 3. $V(G) = \{u, v, y\} \cup X$. Moreover:
P3.1: each vertex of $X \cup \{y\}$ is adjacent to some vertex of $\{u, v\}$,
P3.2: each vertex of $X \cup \{u\}$ is non-adjacent to some vertex of $\{v, y\}$,
P3.3: each vertex of $X \cup \{y\}$ is adjacent to some vertex of $X \cup \{y\}$,
P3.4: each vertex of $X \cup \{u\}$ is non-adjacent to some vertex of $X \cup \{u\}$.

Type 4. $V(G) = \{x, y, u, v\} \cup X$. Moreover:
P4.1: each vertex in $\{x, y\} \cup X$ is adjacent to some vertex of $\{u, v\}$,
P4.2: each vertex in $\{u, v\} \cup X$ is non-adjacent to some vertex of $\{x, y\}$,
P4.3: each vertex in $\{x, y\} \cup X$ is adjacent to some vertex of $\{x, y\} \cup X$,
P4.4: each vertex in $\{u, v\} \cup X$ is non-adjacent to some vertex of $\{u, v\} \cup X$.

Theorem 5.2. *If G be a graph of order $n \geq 2$, then $\gamma_r(G) + \gamma_r(\overline{G}) = 4$ if and only if G or $\overline{G} \in \mathcal{H}$.*

In order to characterize the extremal graphs achieving the upper bound, define $\mathcal{B} = \{P_3, \overline{P}_3\}$, and let $\mathcal{G} = \{G \,|\, G$ or $\overline{G}$ is a galaxy of non-trivial stars$\}$.

Let $\mathcal{S} = \{G \,|\, G$ or $\overline{G} \cong K_1 \cup S$ where S is a star and $|S| \geq 3\}$.

Lastly, let $\mathcal{E} = \mathcal{G} \cup \mathcal{S}$.

Theorem 5.3. *Let $G = (V, E)$ be a graph of order $n \geq 2$ such that $G \notin \mathcal{B}$. Then $\gamma_r(G) + \gamma_r(\overline{G}) \leq n + 2$. Moreover, $\gamma_r(G) + \gamma_r(\overline{G}) = n + 2$ if and only if $G \in \mathcal{E}$.*

6 γ_r-Excellent Trees

In this section, we state a characterization of γ_r-excellent trees, due to Hattingh and Henning [11]. A graph G is called γ_r-excellent if every vertex of G belongs to some $\gamma_r(G)$-set.

Let $G = (V, E)$ be a graph, and let $u \in V$.

Definition 1. *We define the* restrained domination number of G relative to u, *denoted $\gamma_r^u(G)$, as the minimum cardinality of a restrained dominating set in G that contains u. A restrained dominating set of cardinality $\gamma_r^u(G)$ containing u we call a $\gamma_r^u(G)$-set. Hence, the graph G is γ_r-*excellent *if $\gamma_r^u(G) = \gamma_r(G)$ for every vertex $u \in V$.*

Definition 2. *We define a* type-I almost restrained dominating set *(type-I ARDS) of G* relative to u *as a set $S \subseteq V \setminus \{u\}$ such that S dominates $V \setminus \{u\}$ (possibly, S also dominates u) and $G[V \setminus S]$ has no isolated vertex.*

Definition 3. *We define a* type-II almost restrained dominating set *(type-II ARDS) of G* relative to u *as a set $S \subseteq V \setminus \{u\}$ such that S dominates V and $G[V \setminus S]$ has no isolated vertex, except possibly for the vertex u.*

Definition 4. *We define an* almost restrained dominating set *(ARDS) of G* relative to u *as a type-I ARDS or a type-II ARDS of G* relative to u.

Definition 5. *The* almost restrained domination number *of G* relative *to u, denoted $\gamma_r(G;u)$, is the minimum cardinality of an ARDS of G relative to u. An ARDS of G relative to u of cardinality $\gamma_r(G;u)$ we call a $\gamma_r(G;u)$-set. A type-I (respectively, type-II) ARDS of G relative to u of cardinality $\gamma_r(G;u)$ we call a type-I (respectively, type-II) $\gamma_r(G;u)$-set.*

Definition 6. *We say that G is* γ_r-excellent relative to u *if*

- $\gamma_r^u(G) = \gamma_r(G;u) + 1$,
- *there exists both a type-I and a type-II $\gamma_r(G;u)$-set, and*
- *every vertex of G is in some $\gamma_r^u(G)$-set or in some $\gamma_r(G;u)$-set.*

We define the family $\mathcal{F}$ to consist of all trees that can be obtained from a γ_r-excellent tree T' relative to a vertex u of T' by adding a path P_6 to T' and joining u to a central vertex of this path. As an example, the tree T shown in Figure 6.1 and the tree T shown in Figure 6.2 both belong to the family $\mathcal{F}$. (Notice that in both examples, the component T' of $T - uv$ that contains the vertex u is a γ_r-excellent tree relative to a vertex u.)

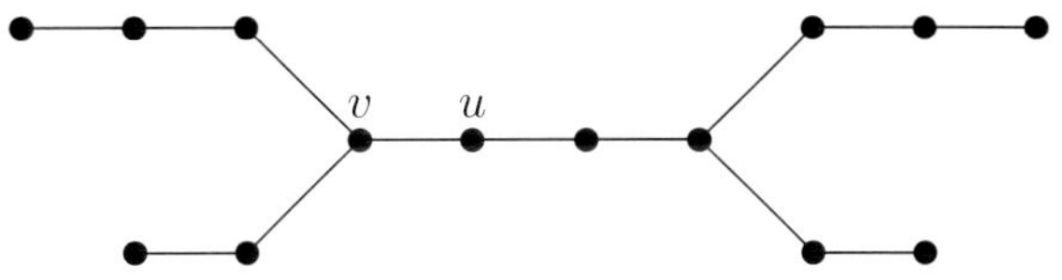

Figure 6.1: A γ_r-excellent tree T in $\mathcal{F}$.

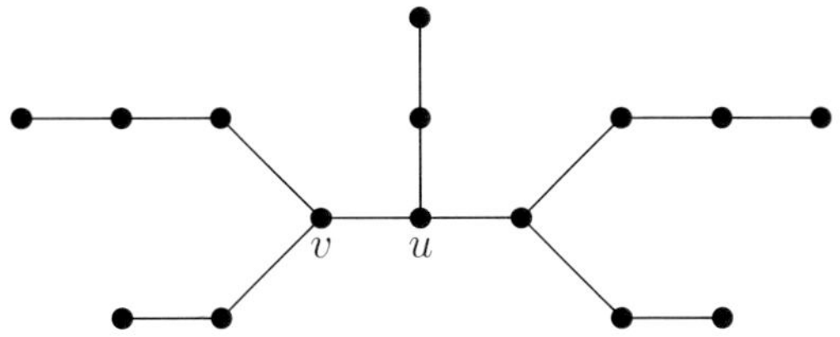

Figure 6.2: A γ_r-excellent tree T in $\mathcal{F}$.

We now construct a family of γ_r-excellent trees. Let $\mathcal{T}$ be the family of trees that contain $\{P_2\} \cup \mathcal{F}$ and is closed under the four operations **O1**, **O2**, **O3** and **O4** listed below, which extend a tree T' by attaching one or more endpaths to a vertex of T', called its *attacher*. We say that a leaf v of a tree T is a *good leaf* if T has order at least 3 and there exists a $\gamma_r(T)$-set that contains $N[w]$ where w is the neighbor of v.

- **Operation O1:** Join a stem of T' to a leaf of P_2.
- **Operation O2:** Join a good leaf of T' to a leaf of P_2.
- **Operation O3:** Join the vertex v on a vxz-endpath of T', where $\deg(v) \geq 2$, to a leaf of a P_2 and a leaf of P_3.
- **Operation O4:** Join the vertex v on a $vxyz$-endpath of T', where $\deg(v) \geq 2$, to a leaf of P_2 and to a leaf of P_3.

Theorem 6.1 (Hattingh and Henning). *A nontrivial tree T is a γ_r-excellent tree if and only if $T \in \mathcal{T}$.*

7 Partitions

A classical result in domination theory is that if S is a minimal dominating set of a graph G without isolates, then $V - S$ is also a dominating set of G. Thus, the vertex set of every graph without any isolates can be partitioned into two dominating sets. However, it is not the case that the vertex set of every graph can be partitioned into two restrained dominating sets. For example, the vertex set of C_5 cannot be partitioned into two total dominating sets.

For a given graph G, Heggernes and Telle [15] established that the decision problem whether there is a partition of $V(G)$ into two restrained dominating sets is NP-complete, even if G is bipartite.

A partition of the vertex set can also be thought of as a coloring. In particular, a partition into two restrained dominating sets is a 2-coloring of the graph such that no vertex has a monochromatic (open) neighborhood. As an example of such a 2-coloring in K_n with $n \geq 4$, take any 2-coloring with at least two vertices of each color, while in $K_{m,n}$ with $m, n \geq 2$ take any 2-coloring where neither partite set is monochromatic.

We now consider the question of how many edges must be added to G to ensure the partition of V into two restrained dominating sets in the resulting graph. We denote this minimum number by $\mathrm{rd}(G)$. It is clear that $\mathrm{rd}(G)$ can only exist for graphs with at least four vertices (we do not allow loops). As observed earlier, $\mathrm{rd}(K_n) = 0$ with $n \geq 4$ and $\mathrm{rd}(K_{m,n}) = 0$ with $m, n \geq 2$.

The follow two results are due to Broere, Dorfling, Goddard, Hattingh, Henning and Ungerer [1].

Lemma 7.1. *(a) For the cycle C_n with $n \geq 4$, $rd(C_n) = 0$ if $4|n$ and 1 otherwise. (b) For the path P_n on $n \geq 4$ vertices, $rd(P_n) = 1$ if $4|n$ and 2 otherwise.*

Theorem 7.2. *Let T be a tree of order at least 4 with ℓ leaves. Then, $rd(T) \leq \ell/2 + 1$.*

The following result is due to Goddard, Hattingh and Henning [10].

Theorem 7.3. *For a graph G on $n \geq 4$ vertices with minimum degree at least 2,*

$$rd(G) \leq \frac{n - 2\sqrt{n}}{4} + c \log n$$

for some constant c.

That this result is best possible by considering the following graph. Start with the complete graph on $2k$ vertices, duplicate each edge, and then subdivide each edge. The resultant graph G^* has order $n = 4k^2$.

8 Closing Remarks

In closing we note that there exists a linear algorithm to compute $\gamma_r(T)$ for a tree T, and that the decision problem for $\gamma_r(G)$ is NP-complete, even when restricted to bipartite and chordal graphs. The details appear in [6].

References

[1] I. Broere, M. Dorfling, W. Goddard, J.H. Hattingh, M.A. Henning and E. Ungerer, Augmenting trees to have two disjoint total dominating sets, *Bulletin of ICA*, **42** (2004), pp. 12–18.

[2] T.A. Burton and D.P. Sumner, Domination dot-critical graphs, manuscript

[3] G. Chartrand and L. Lesniak, *Graphs & Digraphs: Third Edition*, Chapman & Hall, London, (1996).

[4] P. Dankelmann, D. Day, J.H. Hattingh, M.A. Henning, L.R. Markus and H.C. Swart, On equality in an upper bound for the restrained and total domination numbers of a graph, *Discrete Mathematics*, **307** (2007), pp. 2845–2852.

[5] P. Dankelmann, J.H. Hattingh, M.A. Henning and H.C. Swart, Trees with equal domination and restrained domination numbers, *J. Global Optim.*, **34** (2006), pp. 597–607.

[6] G.S. Domke, J.H. Hattingh, S.T. Hedetniemi, R.C. Laskar, and L.R. Markus, Restrained domination in graphs, *Discrete Math.*, **203** (1999), pp. 61–69.

[7] G.S. Domke, J.H. Hattingh, M.A. Henning and L.R. Markus, Restrained domination in trees, *Discrete Math.*, **211** (2000), pp. 1–9.

[8] G.S. Domke, J.H. Hattingh, M.A. Henning and L.R. Markus, Restrained domination in graphs with minimum degree two, *J. Combin. Math. Combin. Comput.*, **35** (2000), pp. 239–254.

[9] G.H. Fricke, T.W. Haynes, S.M. Hedetniemi, S.T. Hedetniemi and R.C. Laskar, Excellent trees, *Bulletin of ICA*, **34** (2002), pp. 27–38.

[10] W. Goddard, J.H. Hattingh and M.A. Henning Augmenting a graph of minimum degree 2 to have two disjoint total dominating sets, *Discrete Math.*, **300** (2005), pp. 82–90.

[11] J.H. Hattingh and M.A. Henning, Restrained domination excellent trees, To appear in *Ars Combin.*

[12] J.H. Hattingh, E. Jonck, E.J. Joubert and A.R. Plummer, Nordhaus-Gaddum results for restrained domination and total restrained domination in graphs, *Discrete Mathematics*, **308** (2008), pp. 1080–1087.

[13] J.H. Hattingh and A.R. Plummer, A note on restrained domination in trees, To appear in *Ars Combin.*

[14] M.A. Henning, Graphs with large restrained domination number, *Discrete Math.*, **197/198** (1999), pp. 415–429.

[15] P. Heggernes and J.A. Telle, Partitioning graphs into generalized dominating sets, *Nordic J. Comput.*, **5(2)** (1998), pp. 128–142.

[16] F. Jaeger and C. Payan, Relations du type Nordhaus-Gaddum pour le nombre d' absorption d' un graphe simple, *C. R. Acad. Sci Ser A*, **274** (1972), pp. 728–730.

[17] W. McCuaig and B. Shepherd, Domination in graphs with minimum degree two, *J. Graph Theory*, **13** (1989), pp. 749–762.

[18] C.M. Mynhardt, H.C. Swart and E. Ungerer, Excellent trees and secure domination, To appear in *Util. Math.*

[19] E.A. Nordhaus and J.W. Gaddum, On complementary graphs, *Amer. Math. Monthly*, **63** (1956), pp. 175–177.

[20] J.A. Telle and A. Proskurowski, Algoritms for vertex partioning problems on partial k-trees. *SIAM J. Discrete Math.* **10** (1997), pp. 529–550.

Contributed Papers

Proc. of ICDM 2006, Pages 227–231

Some Properties of the Clique Graph of a Cograph

S. Aparna Lakshmanan
Email: aparna@cusat.ac.in
A. Vijayakumar
Department of Mathematics, Cochin University of Science and Technology, Cochin-682 022, India. *Email:* vijay@cusat.ac.in

Abstract. In this paper, we prove that the diameter of the clique graph of a cograph cannot exceed 2. A cograph G is clique complete if and only if G has a vertex of full degree. Though the chromatic number of G and $K(G)$ are independent, the number of clique graphs of a cograph with $\chi(K(G)) = s$, where s is a fixed integer is finite. A realization of cographs G and its clique graph $K(G)$ which have specific values for the domination number, the clique transversal number and the clique independence number are given. It is also proved that, cographs G satisfy $\tau_c(K(G)) = \alpha_c(K(G))$, but still are not clique perfect.

1 Introduction

We consider only finite, simple graphs $G = (V, E)$ with $|V| = n$ and $|E| = m$.

A graph G is a cograph if it does not contain P_4 - the path on four vertices, as an induced subgraph [8]. The recursive characterization of cographs [4] is,

(1) K_1 is a cograph

(2) If G is a cograph, so is its complement G^c and

(3) If G and H are cographs, so is their join, $G \vee H$, where $G \vee H$ is defined as $V(G \vee H) = V(G) \cup V(H)$ and
$E(G \vee H) = E(G) \cup E(H) \cup \{uv$, where $u \in V(G)$ and $v \in V(H)\}$.

For a detailed literature on cographs the reader may refer to [3, 4, 7, 8].

A vertex v of a graph G is of full degree if its degree is $n - 1$. A clique of a graph G is a maximal complete subgraph of G. The clique graph of a graph G, denoted by $K(G)$, has cliques of G as its vertices and any two vertices of $K(G)$ are adjacent if the corresponding cliques in G intersects. A graph G is clique complete if the clique graph of G is complete.

F. Larrion et.al, [7] have studied in detail the clique operator on cographs and proved that a cograph is clique convergent if and only if it is clique Helly. A characterization of cographs whose clique graph is also a cograph is obtained. In [1] all graphs G for which $d(K(G)) = d(G) - 1$, $d(K(G)) = d(G)$ and $d(K(G)) = d(G) + 1$ are characterized. The main open question concerning the clique graphs of cographs is a characterization of their clique graphs. The vast literature on clique graphs is very well summarized in [10].

In this paper, we prove that the diameter of the clique graph of a cograph cannot exceed two. Any non-complete graph with prime number of vertices cannot be the clique graph of a cograph. A cograph G is clique complete if and only if G has a vertex of full degree. Though the difference between the chromatic numbers of G and $K(G)$ is arbitrary, the number of clique graphs of a cograph with $\chi(K(G)) = s$, where s is a fixed integer is finite. A realization of cographs G and its clique graph $K(G)$ which have specific values for the domination number, the clique transversal number and the clique independence number are given. It is also proved that, cographs G satisfy $\tau_c(K(G)) = \alpha_c(K(G))$, but still are not clique perfect.

For all graph theoretic notations and preliminaries, we follow [2].

2 The Clique Graph of a Cograph

Theorem 1. *If G is a connected cograph then the diameter of $K(G)$ is less than or equal to 2.*

Proof. Let S_1 and S_2 be any two non-adjacent vertices in $K(G)$. If a vertex in S_1 is adjacent to a vertex in S_2, then there exists a clique S which contains this edge and hence is adjacent to both S_1 and S_2 in $K(G)$. Therefore, $d(S_1, S_2) = 2$. If possible assume that no vertex in S_1 is adjacent to a vertex in S_2. Let $v_1 \in V(S_1)$ and $v_2 \in V(S_2)$. Then, $d(v_1, v_2) = 2$. Hence there exists a vertex v adjacent to both v_1 and v_2. If v_1' is another vertex in $V(S_1)$ then $v_1'v_1vv_2$ should not induce P_4 in G and therefore v_1' is adjacent to v. Since v_1' was arbitrary, every vertex in $V(S_1)$ is adjacent to v. But, this is a contradiction to the maximality of S_1. Hence, for a connected cograph G, diameter of $K(G) \leq 2$.

Theorem 2. *If G is a connected cograph with prime number of cliques, then G is clique complete.*

Proof. Let $G = G_1 \vee G_2$. The number of cliques in G is the product of the number of cliques in G_1 and G_2. But, the number of cliques in G is prime and hence one of the G_i's must have prime number of cliques and other must be complete. Every clique in G is the join of the cliques of G_1 and G_2. Hence any two cliques in G have a non-empty intersection and therefore the clique graph of G is complete.

Corollary 3. *Any graph of prime order, other than K_p, cannot be the clique graph of a cograph.*

Theorem 4. *A cograph is clique complete if and only if there exists a vertex of full degree.*

Proof. If there exists a vertex of full degree in G then that vertex will be present in every clique of G and hence $K(G)$ is complete.

Now, assume that a cograph G is clique complete. Let S be a clique of G with maximum cardinality and ω be its clique number. The proof is by induction on $|V(G) - V(S)|$. If $|V(G) - V(S)| = 0$ then $G(= S)$ itself is complete. If $|V(G) - V(S)| = 1$ then there exist only one vertex v outside S. Since G is connected there exists at least one vertex $u \in S$ which is adjacent to v. Then $\deg(u) = n-1$. Assume that if $|V(G)-V(S)| = k$ then there exists a vertex of full degree in G.

Now, let $|V(G) - V(S)| = k + 1$ and $v_1, v_2, ..., v_{k+1} \in V(G) - V(S)$. Let G_i be the graph obtained by deleting the vertex v_i for $i \in \{1, 2, .., k + 1\}$. Then $|V(G_i)| = n - 1$ and S is a clique in G_i. Also $|V(G_i) - V(S)| = k$. Therefore by the induction assumption, there exists a vertex v_i' of degree $n-2$ in G_i for all i. Then v_i'

belongs to $V(S)$, since it is adjacent to all vertices in G_i and S is maximal complete. If for at least one v_i, v_i is adjacent to v_i', then v_i' will be of full degree in G.

Now, assume that v_i is not adjacent to v_i' for all i and hence $v_i' \neq v_j'$ if $i \neq j$. Consider two arbitrary vertices v_i and v_j where $i \neq j$ and $i, j \in \{1, 2, ..., k+1\}$. If v_i is not adjacent to v_j, then $v_i v_j' v_i' v_j$ is an induced P_4 in G which is a contradiction. Therefore v_i is adjacent to v_j for all $i \neq j$ and $i, j \in \{1, 2, ..., k+1\}$. Hence $\{v_1, v_2, ..., v_{k+1}\}$ induces a complete graph. So there exists a clique in G which contains all the vertices $v_1, v_2, ..., v_{k+1}$. This clique has non-empty intersection with S, since G is clique complete. Therefore there exists $u \in V(S)$ which in adjacent to v_i for all $i \in 1, 2, ..k+1$ and hence u will be a vertex of full degree. The proof now follows by the mathematical induction.

3 The Chromatic Number

The chromatic number, $\chi(G)$ of a graph G is the minimum number of colors required to properly color its vertices.

Note 5. Given any two positive integers $a, b > 1$, there exists a cograph G such that $\chi(G) = a$ and $\chi(K(G)) = b$.

Theorem 6. *The number of clique graphs of a connected cograph G with $\chi(K(G)) = s$ is finite.*

Proof. Let G be a cograph with $\chi(K(G)) = s$. Let $G = G_1 \vee G_2$ be a decomposition of G. Let the number of cliques of G_i be p_i for $i = 1, 2$. If $p_i > s$ for some i, say $i = 1$, then G_1 will have at least $s+1$ cliques, $H_{11}, H_{12}, ..., H_{1,s+1}$. Let H_2 be a clique of G_2. Then $H_{11} \vee H_2, H_{12} \vee H_2, ..., H_{1,s+1} \vee H_2$ are cliques of G which induce K_{s+1} in $K(G)$. But, then $\chi(K(G)) \geq s+1$ which is a contradiction. Therefore each $p_i \leq s$ and hence $|V(K(G))| \leq s^2$. Hence, the number of clique graphs of a connected cograph with $\chi(K(G)) = s$ is finite.

4 Some Graph Parameters

A subset S of V in a graph G is called a dominating set if every vertex $v \in V$ is either an element of S or is adjacent to an element of S. The domination number $\gamma(G)$ is the minimum cardinality of a dominating set in G.

A subset V' of V is called a clique transversal, if it intersects with every clique of G. The clique transversal number $\tau_c(G)$ of a graph G is the minimum cardinality of a clique transversal of G [5]. A clique independent set is a subset of pairwise disjoint cliques of G. The clique independence number $\alpha_c(G)$ of a graph G is the maximum cardinality of a clique independent set of G [6]. A graph G is clique perfect if $\tau_c(G) = \alpha_c(G)$ for G and all its induced subgraphs.

Theorem 7. *There exists a cograph G such that $\gamma(G) = a$ and $\gamma(K(G)) = b$ if and only if*
(1) $a \leq 2$.
(2) $a = 1$ *if and only if* $b = 1$.
(3) $a = 2$ *and* $b \geq a$.

Proof. If G is a cograph then $\gamma(G) \leq 2$ [9]. Therefore (1) holds. If $\gamma(G) = 1$ then G has a vertex of full degree and hence $K(G)$ is complete. Therefore, $a = 1$ implies that $b = 1$. If $b = 1$ then $K(G)$ has a vertex of full degree. Let C be the clique in G which corresponds to this vertex of full degree in $K(G)$. Let $u_1, u_2, ..., u_p \in V(G) - V(C)$. Every clique in G intersects with C and hence u_i s for $i = 1, 2, ..., p$ must be adjacent to at least one vertex of $V(C)$.

Claim. Every u_i is adjacent to a common vertex $v \in V(G)$.

On the contrary, assume that u_1 and u_2 do not have a common neighbor in C. Let u_1 be adjacent to v_1 and u_2 be adjacent to v_2. But, $u_1v_1v_2u_2$ cannot induce a P_4 in G and hence u_1 is adjacent to u_2. Since, u_1 and u_2 have no common neighbors in C, the clique of G which contains the edge u_1u_2 does not intersect C which is a contradiction. Therefore, our claim holds.

Therefore, v is a vertex of full degree in G and hence $a = \gamma(G) = 1$. Hence, (2) holds.

If $a = 2$ then $b \neq 1$ by (2). Therefore, $b \geq a$ and (3) holds.

Conversely, assume that a and b satisfy the given conditions. Let G be the cograph $K_{b,b}$. The clique graph of $K_{b,b}$, $K(K_{b,b}) = K_b \times K_b$. Therefore, $\gamma(K(K_{b,b})) = b$. If $b > 1$ then $\gamma(G) = 2$ and if $b = 1$ then $\gamma(G) = 1$. Hence, G is the required graph.

Theorem 8. *If G is a cograph then $\tau_c(K(G)) = \alpha_c(K(G))$.*

Proof. We use the recursive definition of cographs to prove the theorem. If $G = K_1$, then $K(G) = K_1$ and $\tau_c(K_1) = \alpha_c(K_1) = 1$.

Let G_1 and G_2 be cographs which satisfy $\tau_c(K(G_i)) = \alpha_c(K(G_i))$ for $i = 1, 2$. Let $G = G_1 \cup G_2$. Then, $K(G) = K(G_1) \cup K(G_2)$ and hence $\tau_c(K(G)) = \tau_c(K(G_1)) + \tau_c(K(G_2)) = \alpha_c(K(G_1)) + \alpha_c(K(G_2)) = \alpha_c(K(G))$.

Let $G = G_1 \vee G_2$. Let H_1 be a clique in $K(G_1)$ induced by the vertices corresponding to the cliques $G_{11}, G_{12}, ...G_{1k}$ in G_1. Let $G_{21}, G_{22}, ..., G_{2t}$ be the cliques in G_2. Therefore, $\{G_{1i} \vee G_{2j} : i = 1, 2, ...k \text{ and } j = 1, 2, ..., t\}$ are cliques in $G_1 \vee G_2$ and the vertices corresponding to these cliques induce a clique in $K(G_1 \vee G_2)$. Let this clique be H_1'. Similarly, if H_2 is a clique in $K(G_1)$, then we can find a clique H_2' in $K(G_1 \vee G_2)$. Moreover, if H_1 and H_2 are independent, then H_1' and H_2' are also independent. Therefore, $\alpha_c(K(G_1 \vee G_2)) \geq \alpha_c(K(G_1))$. Similarly we can prove that $\alpha_c(K(G_1 \vee G_2)) \geq \alpha_c(K(G_2))$. Therefore, $\alpha_c(K(G_1 \vee G_2)) \geq max\{\alpha_c(K(G_1)), \alpha_c(K(G_2))\}$. Using similar arguments, we can prove that $\tau_c(K(G_1 \vee G_2)) \leq max\{\tau_c(K(G_1)), \tau_c(K(G_2))\}$. Therefore, $\tau_c(K(G_1) \vee K(G_2)) \leq \alpha_c(K(G_1) \vee K(G_2)$. But, by definition, $\tau_c(K(G_1) \vee K(G_2)) \geq \alpha_c(K(G_1) \vee K(G_2)$. Therefore, $\tau_c(K(G_1) \vee K(G_2)) = \alpha_c(K(G_1) \vee K(G_2)$.

Hence the theorem.

Note 9. Even though $\tau_c(K(G)) = \alpha_c(K(G))$ for a cograph G, $K(G)$ is not necessarily clique perfect. For example, let $G = (K_1 \cup C_4) \vee (2K_1 \cup P_3)$ as in Fig 1. The cliques of G formed by the vertices $\{u_1, v_1\}$, $\{u_1, v_2, v_3\}$, $\{u_4, u_5, v_3, v_4\}$, $\{u_3, u_4, v_5\}$ and $\{u_2, u_3, v_1\}$ induce a C_5 in $K(G)$ and hence $K(G)$ is not clique perfect.

Note 10. There exists a cograph G such that $\alpha_c(G) = \tau_c(G) = a$ and $\alpha_c(K(G)) = \tau_c(K(G)) = b$ if
(1) $a = 1$ if and only if $b = 1$
(2) $a \leq b$.

Acknowledgement: The authors thank Prof. S.B. Rao for his valuable suggestions for the preparation of this paper and the referees for their suggestions and revision.

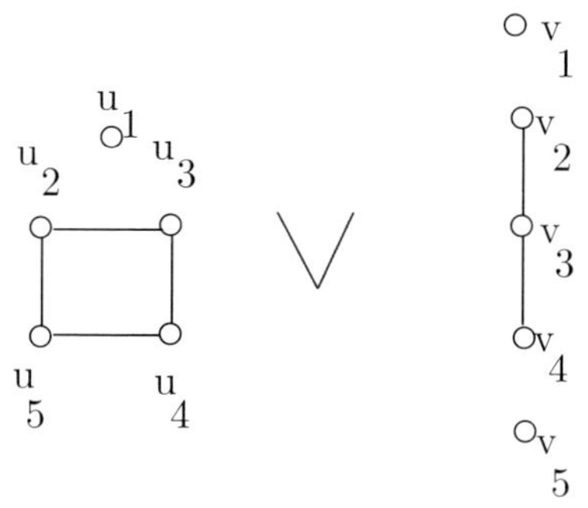

Figure 4.1:

References

[1] R. Balakrishnan and P. Paulraja, Self-clique graphs and diameters of iterated clique graphs, *Util. Math.*, **29** (1986), pp. 263–268.

[2] R. Balakrishnan and K. Ranganathan, *A text book of graph theory*, Springer (1999).

[3] A. Brandstät, V.B. Le and J.P. Spinrad, Graph classes – a survey, *SIAM*, (1999).

[4] D.G. Corneil, H. Lerchs and L.S. Burlington, Complement reducible graphs, *Discrete Appl. Math.*, **3(3)** (1981), pp. 163–174.

[5] P. Erdös, T. Gallai and Z. Tuza, Covering the cliques of a graph with vertices, *Discrete Math.*, **108** (1992), pp. 279–289.

[6] V. Guruswami and C. Pandu Rangan, Algorithmic aspects of clique-transversal and clique-independent sets, *Discrete Appl. Math.*, **100** (2000), pp. 183–202.

[7] F. Larrión, C.P. de Mello, A. Morgana, V. Neumann-Lara and M.A. Pizaña, The clique operator on cographs and serial graphs, *Discrete Math.*, **282** (2004), pp. 183–191.

[8] T.A. Mckee, Dimensions for cographs, *Ars Combin.*, **56** (2000), pp. 85–95.

[9] S.B. Rao, L.S. Aparna and A. Vijayakumar, Cographic and global cographic domination number of a graph, *Ars Combin.*, (to appear).

[10] J.L. Szwarcfiter, A survey on clique graphs, *Recent Advances in Algorithms and Cobinatorics*, (2003), pp. 109–136.

Proc. of ICDM 2006, Pages 233–236

$\hat{S}_k$-Factorization of Symmetric Digraph of Tensor Product of Graphs

P. Hemalatha
Dept. of Mathematics, Kongu Engineering College, Erode 638052
A. Muthusamy
Dept. of Mathematics, Bharathidasan University, Tiruchirappalli 620024.

Abstract. In this paper, a necessary condition and some sufficient conditions are obtained for the existence of an $\hat{S}_k$- factorization of symmetric digraph of tensor product of complete graphs.

Keywords: Star-factorization, Wreath product of graphs, Tensor product of graphs.

1 Introduction

All the graphs considered here are finite. Let $G = (V, E)$ be a graph. Partition of G into edge-disjoint sub graphs $G_1, G_2, \ldots, G_r$ such that $E(G) = E(G_1) \cup E(G_2) \cup \cdots \cup E(G_r)$ is called a *decomposition* of G. Decomposition of the given graph into spanning subgraphs is called a *factorization*. An r-regular spanning subgraph of G is called an r-factor of G. A cycle of length k is denoted by C_k. $\hat{S}_k$ is an $S_k = K_{1,k-1}$ (using vertices from two parts of the m-partite digraph)with arcs oriented from the centre vertex to end vertices. An $\hat{S}_k$*-factor* is a spanning subgraph H of G such that each component of H is an $\hat{S}_k$. Partitioning the edge set of G into $\hat{S}_k$-factors is called an $\hat{S}_k$*-factorization* of G, denoted by $\hat{S}_k \parallel G$. For a graph G and a positive integer r, $r(G)$ denote r disjoint copies of G and $G(r)$ is the graph obtained from G by replacing each of its edges by r edges. G^* is the symmetric digraph obtained from G by replacing every edge by a symmetric pair of arcs. The complete graph on m vertices is denoted by K_m and its complement is denoted by $\overline{K}_m$. The subgraph of G induced by $U \subset V(G)$ is denoted by $G[U]$.
We write $G = H_1 \oplus H_2 \oplus \cdots \oplus H_k = \oplus_{i=1}^{k} H_i$ if $H_1, H_2, \ldots, H_k$ are edge - disjoint subgraphs of G and $E(G) = E(H_1) \cup E(H_2) \cup \cdots \cup E(H_k)$. Graph theoretical terms not defined here can be found in [1].The *tensor product* of two graphs G and H, denoted by $G \times H$, has the vertex set $V(G) \times V(H)$ in which (g_1, h_1) is joined to (g_2, h_2) by an edge whenever $g_1g_2 \in E(G)$ and $h_1h_2 \in E(H)$.

Nowadays, decomposition/factorization of product graphs into cycles, stars, paths etc. have attracted much attention due to their applications in many fields especially in G-designs. Study on factorization of digraphs into stars, paths and cycles are not new (see [2–6]). Ushio [4, 5] has obtained a necessary and some sufficient conditions for the existence of an $\hat{S}_k$-factorization of complete bipartite and tripartite symmetric digraphs. Recently, the authors [2] have obtained a necessary and some sufficient conditions for the existence of $\hat{S}_k$-factorization of complete multipartite symmetric digraphs. In this paper, we have obtained a necessary and some sufficient conditions for the existence of $\hat{S}_k$-factorization in $(K_m \times K_n)^*$.

2 Necessary Condition

Theorem 2.1. *If $(K_m \times K_n)^*$ has an $\hat{S}_k$-factorization, then (i) $mn \equiv 0 (mod\ k)$ and (ii) $n \equiv k(mod\ k(k-1)/d)$ where $d = (m, k)$.* □

Proof follows easily from the definitions of $(K_m \times K_n)^*$ and $\hat{S}_k$-factorization.

3 Sufficient Conditions

Notation. Let $\overrightarrow{K}_{|V_1|,|V_2|}$ denote the complete bipartite digraph with arcs oriented from V_1 to V_2. We denote $\hat{S}_k$ with centre vertex u and end vertices $v_1, v_2, \ldots, v_{k-1}$, by $(u; v_1, v_2, \ldots, v_{k-1})$.

We first prove the following lemmas to prove our main result.

Lemma 3.1. *For $k \geq 2$, $\hat{S}_k \parallel \overrightarrow{K}_{k,k(k-1)}$.* □

Lemma 3.2. *For $k \geq 2$, $\hat{S}_k \parallel (K_2 \times K_k)^*$.* □

Proofs of the Lemmas 3.1 and 3.2 follows easily from the definitions.

Lemma 3.3. *For $k \geq 2$, $\hat{S}_k \parallel (K_2 \times K_{k^2})^*$.*

Proof. Let $G = (K_2 \times K_{k^2})^*$ and V_1 and V_2 be the two partite sets of $V(G)$. Now, divide V_1 and V_2 into k subsets of equal cardinality. Denote the subsets of V_1 by $X_1, X_2, \ldots, X_k$ and the subsets of V_2 by $Y_1, Y_2, \ldots, Y_k$. To find $\hat{S}_k$-factorization of G, consider $G' = G - \bigcup_{i=1}^{k}(E(G[X_i, Y_i]))$. Identify the subsets of V_1 and V_2 in G' into vertices and join two vertices by a symmetric pair of arcs if and only if, the corresponding subsets form a $K^*_{k,k}$ in G'. Now the resulting graph will be isomorphic to $(K_2 \times K_k)^*$. But by Lemma 3.2, $\hat{S}_k \parallel (K_2 \times K_k)^*$. By expanding the vertices, each $\hat{S}_k$ in $(K_2 \times K_k)^*$ will be isomorphic to $\overrightarrow{K}_{k,k(k-1)}$ in G' with arcs oriented from k subset to the $k(k-1)$ subset. i.e.,$\hat{S}_k$-factorization of $(K_2 \times K_k)^*$ implies $\overrightarrow{K}_{k,k(k-1)}$-factorization of G'. By Lemma 3.1,$\hat{S}_k \parallel \overrightarrow{K}_{k,k(k-1)}$ and hence $\hat{S}_k \parallel G'$. To complete the proof, it remains to show that $\hat{S}_k \parallel \bigcup_{i=1}^{k} G[X_i, Y_i]$. By Lemma 3.2, $\hat{S}_k \parallel G[X_i, Y_i]$, since each $G[X_i, Y_i]$, $i = 1, 2, \ldots, k$ is isomorphic to $(K_2 \times K_k)^*$ and hence $\hat{S}_k \parallel \bigcup_{i=1}^{k} G[X_i, Y_i]$. Thus $\hat{S}_k \parallel G$. □

Lemma 3.4. *For $k \geq 2$, and $m > 2$, $\hat{S}_k \parallel (C_m \times K_k)^*$.*

Proof. Let $V_1 = \{1^{(1)}, 2^{(1)}, \ldots, k^{(1)}\}$, $V_2 = \{1^{(2)}, 2^{(2)}, \ldots, k^{(2)}\}$, $\ldots$, $V_m = \{1^{(m)}, 2^{(m)}, \ldots, k^{(m)}\}$ be the partite sets of $(C_m \times K_k)^*$. Now we construct an $\hat{S}_k$-factor of $(C_m \times K_k)^*$ as follows:

$F_i = \left\{\bigcup_{p=1}^{m}(i^{(p)}; (i+1)^{(p+1)}, (i+2)^{(p+1)}, \ldots, (i+k-1)^{(p+1)})\right\}$, $i = 1, 2, \ldots, k$, where the additions are taken modulo k with residues $1, 2, \ldots, k$ and the superscripts are taken modulo m with residues $1, 2, \ldots, m$. Clearly, each $F_i, i = 1, 2, \ldots, k$ is an $\hat{S}_k$-factor of $(C_m \times K_k)^*$. When i varies we have k $\hat{S}_k$-factors of $(C_m \times K_k)^*$. Due to symmetry of arcs we have another k $\hat{S}_k$-factors of $(C_m \times K_k)^*$. All these $2k$ $\hat{S}_k$-factors together form an $\hat{S}_k$-factorization of $(C_m \times K_k)^*$.

Thus $\hat{S}_k \parallel (C_m \times K_k)^*$. □

Lemma 3.5. *For $k \geq 2$, and $m > 2$, $\hat{S}_k \parallel (C_m \times K_{k^2})^*$.*

Proof. Let $(C_m \times K_{k^2})^* = G$ and let $V_1, V_2, \ldots, V_m$ be the partite sets of $V(G)$. Divide each partite set $V_i, i = 1, 2, \ldots, m$ into k subsets $\left\{X_1^{(i)}, X_2^{(i)}, \ldots, X_k^{(i)}\right\}$ such that each subset has k vertices. Let $G' = \bigcup_{i=1}^{k} \left\{\oplus_{i=1}^{m} E(G[X_j^{(i)}, X_j^{(i+1)}])\right\}$, where the superscripts are taken modulo m with residues $1, 2, \ldots, m$. i.e., $G' \cong k(C_m \times K_k)^*$. Let $G'' = G - G'$. Now identify each subset $X_j^{(i)}$ of G'' into a vertex $x_j^{(i)}, i = 1, 2, \ldots, m$ and $j = 1, 2, \ldots, k$ and join two vertices by a symmetric pair of arcs if the corresponding subsets form a $K_{k,k}^*$ in G''. The new graph G''' is isomorphic to $(C_m \times K_k)^*$. By Lemma 3.4, $\hat{S}_k \parallel (C_m \times K_k)^*$. i.e., $\hat{S}_k \parallel G'''$. Now going back to G'', each $\hat{S}_k$ in G''' becomes a graph isomorphic to $\overrightarrow{K}_{k,k(k-1)}$ in G''. i.e.,$\hat{S}_k$-factorization of G''' implies $\overrightarrow{K}_{k,k(k-1)}$-factorization of G''. By Lemma 3.1, $\hat{S}_k \parallel \overrightarrow{K}_{k,k(k-1)}$ and hence $\hat{S}_k \parallel G''$. Also by Lemma 3.4, $\hat{S}_k \parallel G'$. Therefore $\hat{S}_k \parallel (G = G' \oplus G'')$. □

Theorem 3.6. *For $k \geq 2$, and $m > 1$, $\hat{S}_k \parallel (K_m \times K_{k^t})^*$.*

Proof. **Case 1.** m even.

Now $(K_m \times K_{k^t})^* \cong \oplus_{i=1}^{m-1}(F_i \times K_{k^t})^*$, since $K_m = \oplus_{i=1}^{m-1} F_i$ where $F_i's$ are 1-factors of K_m. We know that, each $(F_i \times K_{k^t})^* \cong \frac{m}{2}(K_2 \times K_{k^t})^*$. To complete the proof, it is enough to show that, $\hat{S}_k \parallel (K_2 \times K_{k^t})^*$. We prove this by induction on t. When $t = 1$ by Lemma 3.2, $\hat{S}_k \parallel (K_2 \times K_k)^*$ and when $t = 2$ by Lemma 3.3, $\hat{S}_k \parallel (K_2 \times K_{k^2})^*$. Without loss of generality, assume that $\hat{S}_k \parallel (K_2 \times K_{k^{t-1}})^*$. Let $G = (K_2 \times K_{k^t})^*$. Now divide the partite sets V_1, V_2 of G into k^{t-1} subsets of equal cardinality k. Denote the subsets of V_1 by $X_1, X_2, \ldots, X_{k^{t-1}}$ and the subsets V_2 by $Y_1, Y_2, \ldots, Y_{k^{t-1}}$. Then $G[X_i, Y_i]$, $i = 1, 2, \ldots, k^{t-1}$ is isomorphic to $(K_2 \times K_k)^*$. Let $G' = \oplus_{i=1}^{k^{t-1}} G[X_i, Y_i] \cong k^{t-1}(K_2 \times K_k)^*$. By induction hypothesis, $\hat{S}_k \parallel G'$. Now consider $G'' = G - G'$. To find the $\hat{S}_k$ - factors of G'', identify each X_i, Y_i, $i = 1, 2, \ldots, k^{t-1}$ into a vertex x_i, y_i, and join two vertices by a symmetric pair of arcs if and only if the corresponding subsets form a graph isomorphic to $K_{k,k}^*$ in G''. The resulting graph G''' will be isomorphic to $(K_2 \times K_{k^{t-1}})^*$. By induction hypothesis, $\hat{S}_k \parallel (K_2 \times K_{k^{t-1}})^*$. When we go back to G'' each $\hat{S}_k$ in G''' will be isomorphic to $\overrightarrow{K}_{k,k(k-1)}$ in G''. i.e., $\hat{S}_k$- factorization of G''' implies $\overrightarrow{K}_{k,k(k-1)}$ - factorization of G''. By Lemma 3.1, $\hat{S}_k \parallel \overrightarrow{K}_{k,k(k-1)}$ and hence, $\hat{S}_k \parallel G''$. Thus $\hat{S}_k \parallel (G = G' \oplus G'')$.

Case 2. m odd.

Now $(K_m \times K_{k^t})^* \cong \oplus_{i=1}^{\frac{m-1}{2}} (H_i \times K_{k^t})^*$, where $H_i's$ are are edge disjoint Hamilton cycles of K_m. Also each $(H_i \times K_{k^t})^* \cong (C_m \times K_{k^t})^*$. To complete the proof, we have to show that, $\hat{S}_k \parallel (C_m \times K_{k^t})^*$. We prove this by induction on t. When $t = 1, 2$ $\hat{S}_k \parallel (C_m \times K_{k^t})^*$ by Lemmas 3.4 and 3.5. Assume that $\hat{S}_k \parallel (C_m \times K_{k^r})^*$, for all $r < t$. Now consider $G = (C_m \times K_{k^t})^*$. As in Case 1, divide the m-partite sets $V_1, V_2, \ldots, V_m$ of G into k^{t-1} subsets of cardinality k and denote the subsets of V_i by $X_i^{(1)}, X_i^{(2)}, \ldots, X_i^{(k^{t-1})}$. Let $G' = \bigcup_{j=1}^{k^{t-1}} \{\oplus_{i=1}^{m} G[X_i^{(j)}, X_{i+1}^{(j)}]\}$. For each j, $\oplus_{i=1}^{m} G[X_i^{(j)}, X_{i+1}^{(j)}] \cong (C_m \times K_k)^*$, and hence $G' \cong k^{t-1}(C_m \times K_k)^*$. By induction hypothesis $\hat{S}_k \parallel G'$. Now consider $G'' = G - G'$. To find the $\hat{S}_k$-factorization of G'', identify each $X_i^{(j)}$ into a vertex $x_i^{(j)}$ and join two vertices by a symmetric pair of arcs if and only if the corresponding subsets form a $K_{k,k}^*$ in G''. The resulting graph $G''' \cong (C_m \times K_{k^{t-1}})^*$. By induction hypothesis, $\hat{S}_k \parallel (C_m \times K_{k^{t-1}})^*$. When we go back to G'' each $\hat{S}_k$ in G''' becomes a graph isomorphic to $\overrightarrow{K}_{k,k(k-1)}$ in

G''. By Lemma 3.1, $\hat{S}_k \parallel \overrightarrow{K}^*_{k,k(k-1)}$. Therefore, $\hat{S}_k \parallel G''$ and hence $\hat{S}_k \parallel G$. Thus $\hat{S}_k \parallel (K_m \times K_{k^t})^*$. □

Note. Theorem 3.6 makes clear that the necessary condition given in Section 2 for the existence of $\hat{S}_k$-factorization of $(K_m \times K_n)^*$ is also sufficient when $n = k^t$, $t \geq 1$.

Acknowledgements.
The second author thanks the Department of Science and Technology (DST), New Delhi for its financial support (Project Grant No.SR/S4/MS: 372/06).

References

[1] J.A. Bondy and U.S.R. Murty, *Graph Theory with Applications*, MacMillan, New York (1976).

[2] P. Hemalatha and A. Muthusamy, $\hat{S}_k$-factorization of complete multi-partite symmetric digraphs, Revised version to be submitted to Discrete Mathematics.

[3] K. Ushio and R. Tsuruno, Cyclic S_k-factorization of complete bipartite digraphs, *Graph theory, Combinatorics, Algorithms and Applications, SIAM, Philadelphia*, **PA** (1991), pp. 551–563.

[4] K. Ushio, Star factorizations of symmetric complete bipartite digraphs, *Discrete Math.*, **167/168** (1997), pp. 593–596.

[5] K. Ushio, $\hat{S}_k$-factorizations of symmetric complete tripartite digraphs, *Discrete Math.*, **197/198** (1999), pp. 791–797.

[6] H. Wang, $K_{1,k}$-factorization of a complete bipartite digraph, *Discrete Math.*, **126** (1994), pp. 359–364.

Proc. of ICDM 2006, Pages 237–243

Some New Degree Concepts in Graphs

S.S. KAMATH
Email: shyam.kamath@gmail.com
R.S. BHAT
Department of Mathematical and Computational Sciences,
National Institute of Technology Karnataka, Surathkal,
Srinivasnagar, Mangalore 575 025, India *Email:* rsbhat_2000@yahoo.com

Abstract. Let $G = (V, E)$ be any graph. For any $v \in V$, the set $N(v) = \{u \in V/uv \in E\}$ is the open neighbourhood of the vertex v; while the set $N[v] = N(v) \bigcup \{v\}$ is the closed neighbourhood of v. Similarly, for any edge $x = uv$, $N(x) = \{y \epsilon E/y$ is adjacent to $x\}$ and $N[x] = N(x) \bigcup \{x\}$. Then $\langle N[x] \rangle$ denotes the induced subgraph of G induced by the set $N(v) \bigcup N(u)$. We define three types of degrees: The VE-degree of a vertex $u \epsilon V$, $d_{ve}(u)$ is the number of edges in $\langle N[v] \rangle$. The EV-degree of an edge $x \in E$, $d_{ev}(x)$ is the number of vertices in $N[x]$ and the EE-degree of an edge $d_{ee}(x)$ is the number of edges in $\langle N[x] \rangle$.
In this paper, we obtain an expression for the sum of VE-degrees of all vertices, the sum of EV-degrees of all the edges and the sum of EE-degrees of all the edges, all of which are similar to the handshaking lemma. We have defined three types of domination, viz., Vertex-Edge Domination, Edge-Vertex Domination and Edge-Edge Domination. We also obtain some bounds for the three domination numbers using the new degree concepts.

Keywords: VE-degree, EV-degree, EE-degree
Subject Classification Number: 05C69

1 Introduction

For the usual notations and terminologies, we refer [5]. For any real number r, $\lfloor r \rfloor$ denotes the greatest integer not greater than r; while $\lceil r \rceil = -\lfloor -r \rfloor$.

For any $v \in V$, the set $N(v) = \{u \in V/uv \in E\}$ is the *open neighbourhood* of the vertex v; while the set $N[v] = N(v) \bigcup \{v\}$ is the *closed neighbourhood* of v. Similarly, for any edge $x = uv$, $N(x) = \{y \in E/y$ is adjacent to $x\}$ and $N[x] = N(x) \bigcup \{x\}$. Then $\langle N[v] \rangle$ denotes the subgraph of G induced by the set $N[v]$. For an edge $x = uv$, $\langle N[x] \rangle$ denotes the subgraph of G induced by the set $N(v) \bigcup N(u)$. The *degree of a vertex* v, $d(v) = |N(v)|$ and the *degree of an edge* $x = uv$, $d_e(x) = d(u) + d(v) - 2$. Then $\Delta(G)$ denotes the *maximum degree* of G and $\Delta_e(G)$ denotes the *maximum edge degree* of G. Similarly, $\delta(G)$ and $\delta_e(G)$ denote the *minimum degree* and *minimum edge degree* of G respectively.

The mixed domination number has been defined by E. Sampathkumar and S.S. Kamath [10]. A vertex $v \in V$ *m-dominates* an edge $x \in E$ if $x \in \langle N[v] \rangle$. A set $S \subseteq V$ is a *Vertex-Edge Dominating set* (*VED-set*) if every edge in G is m-dominated by a vertex in S. The *vertex-edge domination number* $\gamma_{ve}(G)$ is the minimum cardinality of a VED-set. An edge x, *m- dominates* a vertex v if $v \in N[x]$. A set $L \subseteq E$ is an *Edge-Vertex Dominating set* (*EVD-set*) if every vertex in G is m-dominated by an edge in L. The *edge-vertex domination number* $\gamma_{ev}(G)$ is the minimum cardinality of an EVD-set. The edge-edge domination is defined by S.S.Kamath and R.S.Bhat [7]. Let $x,\ y \in E$. Then the edge x, *e-dominates* an edge y if $y \in \langle N[x] \rangle$. A set $L \subseteq E$ is an *Edge-Edge Dominating set* (*EED-set*) if every edge in $E - L$ is e-dominated by an edge in L. The *edge-edge domination number* $\gamma_{ee}(G)$is the minimum cardinality of an EED-set. These concepts are further studied in [4, 6, 8, 9].

Handshaking Lemma. We quote here the well known first theorem in Graph Theory for our reference which gives an expression for the sum of all vertex degrees.

Theorem 1.1. *For any (p, q) graph G =(V, E),*

$$\sum_{v \in V} d(v) = 2q.$$

The edge analogue of the handshaking lemma is discussed in [1–3].

Theorem 1.2. *For any (p, q) graph G =(V, E),*

$$\sum_{x \in E} d_e(x) = \sum_{x=uv,\ x \in E} (d(u) + d(v) - 2) = \sum_{u \in V} [d(u)]^2 - 2q.$$

The following corollary is immediate from the above theorem.

Corollary 1.3. *For any (p, q) graph G =(V, E),*

$$\sum_{x=uv} [d(u) + d(v)] = \sum_{u \in V} [d(u)]^2.$$

In our attempts to obtain some bounds for the three domination parameters, we have defined the VE-degree of a vertex, the EV-degree of an edge and the EE-degree of an edge. Naturally, we give the handshaking lemma type results for these new degree parameters. We also obtain some bounds for the three domination parameters using the new degree concepts.

2 VE-Degree

The *VE-degree* (*Vertex-Edge degree*) of a vertex $v \in V$, $d_{ve}(u)$ is the number of edges m-dominated by v or equivalently, $d_{ve}(u)$ is the number of edges in $\langle N[u] \rangle$. Then $\Delta_{ve}(G)$ and $\delta_{ve}(G)$ denote the *maximum* and the *minimum VE-degrees* of G respectively.

Example 1. The vertex labels of G_1 and G_2 in Fig. 1 represent the VE-degree of the corresponding vertices. Here, $\Delta_{ve}(G_1) = 10, \delta_{ve}(G_1) = 5, \Delta_{ve}(G_2) = 7$, $\delta_{ve}(G_2) = 1$. Also, observe that $\gamma_{ve}(G_1) = 2$ and $\gamma_{ve}(G_2) = 3$.

Remark 1. We observe that for any triangle free graph G, $d(u) = d_{ve}(u)$, $u \in V$ and hence, $\Delta(G) = \Delta_{ve}(G)$.

We now obtain an expression for the sum of the VE-degrees of all vertices (a result similar to the Handshaking Lemma).

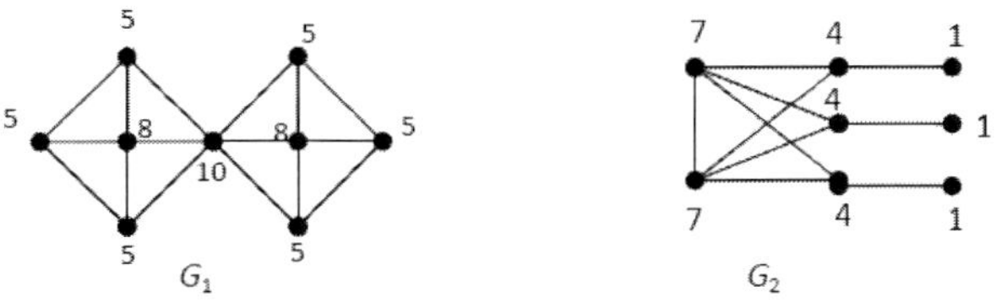

Fig.1.

Proposition 2.1. *For any graph G of order p and size q with t triangles,*

$$\sum_{u \in V} d_{ve}(u) = 2q + 3t \tag{1}$$

Proof. Let m be the number of triangles in G containing the vertex $u \in V$. We first observe that $d_{ve}(u) = d(u) + m$, since in counting the VE-degree of the vertex u, we count $d(u)$ edges and one edge in each of the m triangles which is not incident on u is also counted. Further, whenever a triangle exists, the three vertices of the triangle count a total of 3 to the sum and hence, we have $\sum_{u \in V} d_{ve}(u) = \sum_{u \in V}[d(u) + m] = 2q + 3t$ where t is the total number of triangles in G.

Corollary 2.2. *For any triangle free graph G of order p and size q,*

$$\sum_{u \in V} d_{ve}(u) = 2q.$$

Example 2. For the graph G_1 in Fig. 1, the number of triangles $t = 8$ and $q = 16$. Then $\sum\limits_{v \in V} d_{ve}(u)$ = sum of all the vertex labels $= 56 = 2 \times 16 + 3 \times 8 = 2q + 3t$.

3 EV-Degree

The *EV-degree* (*Edge-Vertex degree*) $d_{ev}(x)$, $x \in E$, is the number of vertices m-dominated by x, or equivalently, $d_{ev}(x)$ is the number of vertices in $N[x]$. Then $\Delta_{ev}(G)$, $\delta_{ev}(G)$ denote the maximum and minimum EV-degrees of G respectively.

Example 3.

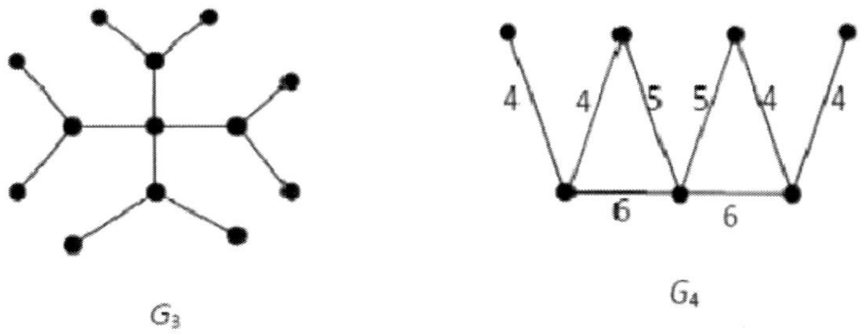

Fig.2

In G_3 of Fig. 2, for every pendant edge x, $d_{ev}(x) = 4$. For every non-pendant edge x, $d_{ev}(x) = 7$. Further, $\Delta_{ev}(G_3) = 7$, $\delta_{ev}(G_3) = 4$. The edge labels in G_4

represent the EV-degree of each edge and $\Delta_{ev}(G_4) = 6$, $\delta_{ev}(G_4) = 4$. Also, observe that $\gamma_{ev}(G_3) = 4$ and $\gamma_{ev}(G_4) = 2$.

Remark 2. We observe that for any triangle free graph G, $d_{ev}(x) = d_e(x) + 2$, for every $x \in E$ and hence, $\Delta_{ev}(G) = \Delta_e(G) + 2$.

We now obtain an expression for the sum of the EV-degrees of all the edges.

Proposition 3.1. *For any $(p,\ q)$ graph G with t triangles,*

$$\sum_{x\in E} d_{ev}(x) = \sum_{u\in V} [d(u)]^2 - 3t \tag{2}$$

Proof. We first note that for any edge $x = uv$, which is contained in m triangles $d_{ev}(x) = d(v)+d(u)-m$ because if x is in a triangle then the vertex which is not incident on the edge x is counted twice once in $d(u)$ and once in $d(v)$ and hence we have to subtract 1 for each of the m triangles from the sum $d(v) + d(u)$. Further, whenever a triangle exists in G, the three edges of the triangle reduce the sum by 3 and hence the t triangles in G reduce the sum by $3t$ and thus, we have $\sum_{x\in E} d_{ev}(x) = \sum_{x=uv}[d(u) + d(v) - m] = \sum_{x=uv}[d(u) + d(v)] - 3t = \sum_{u\in V}[d(u)]^2 - 3t$ (using corollary 1.3).

Corollary 3.2. *For any $(p,\ q)$ graph G without triangles,*

$$\sum_{x\in E} d_{ev}(x) = \sum_{u\in V} [d(u)]^2.$$

Example 4. For the graph G_4 in Fig. 2, the number of triangles = 2 and $\sum_{u\in V}[d(u)]^2 = 44$. Therefore, $\sum_{x\in E} d_{ev}(x) =$ sum of all the edge labels $= 38 = 44 - 3 \times 2 = \sum_{u\in V}[d(u)]^2 - 3t$.

4 EE-Degree

The *EE-degree* (*Edge-Edge degree*) $d_{ee}(x)$ of an edge $x \in E$, is the number of edges e-dominated by x together with x or equivalently, it is the number of edges in $\langle N[x]\rangle$. Then $\Delta_{ee}(G)$ and $\delta_{ee}(G)$ denote the maximum and minimum EE-degrees of G.

Example 5.

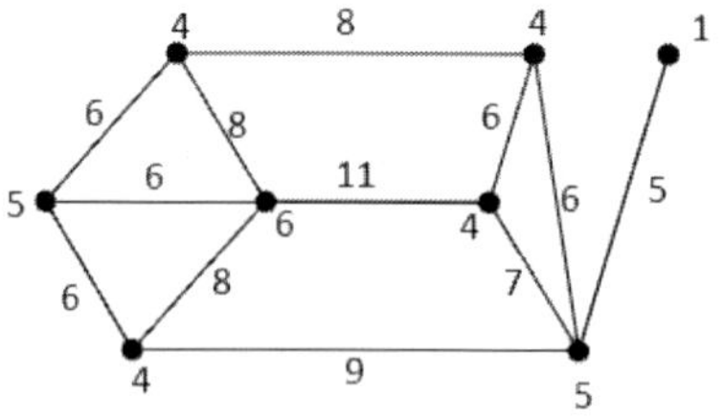

Fig .3 G_5

The edge labels represent the EE-degree of each edge and $\Delta_{ee}(G_5) = 11$ and $\delta_{ee}(G_5) = 5$. Also, the vertex labels represent the VE-degree of each vertex. We observe that $\gamma_{ee}(G_5) = 2$.

Remark 3. If G is a graph without triangles and quadrilaterals (cycle C_4), then $d_{ee}(x) = d_e(x) + 1$. Thus, if G is any tree or a bipartite graph without C_4, then $\Delta_{ee}(G) = \Delta_e(G) + 1$ and $\delta_{ee}(G) = \delta_e(G) + 1$.

Lemma 4.1.

$$\sum_{x=uv,\ x\in E} [d_{ve}(u) + d_{ve}(v)] = \sum_{u\in V} [d_{ve}(u)d(u)]$$

.

Proof. In counting the VE-degrees at both the vertices of every edge, the VE-degree of every vertex u is counted $d(u)$ times. Hence the result follows.

Theorem 4.2. *Let G be a graph of order p and size q. Let t_1 be the number of triangles in G and t_2 be the number of quadrilaterals without induced triangles. Then*

$$\sum_{x\in E} d_{ee}(x) = \sum_{u\in V} [d_{ve}(u)d(u)] - q - 6t_1 + 4t_2. \tag{3}$$

Proof. By definition, for any $x \in E$, $d_{ee}(x)$ is the number of edges in $\langle N[x]\rangle$. We first observe that the contribution of an edge x to varies in the following cases:

1. An edge x is not contained in any triangle or quadrilateral.

2. An edge x is contained in a triangle.

3. An edge x is contained in a quadrilateral which does not contain any induced triangle.

4. An edge x is a common edge of a triangle and a quadrilateral.

We now discuss the contribution of an edge to the sum in each case.

Case 1. If an edge $x = uv$ is not contained in any triangle or quadrilateral. Then $d_{ee}(x) = d_{ve}(u) + d_{ve}(v) - 1$. Since in counting $d_{ve}(u)$ and $d_{ve}(v)$ the edge x is counted twice. Hence we subtract 1 from the sum $d_{ve}(u) + d_{ve}(v)$.

Case 2. If x is contained in a triangle. Let m_1 be the number of triangles containing the edge $x = uv$ and (u, v, w) be a triangle containing the edge x. Then in counting $d_{ve}(u)$ and $d_{ve}(v)$ the edges uw and vw are counted twice and hence we have to subtract 2 from the sum $d_{ve}(u) + d_{ve}(v) - 1$ and this is true for all the m_1 triangles containing the edge x. Therefore, we have $d_{ee}(x) = d_{ve}(u) + d_{ve}(v) - 1 - 2m_1$.

Case 3. If x is contained in a quadrilateral which does not contain any induced triangle. Let m_2 be the number of quadrilaterals containing the edge $x = uv$ which does not contain any induced triangle. Let (u, v, w, z) be such a quadrilateral. Then in counting $d_{ve}(u)$ and $d_{ve}(v)$ the edge wz is not at all counted. Hence we have to add 1 to the sum $d_{ve}(u)+d_{ve}(v)-1$. Further, this is true for all the m_2 quadrilaterals containing the edge x and therefore, we have $d_{ee}(x) = d_{ve}(u) + d_{ve}(v) - 1 + m_2$.

Case 4. If x is a common edge of a triangle and a quadrilateral. Since x is an edge of a triangle, this case is same as case 2.

Combining all the three distinct cases, in general we have $d_{ee}(x) = d_{ve}(u) + d_{ve}(v) - 1 - 2m_1 + m_2$. Further, whenever a triangle exists in G, each edge reduces the sum by 2 (from case 2); therefore, the three edges of the triangle reduce the sum by 6. Thus, the t_1 triangles reduce the sum by $6t_1$. Moreover, whenever a quadrilateral without induced triangle exists each edge counts 1 to the sum (from case 3), then the four edges of the quadrilateral count a total of 4 to the sum. Hence the t_2 quadrilaterals without induced triangles count a total of $4t_2$ to the sum. Thus, we have $\sum_{x\in E} d_{ee}(x) = \sum_{x=uv_{x\in E}} [d_{ve}(u) + d_{ve}(v) - 1 - 2m_1 + m_2] = \sum_{x=uv_{x\in E}} [d_{ve}(u) + d_{ve}(v)] - q - 6t_1 + 4t_2 = \sum_{u\in V} [d_{ve}(u)d(u)] - q - 6t_1 + 4t_2$ using lemma 4.1. This completes the proof.

Corollary 4.3. *If G is a graph which is free from triangles and quadrilaterals, then*

$$\sum_{x\in E} d_{ee}(x) = \sum_{u\in V} [d(u)]^2 - q.$$

Example 6. The above result is illustrated for the graph G_5 given in Fig. 3. We observe that $\sum_{x\in E} d_{ee}(x)$ = sum of all the edge labels = 86. In the graph G_5, $q = 12$, the number of triangles = $t_1 = 3$, the number of quadrilaterals without induced triangles = $t_2 = 2$ and hence, $\sum_{u\in V}[(d_{ve}(u)d(u)) - q - 6t_1 + 4t_2] = 108 - 12 - 18 + 8 = 86$.

Using the above degree concepts the following bounds are obtained for the three domination numbers.

Proposition 4.4. *For any $(p,\ q)$ graph G,*

$$\lceil \frac{q}{\Delta_{ve}} \rceil \leq \gamma_{ve} \leq p - \Delta \tag{4}$$

$$\lceil \frac{p}{\Delta_{ev}} \rceil \leq \gamma_{ev} \leq p - \Delta_{ev} + 1 \tag{5}$$

$$\lceil \frac{q}{\Delta_{ee}} \rceil \leq \gamma_{ee} \leq p - \Delta_{ee} + 1 \tag{6}$$

Further, these bounds are sharp.

Proof. Since a vertex can m-dominate at most $\frac{q}{\Delta_{ve}}$ edges. This implies the lower bound in (4). If v is a vertex of maximum degree then $V - N(v)$ is a VED-set of G. Hence $\gamma_{ve} \leq |V - N(v)| = p - \Delta$. Since any edge can m-dominate at most Δ_{ev} vertices, and we have to exhaust all the p vertices, we need at least $\frac{p}{\Delta_{ev}}$ edges to m-dominate all the vertices. This implies the lower bound in (5). Let D be the set of vertices that are m-dominated by the edge x, so that $|D| = \Delta_{ev}$. Let $V - D = \{u_1, u_2, ..., u_k\}$. Choose an edge y_i, $1 \leq i \leq k$, such that y_i is incident on the vertex u_i. Then the set $L = \{y_1, y_2, ..., y_k\}$ m-dominates all the vertices in $V - D$. Hence $L \bigcup \{x\}$ is an EVD-set of G. Therefore, $\gamma_{ev} \leq |L \bigcup \{x\}| = |L| + 1 = |V - D| + 1 = p - \Delta_{ev} + 1$. Hence the upper bound in (5) follows.

Let x be an edge of degree $\Delta_{ee}(G)$. Then x e-dominates all the edges in $\langle N[x]\rangle$. Further, all the edges in $E - \langle N[x]\rangle$ are e-dominated by themselves. Therefore, $E - \langle N[x]\rangle \bigcup \{x\}$ is an EED-set of G. Hence $\gamma_{ee} \leq q - \Delta_{ee} + 1$. Since an edge in G can e-dominate at most Δ_{ee} edges and we have to exhaust all the q edges of G, we need at least $\frac{q}{\Delta_{ee}}$ edges to e-dominate all the edges. This implies the lower bound in (6). That the bounds are sharp, is evident as the complete bipartite graph $K_{m,n}$ attains the upper and the lower bounds in (4)–(6).

Acknowledgement: The authors are grateful to the referees for their valuable comments/ suggestions.

References

[1] B.D. Acharya, Sum-degrees and difference degrees of edges in a graph, in *Proc. 2nd International Conference on "Matrix Theory and Number Theory" (CFTRI, Mysore: March, 1972).*

[2] B.D. Acharya and M.N. Vartak, On the construction of graphs with given constant valence-difference(s) on each of their lines, *Wiss. Z., TH Ilmenau*, **23(6)** (1977), pp. 33–60. MR-05C99.

[3] R. Balakrishnan and K. Ranganathan, *A text book of Graph Theory*, Springer, Special Price Edition (2003).

[4] R.S. Bhat, *A study of strong (weak) domination parameters*, Ph.D. Thesis, National Institute of Technology Karnataka, Surathkal, India (2007).

[5] F. Harary, *Graph Theory*, Addison Wesley, (1969).

[6] S.S. Kamath, *Some studies in Graph Theory (Domination Parameters)*, Ph.D. Thesis, Karnatak University, Dharwad, India (1991).

[7] S.S. Kamath and R.S. Bhat, Strong (Weak) Edge-Edge Domination Number of a Graph, *Indian Journal of Mathematics*, (submitted).

[8] S.S. Kamath and R.S. Bhat, Strong (Weak) Edge-Vertex Domination Number of a Graph, *Ind. J. of Pure and Appl. Math.*, (submitted).

[9] S.S. Kamath and R.S. Bhat, Strong/Weak Neighbourhood Number of a graph (K_3-coverings), in *Proc. International Conference on Discrete Mathematics, Editor: Sethumadhavan* Narosa Pub., New Delhi, India (2004), pp. 142–150.

[10] E. Sampathkumar and S.S. Kamath, Mixed Domination in Graphs, *Sänkhya*, (Special Volume) **54** (1992), pp. 399–402.

Proc. of ICDM 2006, Pages 245–248

The Edge C_4 Graph of a Graph

Manju K. Menon
Email: manjukmenon@cusat.ac.in
A. Vijayakumar
Department of Mathematics, Cochin University of Science and Technology, Cochin-682 022, India. *Email:* vijay@cusat.ac.in

Abstract. The edge C_4 graph $E_4(G)$ of a graph G has all the edges of G as its vertices, two vertices in $E_4(G)$ are adjacent if their corresponding edges in G are either incident or are opposite edges of some C_4. In this paper, characterizations for $E_4(G)$ being connected, complete, bipartite, tree etc are given. We have also proved that $E_4(G)$ has no forbidden subgraph characterization. Some dynamical behaviour such as convergence, mortality and touching number are also studied.

1 Introduction

The study of graph dynamics has been receiving wide attention since Ore's work on the line graph operator $L(G)$, see [5,6]. Of concern in this paper is the notion of edge C_4 graph $E_4(G)$ of a graph G. The vertices of $E_4(G)$ are in one one correspondence with the edges in G and two vertices in $E_4(G)$ are adjacent if their corresponding edges in G either intersect or are opposite edges of some C_4 in G. So two vertices are adjacent vertices in $E_4(G)$ if the union of the corresponding edges in G induces any one of the graphs P_3, C_3, C_4, $K_4 - \{e\}$, K_4. This edge C_4 graph is the edge graph mentioned in [6].

Clearly the edge C_4 graph coincides with the line graph for any acyclic graph. But they differ in many properties. As a case, for a connected graph G, $E_4(G) = G$ if and only if $G = C_n$, $n \neq 4$. Then we say that C_n, $n \neq 4$ is fixed under E_4. Also Beineke has proved in [2] that the line graph has nine forbidden subgraphs. In this paper we see that $E_4(G)$ has no forbidden subgraphs.

In [1], Bandelt and others proved that a bipartite graph is dismantlable if and only if its edge C_4 graph is dismantlable and a bipartite graph is neighbourhood-Helly if and only if its edge C_4 graph is neighbourhood-Helly.

All the graphs considered here are finite, undirected and simple. We denote by P_n (respectively C_n), a path (respectively cycle) on n vertices. The graph obtained by deleting any edge of K_n is denoted by $K_n - \{e\}$. A triangle with a pendant edge attached to any one of its vertices is called a 'paw'. A graph G is H-free if G does not contain H as an induced subgraph. P_4-free graphs are called cographs [3, 4]. A graph H is a forbidden subgraph for a property P of graphs if no graph having property P contains an induced subgraph isomorphic to H. The cross product $G_1 \times G_2$ of two graphs G_1 and G_2 is a simple graph with $V(G_1) \times V(G_2)$ as its vertex set and two vertices (u_1, v_1) and (u_2, v_2) are adjacent in $G_1 \times G_2$ if and only if either $u_1 = u_2$ and v_1 is adjacent to v_2 in G_2, or u_1 is adjacent to u_2 in G_1

and $v_1 = v_2$. For all basic concepts and notations not mentioned in this paper we refer [7].

In this paper, characterizations for the edge C_4 graph of a graph G being connected, complete, bipartite etc are obtained. We have also proved that the edge C_4 graph has no forbidden subgraph characterization. The dynamical behaviour such as convergence, periodicity, mortality, transition number and touching number of $E_4(G)$ are also studied.

2 The Edge C_4 Graph of a Graph

Consider the edge C_4 graph $E_4(G)$ of a graph G . If $a_1 - a_2$ is an edge in G, the corresponding vertex in $E_4(G)$ is denoted by a_1a_2.

Theorem 1. *$E_4(G)$ is connected if and only if exactly one component of G contains edges.*

Theorem 2. *For a connected graph G, $E_4(G)$ is complete if and only if G is a complete multipartite graph.*

Proof. Let G be a connected graph such that $E_4(G)$ is complete. We shall first show that G is a cograph and is paw-free. If G contains an induced P_4 then the first and the third edges in P_4 correspond to two non adjacent vertices in $E_4(G)$, and $E_4(G)$ is not complete. Thus G must be a cograph. Further if G contains a paw as an induced subgraph then the pendant edge and the edge in the triangle of the paw to which the pendant edge is not adjacent correspond to non adjacent vertices in $E_4(G)$. Hence G is also paw-free. Claim: G is a complete multipartite graph. If not, $\overline{G}$ is not a union of complete graphs. Then $\overline{G}$ contains an induced P_3. But, since G is a connected cograph, $\overline{G}$ is disconnected. Hence $\overline{G}$ is a disconnected graph containing an induced P_3 and so G has a paw, giving a contradiction. This proves the claim.

Conversely suppose that G is a complete multipartite graph. Let e_1 and e_2 be any two edges in G. If they are not adjacent, then since G is complete multipartite, they are opposite edges of some C_4 in G. Hence $E_4(G)$ is a complete graph.

Theorem 3. *There is no forbidden subgraph characterization for $E_4(G)$.*

Proof. We shall prove that given any graph G, we can find a graph H such that G is an induced subgraph of $E_4(H)$. For any graph G, let $H = G \times K_2$. Then in $E_4(H)$, all the vertices of the form uu' where u is a vertex in G and u' is the corresponding vertex in the copy of G used in the construction of $G \times K_2$ will induce G. For, if u and v are any two adjacent vertices in G, uu' and vv' correspond to adjacent vertices in $E_4(H)$ as $uu'v'v$ forms a C_4 in H. If u and v are any two non adjacent vertices in G then uu' and vv' are non adjacent vertices in $E_4(H)$. □

Theorem 4. *For a connected graph G, $E_4(G)$ is bipartite if and only if G is either a path or an even cycle of length greater than five.*

Corollary 1. *For a connected graph G, $E_4(G)$ is a tree if and only if G is a path.*

3 Dynamical Properties

We shall first list some graph dynamical terminologies from [6].
Let G be any graph. The nth iterated graph is iteratively defined as $E_4^2(G) = E_4(E_4(G))$ and $E_4^n(G) = E_4(E_4^{n-1}(G))$ for $n > 2$. We say that G is convergent

under E_4 if $\{E_4^n(G),\ n \in N\}$ is finite. If G is not convergent under E_4, then G is divergent under E_4. A graph G is periodic if there is some natural number n with $G = E_4^n(G)$. The smallest such number is called the period of G. The transition number $t(x)$ of a convergent graph G is defined as zero if G is periodic and as the smallest number n such that $E_4^n(G)$ is periodic. A graph G is mortal if for some $n \in N$, $E_4^n(G) = \phi$, the empty graph. The touching number of a cycle is the cardinality of the set of all edges having exactly one of its vertices on the cycle. For every integer $n \geq 3$, the n-touching number $t_n(G)$ of a graph G is the supremum of all touching numbers of C_n, provided G contains some C_n. If not, $t_n(G)$ is undefined.

Theorem 5. *The graphs P_n, $K_{1,3}$, $C_n(n \neq 4)$ are the only E_4 convergent graphs.*

Proof. If G contains a vertex of degree > 3, then $E_4(G)$ contains K_4. In the subsequent iterations the clique size goes on increasing and hence G diverges. So, for convergent graphs $\Delta(G) \leq 3$. If G is a tree which is neither P_n nor $K_{1,3}$, then K_4 is contained at least in the third iterated graph and hence G cannot converge. The paths P_n converge to ϕ and $K_{1,3}$ converges to the triangle. □

Consider the graphs which are not trees. If G is not a cycle, then G contains a cycle with a pendant edge as a subgraph (need not be induced). Then K_4 is a subgraph at least in the second iteration and hence in the subsequent iterations the clique size will go on increasing and hence cannot converge. All cycles except C_4 are fixed under E_4 and C_4 is not convergent.

Corollary 2. *For $E_4(G)$, the only periodic graphs are the cycles $C_n, n \neq 4$ and they have period one.*

Proof. Proof is clear from the remark in [6] that a graph G is convergent if and only if G is either periodic or there is some positive integer n with $E_4^n(G)$ periodic. □

Corollary 3. *The transition number $t(K_{1,3}) = 1$ and $t(C_n)$, $n \neq 4 = 0$.*

Corollary 4. *For $E_4(G)$, the paths are the only mortal graphs.*

Proof. Among the convergent graphs, cycles other than C_4 are fixed and $K_{1,3}$ converges to K_3. The paths are the only graphs converging to ϕ. □

In the following theorem, we consider only the graphs G for which the touching number $t_n(G)$ is defined.

Theorem 6. *For any graph G, $t_n(E_4(G)) \geq 2\ t_n(G)$. Further if G contains C_4 as a subgraph where either an edge or two consecutive edges of C_4 are the edges of the C_n which determines the touching number then $t_n(E_4(G) > 2t_n(G)$.*

Proof. Let the n-cycle in G be $x_1, x_2, \ldots, x_n, x_1$. Then x_1x_2, x_2x_3, $\ldots$, x_nx_1 is an n-cycle in $E_4(G)$. If yx_i is a touching edge in G, then $E_4(G)$ contains two touching edges say $yx_i - x_ix_{i+1}$ and $yx_i - x_{i-1}x_i$.

Let $C_4 = a_1a_2a_3a_4$ be a subgraph of G. Further, suppose that the edge a_3a_4 is a touching edge in G . Then a_1a_4, a_2a_3, a_1a_2 are touching edges in $E_4(G)$. Hence $t_n(E_4(G)) > 2\ t_n(G)$. The proof is similar for the case when any two consecutive edges of C_4 are the edges on the C_n mentioned above. □

Acknowledgement: The authors thank Prof. S.B. Rao, ISI, Kolkata for his help during the preparation of this paper and the referee for his valuable suggestions for the improvement of the paper.

References

[1] H.J. Bandelt, M. Farber and P. Hell, Absolute reflexive retracts and absolute bipartite retracts, *Discrete Appl. Math.*, **44** (1993), pp. 9–20.

[2] Beineke, On derived graphs and digraphs, in *Beiträge zur Graphentheorie* (1968), pp. 17–23.

[3] A. Brandstädt, V.B. Le and J.P. Spinrad, Graph Classes, *SIAM*, (1999).

[4] D.G. Corneil, Y. Perl and I.K. Stewart, A linear recognition algorithm for cographs, *SIAM J. Comput.*, **14** (1985), pp. 926–934.

[5] O. Ore, Theory of Graphs, *Amer. Math. Soc. Coll. Publ.*, **38** Providence (R.I) (1962) p. 21.

[6] E. Prisner, *Graph Dynamics*, Longman (1995).

[7] D.B. West, Introduction to Graph Theory, Prentice Hall of India (2003).

Proc. of ICDM 2006, Pages 249–254

Topological Properties of the Set of All Minimal Dominating Functions of Graphs

K. Reji Kumar
Department of Mathematics, N.S.S College, Pandalam - 689 501, India
Email: rkkmaths@yahoo.co.in

Abstract. A dominating function (DF) of a graph $G = (V, E)$ is a function $f : V \to [0, 1]$ such that for all $v \in V$, sum of the function values over the closed neighbourhood of v is at least one. The dominating function f is called a minimal dominating function (MDF), if f doesn't remain a DF when the function value at some vertex is reduced. In this paper we study some topological properties of the set of all MDFs ($\mathfrak{F}_S$) of certain classes of graphs.

2000 Mathematics Subject Classification: Primary 05C 69;
Secondary 05C 35, 05C 10

Keywords : Minimal dominating function, Convex combination of minimal dominating functions, Basic minimal dominating function, Simplex, Simplicial complex.

1 Introduction

In this paper we consider only finite, simple undirected connected graphs. Exceptional cases will be stated clearly. The terminology is same as that of Harary [5]. The concepts of dominating function and minimal dominating function were introduced by Cockayne et al. [3]. A *dominating function* (DF) of a graph $G = (V, E)$ is a function $f : V \to [0, 1]$ such that $\sum_{x \in N[v]} f(x) \geq 1$ for all $v \in V$, where $N[v]$ is the closed neighborhood of v. A *minimal dominating function* (MDF) is a DF f such that it does not remain a DF if for any $v \in V$, the value of $f(v)$ is decreased. A *dominating set* of $G = (V, E)$ is a subset X of V such that every vertex of $V - X$ is adjacent to some vertex in X. Smallest such set is called a *minimal dominating set.* The set of all minimal dominating functions of a graph G is denoted by $\mathfrak{F}_S(G)$. It is interesting to note that $0-1$ valued DFs and MDFs are characteristic functions of dominating sets and minimal dominating sets respectively. For more details and base papers of MDFs the reader is referred to [4, 6, 7], and [8].

For a DF f of the graph $G = (V, E)$, $f(N[v])$ is used to denote

$$\sum_{x \in N[v]} f(x).$$

The *boundary* of f or B_f is the set $\{v \in V / \sum_{x \in N[v]} f(x) = 1\}$. The *positive set* of f or P_f is $\{v \in V / f(v) > 0\}$. For subsets A and B of V, we write $A \to B$ if every vertex in $B - A$ is adjacent to some vertex in A. For two MDFs f and g of G, the convex combination of f and g is defined by $h_\lambda = \lambda f + (1 - \lambda)g$ where $0 < \lambda < 1$. Convexity of the set of MDFs of a graph was first studied by B.D. Acharya [1]. Since any convex combination of dominating functions is again a dominating function, it follows that the set of all DFs forms a convex set. However it is evident from the following theorems that the convex combination of two MDFs need not be an MDF always.

Theorem 1.1 **([3]).** *A dominating function f of the graph G is minimal iff $B_f \to P_f$.*

Cockayne et al. [3] obtained a necessary and sufficient condition for the convex combination of two MDFs to be again an MDF.

Theorem 1.2 **([3]).** *Let f and g be two MDFs of $G = (V, E)$, then h_λ is minimal iff $B_f \bigcap B_g \to P_f \bigcup P_g$.*

An MDF is called a *basic minimal dominating function* (BMDF), if it cannot be expressed as a proper convex combination of two distinct MDFs. Set of all BMDFs of G is denoted by $\mathfrak{F}_{BS}$. K. Reji kumar et al. [9] presented a necessary and sufficient condition for an MDF to be a basic MDF.

Theorem 1.3 **([9]).** *Let f be an MDF. Then f is a BMDF if and only if there does not exist another MDF g such that $B_f = B_g$ and $P_f = P_g$.*

Corollary 1.4 **([10]).** *The number of BMDFs of a graph G is finite.*

Next theorem provide an easy way to check whether a given MDF is basic or not. This has reduced the difficulty in finding all basic minimal dominating functions of a given graph to a great extent.

Theorem 1.5 **([9]).** *Let f be an MDF of a graph $G = (V, E)$ with $B_f = \{v_1, v_2, \ldots, v_m\}$ and $P'_f = \{u \in V / 0 < f(u) < 1\} = \{u_1, u_2, \ldots, u_n\}$. Let $A = (a_{ij})$ be an $m \times n$ matrix defined by*

$$a_{ij} = \begin{cases} 1 & \text{if } v_i \text{ is adjacent to } u_j \text{ or } v_i = u_j \\ 0 & \text{otherwise} \end{cases}$$

Consider the system of linear equations given by

$$\sum_{j=1}^{n} a_{ij} x_j = 0, \quad \text{where } 1 \leq i \leq m. \tag{1}$$

Then f is a BMDF if and only if (1.1) does not have a non-trivial solution.

Corollary 1.6 **([10]).** *Every $0 - 1$ MDF is a BMDF.*

Theorem 1.7 **([10]).** *If the convex combination of a collection of MDFs $f_1, f_2, \ldots, f_n$ is minimal, then the convex combination of any proper sub collection of these MDFs is an MDF.*

An $n-1$ *simplex* in the Euclidean space is the convex combination of n standard basis vectors. A *simplicial complex* consists of a set S of vertices and a set $\mathfrak{T}$ of simplices, such that the following axioms are satisfied.

(i) Each simplex is a finite subset of S.

(ii) Each subset of a simplex is a simplex.

(iii) The union of all simplices equals S.

For further study of simplices and simplicial complexes, the reader is refered to [2] and [11].

2 Topological Properties of the Set of all Minimal Dominating Functions

There is a bijection from the set of all function $f: V \to [0,1]$ of the graph $G = (V, E)$ to the n dimensional cube I^n. Hence the set of all DFs is a subset of I^n. Only a finite number of MDFs are basic (by Corollary 1.4), and all other MDFs can be expressed as the convex combination of a collection of BMDFs. Motivated by all these observations, and Theorem 1.7, we get the following result.

Theorem 2.1. *The set of all MDFs ($\mathfrak{F}_S$) of a graph G is a simplicial complex with maximum dimension $n-1$, where $|\mathfrak{F}_{BS}| = n$.*

Proof. Suppose $\mathfrak{F}_{BS} = \{f_1, f_2, \ldots f_n\}$. We have to consider two cases.

Case (i) : The convex combination of all n BMDFs is minimal. Then $\mathfrak{F}_S$ is an $n-1$ simplex.

Case (ii) : Otherwise proper subsets of $\mathfrak{F}_{BS}$ make convex combinations, which are minimal. Each convex combination is an m simplex, where $m < (n-1)$. Thus $\mathfrak{F}_S$ is the union of a set of simplexes having dimension $\leq (n-2)$. Hence $\mathfrak{F}_S$ is a simplicial complex with dimension at most $n-2$. □

Theorem 2.2. *The set of all MDFs of $G = K_n$ is an $n-1$ simplex.*

Proof. Let $V(K_n) = \{v_1, v_2, \ldots v_n\}$. Define $f_i : V \to [0,1]$ by

$$f_i(v) = \begin{cases} 1 & \text{if } v = v_i \\ 0 & \text{otherwise} \end{cases}$$

Clearly, f_i, where $i = 1, 2, \ldots, n$ are BMDFs, by Corollary 1.6. Let g be any MDF of K_n and let $g(v_i) = \lambda_i$, $1 \leq i \leq n$. Since $N[v_i] = V$ for all i and g is an MDF, we have $\sum_i g(v_i) = \sum_i \lambda_i = 1$. Further $0 \leq \lambda_i \leq 1$, so that any MDF g of K_n is a convex combination of $f_1, f_2, \ldots f_n$. Hence the result. □

Theorem 2.3. *If $G = K_{(1,n)}$ where $n \geq 1$, then $\mathfrak{F}_S = K_2$.*

Proof. Let $V(K_{(1,n)}) = \{v, v_1, v_2, \ldots, v_n\}$ with $d(v) = n$ and $d(v_i) = 1$ for all $i = 1, 2, \ldots, n$. Let $f_1, f_2 : V \to [0,1]$ be defined by,

$$f_1(x) = \begin{cases} 1 & \text{if } x = v \\ 0 & \text{otherwise} \end{cases}$$

$$f_2(x) = \begin{cases} 0 & \text{if } x = v \\ 1 & \text{otherwise} \end{cases}$$

Clearly f_1 and f_2 are BMDFs of G. Let g be an arbitrary MDF of $K_{(1,n)}$. We claim that g is the convex combination of f_1 and f_2. If $g(v) = 1$, then $g(v_i) = 0$ for all i. Hence $g = f_1$. Also if $g(v) = 0$, then $g(v_i) = 1$ for all i. Hence $g = f_2$. Suppose $g(v) = \lambda$, where $0 < \lambda < 1$. We claim that $v \notin B_g$. Suppose $v \in B_g$. Then

$$g(N[v]) = g(v) + \sum_{i=1}^{n} g(v_i) = 1.$$

Hence

$$\sum_{i=1}^{n} g(v_i) = (1 - \lambda) > 0.$$

So $g(v_i) > 0$ for at least one i. Let $g(v_1) > 0$. Then $g(v_2) \leq 1-\lambda-g(v_1)$. Hence $g(N[v_2]) = g(v) + g(v_2) \leq \lambda + 1 - \lambda - g(v_1) = 1 - g(v_1)$, which is a contradiction. Thus $v \notin B_g$. Since $g(v_i) \geq (1-\lambda)$ for all i and $v \notin B_g$, it follows that $v_i \in B_g$ for all i. Thus $g(v_i) = (1-\lambda)$ for all i. Hence g is a convex combination of f_1 and f_2 and the proof is complete. □

Now we proceed to prove the converse of the above theorem. Before that, we prove four lemmas which are used in the final step.

Lemma 2.4. *If a graph contains a triangle, then it has at least three* $0-1$ *BMDFs.*

Proof. Let v_1, v_2, v_3, v_1 be the triangle in the graph G. We build a set of vertices S_1 such that, $v_1 \in S_1$ and $N(v_1) \bigcap S_1 = \phi$. Choose other vertices so that all vertices in $(V - S_1) - N(v_1)$ are adjacent to at least one vertex in $S_1 - \{v_1\}$ and the set S_1 is minimal. Let $f_1 = \chi_{S_1}$. Similarly, f_2 and f_3 are defined using the vertices v_2 and v_3 respectively. By Theorem 1.6, f_1, f_2 and f_3 are BMDFs. □

Lemma 2.5. *If the graph G contains at least one path v_1, v_2, v_3 such that $N(v_1) \neq N(v_3)$, then G has at least three* $0-1$ *BMDFs.*

Proof. We can construct a minimal dominating set S_1 such that, $v_1 \in S_1$ and $N(v_1) \bigcap S_1 = \phi$. The vertex $v_3 \notin S_1$ and all vertices in $(V - S_1) - N(v_1)$ are adjacent to at least one vertex in $S_1 - \{v_1\}$. Define, $f_1 = \chi_{S_1}$.

The minimal dominating set S_2 is constructed such that, $v_2 \in S_2$ and $N(v_2) \bigcap S_2 = \phi$. Other vertices of S_2 are chosen so that all vertices in $(V - S_2) - N(v_2)$ are adjacent to at least one vertex in $S_2 - \{v_2\}$. Let, $f_2 = \chi_{S_2}$.

Finally the MDs S_3 is taken such that S_3 contains the vertex v_3, and $N(v_3) \bigcap S_3 = \phi$. Also all vertices in $(V - S_3) - N(v_3)$ are adjacent to at least one vertex in $S_3 - \{v_3\}$. Define, $f_3 = \chi_{S_3}$.

Clearly, $f_1(v_1) = 1, f_1(v_2) = 0, f_1(v_3) = 0, f_2(v_1) = 0, f_2(v_2) = 1, f_2(v_3) = 0, f_3(v_1) = 1$ or $0, f_3(v_2) = 0$ and $f_3(v_3) = 1$. Also f_1, f_2 and f_3 are BMDFs. □

Lemma 2.6. *If for all paths v_1, v_2, v_3 in the graph G, $N(v_1) = N(v_3)$ and G contains no triangle, then G is a complete bipartite graph.*

Proof. Since G contains no triangle, v_1 and v_3 are not adjacent. Let $N(v_1) = N(v_3) = \{x_1, x_2, \ldots, x_n\}$. Suppose $V' = V - \{v_1, v_3, x_1, x_2, \ldots, x_n\} \neq \phi$. Since the graph is connected, there exists a vertex $u \in V'$ such that u is adjacent to some vertex in $N(v_1)$. Then by the given condition, we get $N(u) = N(v_1)$. This is true for all vertices in V'. Clearly, $A = N(v_1)$ and $B = (V - N(v_1))$ give the required bipartition. Completeness follows from the condition given in the lemma. □

Lemma 2.7. *If G is a complete bipartite graph and $G \neq K_{(1,n)}$ where $n \geq 1$, then G has more than two* $0-1$ *BMDFs.*

Proof. The graph is $G = K_{(m,n)}$ where $m, n \geq 2$. Let the bipartition of G be $A = \{x_1, x_2, \ldots, x_m\}$ and $B = \{y_1, y_2, \ldots, y_n\}$. Define $f_1, f_2, f_3 : V \rightarrow [0, 1]$ as follows,

$$f_1(v) = \begin{cases} 1 & \text{if } v \in A \\ 0 & \text{otherwise} \end{cases}$$

$$f_2(v) = \begin{cases} 1 & \text{if } v \in B \\ 0 & \text{otherwise} \end{cases}$$

$$f_3(v) = \begin{cases} 1 & \text{if } v = x_1 \text{ or } y_1 \\ 0 & \text{otherwise} \end{cases}$$

The functions f_1, f_2 and f_3 are different BMDFs. □

Theorem 2.8. *If $\mathfrak{F}_S(G)$ is a one simplex, then $G = K_{(1,n)}$, where $n \geq 1$.*

Proof. If the graph G contains either a triangle or a path v_1, v_2, v_3 such that $N(v_1) \neq N(v_3)$ or both, then by Lemma 2.4 and Lemma 2.5, G has at least three $0-1$ BMDFs. Next suppose that G does not contain a triangle and a path with the prescribed properties, then by Lemma 2.6, G is a complete bipartite graph. Now by Lemma 2.7, it is clear that the given graph is $K_{(1,n)}$ where $n \geq 1$. □

It is very interesting to search weather $\mathfrak{F}_S(G)$ is connected or disconnected. In all the known cases, the set is connected. So it is conjectured that $\mathfrak{F}_S$ of any graph is connected.

Conjecture 2.9. *Set of all minimal dominating functions ($\mathfrak{F}_S$) of any graph is connected.*

3 Conclusion and Scope

In this paper we have characterized the graphs, whose minimal dominating set is one simplex. Characterization of graphs having higher dimensional simplices is an interesting open problem. Proving or disproving the conjecture is another area of research. Minimal dominating set ($\mathfrak{F}_S$) of a number of families of graphs are still unknown. This is another area which will give good results.

References

[1] B.D. Acharya, On minimal dominating functions of a graph, *IMS Mathematics Student*, **64 (1-4)** (1995), pp. 170–174, MR-05C35.

[2] Allen Hatcher, *Algebraic Topology*, Cambridge University Press, (2002).

[3] E.J. Cockayne, G. Fricke, S.T. Hedetniemi and C.M. Mynhardt, Properties of minimal dominating functions of graphs, *ARS. Combinatoria.*, **41** (1995), pp. 107–115.

[4] E.J. Cockayne, and C.M. Mynhardt, Minimality and convexity of dominating and related functions in graphs: A unifying theory, *Utilitas Mathematica.*, **51** (1997), pp. 145–163.

[5] F. Harary, *Graph Theory*, Addison-Wesley, Reading, Mass, (1972).

[6] T. W. Haynes, S. T. Hedetniemi and P. J. Slater, *Fundamentals of Domination in Graphs*, Marcel Dekker, Inc., New York, (1998).

[7] T. W. Haynes, S. T. Hedetniemi and P. J. Slater, *Domination in Graphs–Advanced Topics*, Marcel Dekker, Inc., New York, (1998).

[8] G. MacGillivray, E.J. Cockayne and C.M. Mynhardt, Convexity of minimal dominating functions and universal in graphs, *Bull. Inst. Combin. Appl.*, **5**(1992), pp. 37–48.

[9] K. Reji Kumar and S. Arumugam, Basic minimal dominating functions, To appear in Utilitas Mathematica.

[10] K. Reji Kumar, Studies in Graph Theory-Dominating functions, Ph.D thesis, Manonmaniam Sundaranar University, Tirunelveli, India, (2004).

[11] M. L. J Van De Vel, *Theory of Convex Structures*, North-Holland, (1993)

Proc. of ICDM 2006, Pages 255–262

γ_t-Graph of a Graph G

S. Amutha
N. Sridharan*
Alagappa University, Karaikudi 630 003, India.
**Email:* math_sridhar@yahoo.co.in.

Abstract. *A* subset D of vertices of a graph G is said to be a total dominating set of G if each vertex $u \in V(G)$ is adjacent to at least one vertex $v \in D$. The minimum cardinality of a total dominating set of G is called the total domination number of G and is denoted by $\gamma_t(G)$. A total dominating set D of G with cardinality $\gamma_t(G)$ is called a γ_t- set of G. With the graph G, we can associate a graph denoted by $\gamma_t \cdot G$ which is called γ_t - graph of G as follows:
$V(\gamma_t \cdot G)$ is the set of all γ_t - sets of G and two γ_t-sets S and D of G are adjacent in $\gamma_t \cdot G$ if and only if $|S \cap D| = \gamma_t(G) - 1$. In this paper, we initiate a study on γ_t - graph of G.

Keywords: Total dominating set of a graph, total domination number - γ_t excellent graphs.

1 Introduction

We consider only simple graphs. *A* subset D of vertex set $V(G)$ of a graph G is said to be a total dominating set of G if each vertex $u \in V(G)$ is adjacent to at least one vertex $v \in D$. The minimum cardinality of a total dominating set of G is called the total dominating number and is denoted by $\gamma_t(G)$. A total dominating set with cardinality $\gamma_t(G)$ is called a γ_t- set of G. *A* vertex u of G is said to be γ_t-good, if it belongs to some γ_t- set of G and γ_t- bad otherwise. A graph G is called γ_t- excellent if every vertex of G is γ_t-good. Henning et. al [2], have provided a constructive characterization of γ_t- excellent trees. If D_1 and D_2 are two distinct γ_t- sets of G, then $0 \leq |D_1 \cap D_2| \leq \gamma_t - 1$. Yamuna [3] introduced a new class of graphs called just total excellent graphs and initiated a detailed study on just total excellent graph. *A* graph G is said to be just total excellent graph if to each vertex $u \in V(G)$, there is a unique γ_t- set of G containing u. If D_1 and D_2 are two distinct γ_t- sets of a just total excellent graph, then $D_1 \cap D_2 = \phi$. There are graphs such that for every γ_t- set D of G, there exists (at least) one γ_t- set S satisfying $|D \cap S| = \gamma_t - 1$. i.e., there exist $a \in D$ and $b \notin D$ such that $S = (S \backslash \{a\}) \cup \{b\}$ is a γ_t-set of G. Two γ_t-sets D and S of G are said to be exchangable if $|D \cap S| = \gamma_t - 1$. If G is just total excellent graph, there is no exchangable pair of γ_t-sets of G. On the other extreme, there are graphs G such that given any two γ_t-sets D and S of G, there exists a finite sequence $D = D_1, D_2, \ldots, D_k = S$ of γ_t-sets of G with D_i and D_{i+1} are exchangable, for $1 \leq i \leq k-1$. In order to study these graphs, for a

*Supported by UGC assistance F.No.-31-85/2005(SR)

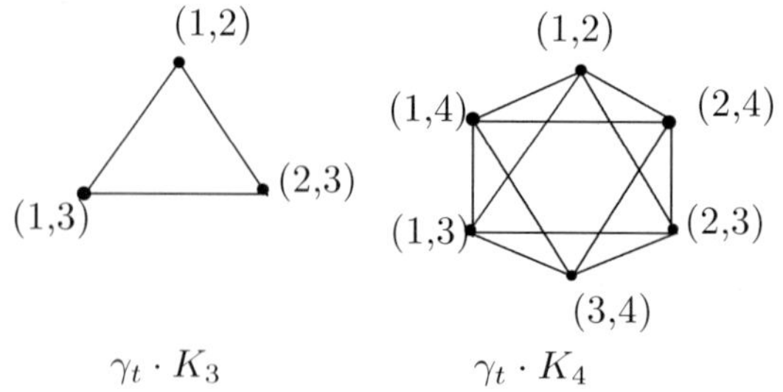

Figure 2.1:

given graph G we associate a graph denoted by $\gamma_t \cdot G$ and called γ_t- graph of G, whose vertex set is the set of all γ_t - sets of G.

Definition (γ_t-graph of G). *If G is a graph, define a new graph denoted by $\gamma_t \cdot G$ as follows:*
$V(\gamma_t \cdot G)$ is the set of all γ_t- sets of G and two γ_t-sets S and D of G are adjacent in $\gamma_t \cdot G$ if and only if $|S \cap D| = \gamma_t(G) - 1$. The graph $\gamma_t \cdot G$ is called the γ_t- graph of G. A graph G is said to be γ_t- connected if $\gamma_t \cdot G$ is connected. G is said to be extremely total excellent if G is γ_t-excellent and γ_t-connected.

Remark. If G is γ_t-connected, then for any two distinct γ_t- sets S_1 and S_2 of G, there is a finite sequence of γ_t - sets $D_1, D_2, \ldots, D_m$ of G such that $S_1 = D_1$; $S_2 = D_m$ and $|D_i \cap D_{i+1}| = \gamma_t(G) - 1, \forall\, i = 1, 2, ...m - 1$.
If D is a γ_t- set of G and $u \in D$, we define $PN(u, D) = \{v \in V(G) | N[v] \cap D = \{u\}\}$, where $N[v]$ is the closed neighbourhood of v in G.

In this paper, we initiate a study on γ_t-graphs. Our notation and terminology are as in [1].

2 γ_t-Graphs of some Standard Graphs

1. $\gamma_t \cdot K_n$: let $n \geq 2$. If $V(K_n) = \{v_i | 1 \leq i \leq n\}$, then for all $i \neq j \in \{1, 2, \ldots, n\}, \{v_i, v_j\}$ is a γ_t- set of K_n. So $\gamma_t \cdot K_n$ has $\frac{n(n-1)}{2}$ vertices and each vertex $D = \{v_i, v_j\}$ of $\gamma_t \cdot K_n$ is adjacent to $2(n-2)$ vertices. Thus $\gamma_t \cdot K_n$ is $2(n-2)$- regular graph.

   Diameter of the graph $\gamma_t \cdot K_n$ is one if $n = 3$ and is 2 for all $n \geq 4$. So K_n is γ_t- connected and hence extremely total excellent. The graphs $\gamma_t \cdot K_3$ and $\gamma_t \cdot K_4$ are shown in Figure 2.1. Each γ_t-set $\{v_i, v_j\}$ is shown as (i, j).

2. $\gamma_t \cdot K_{m,n}$: (Where $m, n \geq 2$.) If $X = \{u_1, u_2, \ldots, u_m\}$ and $Y = \{v_1, v_2, \ldots, v_n\}$ is the partition of the bipartite graph $K_{m,n}$, then each γ_t- set D of $K_{m,n}$ is $\{u_i, v_j\} forsome\ i \in \{1, 2, \ldots, m\}$ and $j \in \{1, 2, \ldots, n\}$ In the graph $\gamma_t \cdot K_{m,n}$, we denote the vertex $\{u_i, v_j\}$ by the ordered pair (i, j). Then $\gamma_t \cdot K_{m,n}$ is the product $K_m \times K_n$, where $K_m = \{1, 2, \ldots, m\}$ and $K_n = \{1, 2, \ldots, n\}$ are complete graphs, and hence $K_{m,n}$ is extremely total excellent.

3. The graphs $\gamma_t \cdot G$, where $G = C_3, C_4$ are shown in Figure 2.2.

4. $\gamma_t \cdot G$ - graph where $G = C_n (n \geq 5)$

If $G = C_n$, let $V(G) = \{v_0, v_1, v_2, ..., v_{n-1}\}$ and $E(G) = \{v_i v_{i+1} \mid 0 \leq i \leq n-1\}$ where addition $i + 1$ is taken under modulo n. If D is a γ_t-set of G, then each

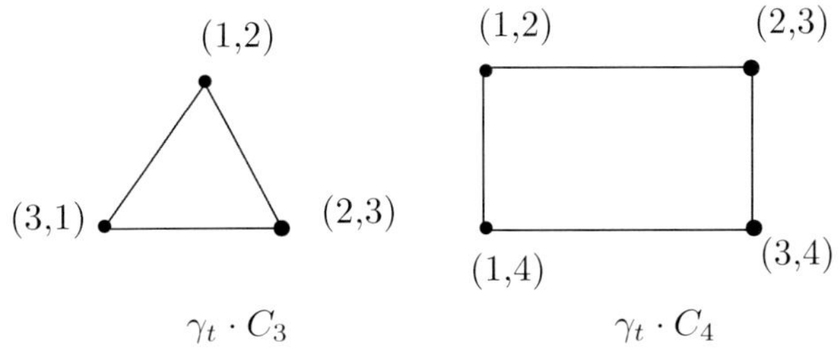

Figure 2.2:

connected component of $G - D$ is either K_1 or K_2. Each connected component of the induced subgraph $< D >$ is P_s for some $s \in \{2, 3, 4\}$. We consider different cases as follows:

Case (i). Let $n = 4k + 3$, where $k \geq 1$.
If D is a γ_t−set of C_n, as $|D| = 2(k + 1), G - D$ has atmost $k + 1$ connected components. As $G - D$ has $2k + 1$ vertices and as each component is either K_1 or $K_2, G - D$ has exactly $k + 1$ components (one of the component is K_1 and all other $k-$ components are K_2). This shows that $\langle D \rangle$ also has $k + 1$ components and each component of $\langle D \rangle$ is K_2. So if D is a γ_t- set of C_n, there is a unique i such that $v_i \notin D$, but both v_{i-1} and $v_{i+1} \in D$, and $D = \{v_j | j = i + 4l + 1$ or $i + 4l + 2, 0 \leq l \leq k\}$. We denote this γ_t- set D by D_i.

Each γ_t- set D is some D_i. If $j = i + 4$, then $D_j = (D_i - \{v_{i+1}\}) \cup \{v_{i+3}\}$ and if $j = i - 4$, then $D_j = (D_i - \{v_{i-1}\}) \cup \{v_{i-3}\}$. Thus if $j = i \pm 4, D_iD_j$ is an edge in $\gamma_t \cdot G$. Let $j \notin \{i, i \pm 4\}$, if $v_i \in D_j$ then either $v_{i+1}, v_{i+2} \notin D_j$ or $v_{i-1}, v_{i-2} \notin D_j$. So $|D_i \cap D_j| \leq \gamma_t(G) - 2$, and D_iD_j is not an edge in $\gamma_t \cdot G$.
If $v_i \notin D_j$, then either $v_{i+1}, v_{i+4} \notin D_j$ and $v_{i+2}, v_{i+3} \in D_j, (\text{as} j \neq i + 4, v_{i+5} \notin D_j)$ or $v_{i-1}, v_{i-4} \notin D_j, v_{i-2}, v_{i-3} \in D_j($ as $j \neq i - 4, v_{i-5} \notin D_j)$.
Thus if $j \neq i, i \pm 4, |D_i \cap D_j| \neq \gamma_t(G) - 1$. So $|D_i \cap D_j| = \gamma_t(G) - 1$ if and only if $j = i \pm 4$.
Thus $V(\gamma_t \cdot G) = \{D_i | 0 \leq i \leq 4k + 2\}$ and $E(\gamma_t \cdot G) = \{D_iD_{i+4} | 0 \leq i \leq 4k + 2\}$.
As $|V|$ is odd, it follows that $\gamma_t \cdot G$ is the cycle C_{4n+3}.

Case (ii). $n = 4k + 2$ for some $k \geq 1$.
If D is a γ_t-set of C_n, then $G - D$ has at the most $k + 1$ components. As $G - D$ has $2k$ vertices and each of its component is either K_1 or K_2 either $G - D$ has exactly k components (in this case each component is K_2) or $k + 1$ components (in this case two components are K_1 and other $k - 1$ components are K_2). If $G - D$ has exactly $k+$ 1 components, then $\langle D \rangle$ also has $k + 1$ components and each component of $\langle D \rangle$ is P_2. If $G - D$ has exactly k components, $\langle D \rangle$ has exactly k-components such that either
(i) two components are P_3 and all other $(k - 2)$ components are P_2 or
(ii) one component is P_4 and all other $(k - 1)$ components are P_2.
There are three types of γ_t-sets of C_n.

Type(1) D_{ij}; where $j = i + 4r + 3$ for some $r = 0, 1, 2, \ldots, (k - 1)$.

$$D_{ij} = \{v_{i+4l+1}, v_{i+4l+2} | 0 \leq l \leq r\} \cup \{v_{i+4l}, v_{i+4l+1} | r + 1 \leq l \leq k\}.$$

Total number of such γ_t-sets $= \frac{k(4k+2)}{2} = 2k^2 + k$.

Type(2) S_{ij}, where $j = i + 4r + 1, (1 \leq r \leq k - 1)$.

$S_{ij} = \{v_{i-1}, v_{i+4r}, v_{i+4r+1} | 0 \leq l \leq r\} \cup \{v_{j+1} \cup \{v_{i+4r+1}, v_{i+4r+2} | r+1 \leq l \leq k-1\}$.
There are $\frac{(k-1)(4k+2)}{2} = (k-1)(2k+1)$ such γ_t- sets of C_n.

Type (3) D_i, where $0 \leq i \leq 4k+1$.
$D_i = \{v_i, v_{i+1}\} \cup \{v_{i+4r+2}, v_{i+4r+3} | 0 \leq r \leq k-1\}$
There are $4k+2$ γ_t- sets of this type.
Thus (1) $|V(\gamma_t \cdot G)| = k(2k+1) + (k-1)(2k+1) + 2(2k+1) = (2k+1)^2$.
(i) each γ_t-set D_{ij}(of type 1) is adjacent in $\gamma_t \cdot C_n$ exactly to the following: $(D_{ij} - \{v_i - 1\} \cup \{v_{i-3}\}, (D_{ij} - v_{i+1}) \cup \{v_{i+3}\}, (D_{ij} - v_{j-1}) \cup \{v_{j-3}\}$ and $(D_{ij} - v_{j+1}) \cup \{v_{j+3}\}$.
If $j \neq i \pm 3$, then $D_{i-4,j}$, $D_{i+4,j}$, $D_{i,j-4}$, $D_{i,j+4}$ are adjacent to D_{ij} in $\gamma_t \cdot C_n$.
If $j = i+3$, the adjacent vertices of D_{ij} are $D_{i+2}, D_{i-2}, D_{i-4,j}, D_{i,j+4}$.
If $j = i-3$, then D_{j+1}, D_{j-2}, $D_{j-4,i}$, $D_{j,i+4}$ are adjacent to D_{ij}.

(ii) each γ_t-set S_{ij}(of type 2) is adjacent to $(S_{ij} - v_{j-1}) \cup \{v_{j-3}\}, (S_{ij} - v_{j+1}) \cup \{v_{j+3}\}, (S_{ij} - v_{i-1}) \cup \{v_{i-3}\}, (S_{ij} - v_{i+1}) \cup \{v_{i+3}\}$ and no other γ_t- sets of C_n. If $j \neq i \pm 5$, these γ_t-sets are $S_{i+4,j}; S_{i-4,j}; S_{i,j+4}; S_{i,j-4}$. If $j = i+5$, they are $D_{i-1}; S_{i,j+4}; S_{i-4,j}; D_{i+3}$. If $j = i-5$, they are D_{j-1}, $S_{i-4,j}; S_{i,j-4}; D_{j+3}$.

(iii) Every γ_t-set $D_i (0 \leq i \leq 4k+1)$ is adjacent to $D_{i+2,i+5}, S_{i+1,i+6}, D_{i-2,i+1}, S_{i-3,i+2}$. Thus $\gamma_t \cdot G$ is 4-regular graph on $(2k+1)^2$-vertices.

Case (iii). Let $n = 4k+1$ where $k \geq l$.
Then $\gamma_t(C_n) = 2k+1$ and if D is a γ_t-set of C_n, then $G - D$ contains $2k$ vertices. So both $< D >$and $< G - D >$ have exactly k components. One of the component of $< D >$ is P_3 and all other components are P_2. Every component of $< G - D >$ is P_2. So every γ_t-set D is of the form $D_i, (0 \leq i \leq 4k)$ where $D_i = \{v_{i-1}\} \cup \{v_{i+4r}, v_{i+4r+1} | 0 \leq r \leq (k-1)\}$. Each D_i is adjacent to D_j iff $j = i \pm 4$. Hence $\gamma_t \cdot C_n$ is the cycle.
$v_0, v_4, v_8, \ldots, v_{4k}, v_3, v_7, \ldots v_{4k-1}, v_2, v_6, \ldots, v_{4k-2}, v_1, v_5, \ldots, v_{4k-3}, v_0$.

Case (iv). If $n = 4k$, where $k \geq 2$, the γ_t-sets of C_n are

$$D_i = \{v_{i+4r}, v_{i+4r+1} | 0 \leq r \leq (k-1)\}$$

for $i = 0, 1, 2, 3. |D_i \cap D_j| = k$ or 0 for $i \neq j \in \{0, 1, 2, 3\}$. So $\gamma_t \cdot C_n = \overline{K_4}$
(5) $\gamma_t \cdot P_n$ for $n \geq 2$.
For every P_n, we label the vertex by $1, 2, \ldots, n$. such that $1, 2, ..., n$ is a path.

(i) Note that P_{4k} has a unique γ_t-set. Therefore $\gamma_t \cdot P_{4k} = k_1$ for $k \geq 1$.

(ii) If D is a γ_t-set of P_{4k+1}, then note that $1 \notin D, 4k+1 \notin D$. Let $D_i = \{4t+2,\ 4t+3 | t = 0, \ldots, (i-1)\} \cup \{4i\} \cup \{4t+3, 4t+4 | t = i, \ldots, k-1\}$ for $i = 1, 2, \ldots, k$. These are the only γ_t-sets of P_{4k+1} and $\gamma_t \cdot P_{4k+1}$ is the path $D_1, D_2, \ldots, D_k$.

(iii) We claim that $\gamma_t \cdot P_{4k+3}$ is the path P_{k+2}. We prove this by induction on k. If $k = 0$ or 1, the result is true. Let us assume that the result is true for some $k \geq 2$. Then $\gamma_t \cdot P_{4k+3}$ is the path P_{k+2} and P_{4k+3} has exactly $(k+2)\gamma_t$-sets. The γ_t-set $D_{k+2} = \{4t+2, 4t+3 | t = 0, 1, \ldots, k\}$ is a pendant vertex of $\gamma_t \cdot P_{4k+3}$. If D is a γ_t-set of P_{4k+3} then $D \cup \{4(k+1)+1, 4(k+1)+2\}$ is a γ_t-set of $P_{4(k+1)+3}$. Apart from these, the only other γ_t-set of $P_{4(4+1)+3}$ is $\{4t+2, 4t+3 | t = 0, ..., (k+1)\}$ and in $\gamma_t \cdot P_{4(k+1)+3}$, it is adjacent only to D_{k+3}. Thus $\gamma_t \cdot P_{4(k+1)+3}$ is the path $P_{(k+1)+2}$.

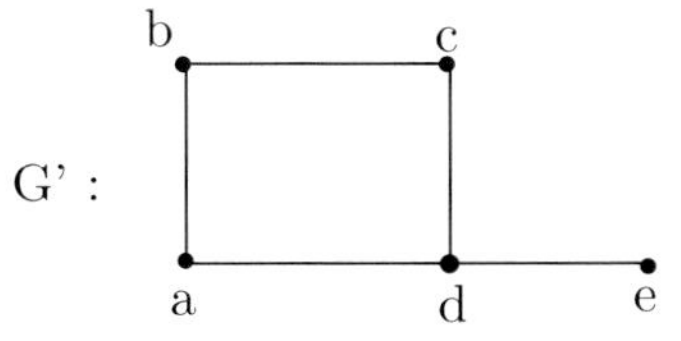

Figure 3.1:

(iv) Now we consider the path P_{4k+2}. There are $(k+1)\gamma_t$-sets containing 1 and 2. They are $S_i = \{4t+1, 4t+2 | 0 \leq t \leq i\} \cup \{4t, 4t+1 | i < t \leq k-1\}$ where $i = 0, 1, \ldots.., k$.

Now we can list all other γ_t-sets of P_{4k+2} as follows.
$S_{i+m} = [S_i - \{4t+1 | 0 \leq t < m-1\}] \cup \{4t+3 | 0 \leq t < m-1\}$ for $1 \leq m \leq k, 0 \leq i \leq k$.
Note that in $\gamma_t \cdot P_{4k+2}$, for $0 \leq i \leq k$, S_i is adjacent to S_{i+1} and $S_{i,j}$ for $1 \leq m \leq k$ and $0 \leq i \leq k$, $S_{i,m}$ is adjacent to $S_{i+1,m}$ and $S_{i,m+1}$.

Thus deg $(S_0) = d(S_k) = d(S_{o,k}) = d(S_{k,k}) = 2$. deg $(S_{o,m}) = d(S_{k,m}) = 3$ for all $0 < m < k$. deg $(S_i) = d(S_{i,k}) = 3$ for all $1 \leq i \leq k-1.d(S_{i,m}) = 4$ for all $1 \leq i \leq k-1, 1 \leq m \leq k-1$.

3 γ_t-Graphs

A graph H is said to be a γ_t-graph, if $H = \gamma_t \cdot G$ for some graph G. Example: Every path P_n is a γ_t-graph, as $\gamma_t \cdot P_{4n+1} = P_n$ for all $n \geq 1$.
We now show that every tree T is a γ_t-graph.

Theorem 1. *Given a tree T, there is a graph G such that,*

(i) $\gamma_t \cdot G = T$,

(ii) For every γ_t-set D of $G, PN(v, D) \neq \phi$ for every $v \in D$,

(iii) For every vertex u of $G, N_G[u]$ contains at least one t-bad vertex.

Proof. We prove this by induction on the order of T (i.e) on the number of vertices of T. If $T = K_2$, consider the following graph G' are shown in Figure 3.1. The γ_t-sets of G' are $S_1 = \{a, d\}; S_2 = \{c, d\}$. As $S_1 \cap S_2 = \{\gamma_t(G) - 1\}, \gamma_t(G')$ is $K_2(withV(K_2) = S_1, S_2)$. Note that $PN(a, S_1) = \{b\}; PN(d, S_1)$ $= \{e, c\}; PN(c, S_2) = \{b\}; PN(d, S_2) = \{a, e\}$. The $t-$ good vertices of G' are a, d, c. So for each $u \in V(G'), N_G[u]$ contains a t-bad vertex. Now assume that the result be true for all trees T with order less than n (for some $n \geq 3$). Let T be a tree with order n. Let u be a pendant vertex of T and v be the vertex of T suchthat $N(u) = \{v\}$. Then $T - u$ is a tree with order $n - 1$.
By our assumption there is a graph G_0 such that,

(i) $\gamma_t \cdot (G_0) = T - u$

(ii) For every γ_t-set D_0 of $G_0, PN(v, D_0) \neq \phi$ for all $v \in D_0$.

(iii) For each vertex $u, N_{G_0}[u]$ contains at least one t-bad vertex of G_0.

Each vertex of $T - u$ can be associated with a γ_t-set of G_0. So the vertex v of $T - u$ represents a γ_t-set of G_0 say v represents the γ_t-set $D^* = \{v_1, v_2, \ldots, v_m\}$ of G_0.

We now obtain the required graph G from G' as follows.
$V(G) = V(G_0) \cup \{x_i, y_i | i = 1, 2, \ldots, m\} \cup \{a, b, c, d, e\}$
$E(G) = E(G_0) \cup \{v_i x_i, v_i y_i, a x_i, a y_i | i = 1, 2, \ldots, m\} \cup \{ab, bc, cd, da, de\}$. Clearly for every γ_t-set D of $G_0, D \cup \{a, d\}$ is a γ_t-set of G and $\gamma_t(G) = \gamma_t(G_0) + 2$. Also $\{v_1, v_2, ...v_m\} \cup \{c, d\}$ is a γ_t-set of G. We claim that these are the only γ_t-sets of G. Let S be a γ_t-set of G. Clearly either $\{a, d\} \subseteq S$ or $\{c, d\} \subseteq S$. Let $S_0 = S \setminus \{a, d, c\}$. Then $|S_0| = \gamma_t(G) - 2 = \gamma_t(G_0)$ and S_0 dominates every vertex of G_0.

Case (i). If $a \notin S$, then S_0 should be a total dominating set for $V(G_0) \cup \{x_i, y_i | i = 1, 2, ...m\}$. As $a \notin S$, it follows that $v_i \in S_0$ and $\{x_i, y_i\} \cap S = \phi$ for every $i = 1, 2, ...m$. Thus $S_0 = D^*$ and $S = D^* \cup \{c, d\}$.

Case (ii). If $a \in S$, then$S_0 = S \setminus \{a, d\}$. If $x_i, y_i \notin S$ for all i, then S_0 is a γ_t-set D of G_0 and$S = D \cup \{a, d\}$. We claim that $x_i, y_i \notin S$ for all i. If possible assume that for some $i, S_0 \cap \{x_i, y_i\} \neq \phi$. Let $A = \{i |$ either $x_i \in S_0$ or $y_i \in S_0\}$. For each $i \in A, N_{G_0}(v_i) \cap S = \phi$. Otherwise $S - \{x_i, y_i\}$ is a γ_t-set of G with less than $|S|$ elements, a contradiction.

Note that every $w_i \in N_{G_0}(v_i)$ is dominated by at least one member of $S_0 - \{x_i, y_i\}$. If $i, j \in A$ and $i \neq j$, then $N_{G_0}(v_i) \cap N_{G_0}(v_j) = \phi$; for if $w \in N_{G_0}(v_i) \cap N_{G_0}(v_j)$, clearly $w \notin S$, as $N_{G_0}(v_i) \cap S = \phi$. So, w is dominated by some $z \in S, z \neq w, x_i, y_i, x_j, y_j$. Now $(S \setminus \{x_i, y_i, x_j, y_j\}) \cup \{w\}$ is the total dominating set for G , which is a contradiction. Therefore $N_{G_0}(v_i) \cap N_{G_0}(v_j) = \phi$ for all $i \neq j \in A$.

If $i \in A$, let w_i be any vertex in $N_{G_0}(v_i)$. Then $S' = (S_0 \cap V(G_0)) \cup \{w_i | i \in A\}$ is a γ_t-set of G_0 and hence each vertex in $N_{G_0}(v_i)$ is t-good vertex of G_0. As v_i is also t-good, every vertex in $N_{G_0}[v_i]$ is t-good, which is a contradiction, as by our assumption of the graph $G_0, N_{G_0}[u]$ contains at least one t-bad vertex.

Hence $A = \phi$ and $S_0 \subseteq V(G_0)$ is a γ_t-set D of G_0. So $S = D \cup \{a, d\}$ for some γ_t-set of G. Thus S is either $\{c, d\} \cup D^*$or $\{a, d\} \cup D$, where D is any γ_t-set of G_0. We also observe that

1. a vertex $u \in V(G_0)$ is t-bad in G_0 iff u is t-bad in G.
2. the vertices $x_i, y_i (1 \leq i \leq m)$ are t-bad in G.
3. the vertices {b,e} are t-bad vertices of G.
   So for each vertex $u, N_G[u]$ contains at least one t-bad vertex of G.
4. If D is any γ_t-set of G, then $PN(v, D) \neq \phi$ for every $v \in D$
5. $\gamma_t \cdot G$ is the given tree T.

From the proof of the above theorem we obtain the following theorem.

Theorem 2. *If a graph H is a γ_t-graph and a new pendant vertex u is attached to a vertex of H, then the new graph is also a γ_t-graph.*

Theorem 3. *Every cycle is a γ_t-graph. i.e. If C_n is a cycle, there is a graph H such that $C_n \cong \gamma_t \cdot H$.*

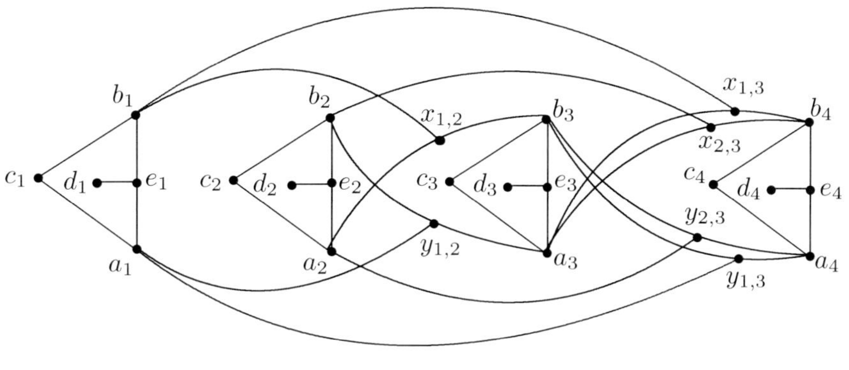

Figure 3.2:

Proof. As $\gamma_t \cdot C_{2n+1}$ is C_{2n+1} for all $n \geq 1$, it is enough to show that every even cycle C_{2n} is γ_t-graph. As $\gamma_t \cdot C_4 = C_4$, it is enough to consider C_k, where $k = 2n$, for all $n \geq 3$.

Let $k = 2n$ for some $n \geq 3$. We construct a graph H as shown in Figure 3.2.

Let $V(H) = \{a_i, b_i, c_i, d_i, e_i | i = 1, 2, ..., n\} \cup \{x_{ij}; y_{ij} | i < j \text{ and } j = 2, ...n - 1\}$

$E(H) = \{a_ie_i, e_id_i, e_ib_i, b_ic_i, c_ia_i | i = 1, 2, ...n\} \cup \{a_iy_{ij}, b_ix_{ij} | i < j, 1 \leq i \leq n-1\} \cup \{a_jx_{ij}, b_jy_{ij} | i < j, 1 < j < n\} \cup \{b_{j+1}x_{ij}; a_{j+1}y_{ij} | i < j, 1 < j < n\}$. Let $A_i = \{a_i, b_i, c_i, d_i, e_i\}$ for all i. As $\{a_i, e_i | 1 \leq i \leq n\}$ and $\{b_i, e_i | 1 \leq i \leq n\}$ are total dominating sets of H, it follows that $\gamma_t(H) \leq 2n$.

If D is any γ_t-set of H, then $e_i \in D$ and $|D \cap A_i| \geq 2$ for each i. So $|D| \geq 2n$. Thus $\gamma_t(H) = 2n$ and $|D \cap A_i| = 2$ for each i. As each $e_i \in D$ and $|D \cap A_i| = 2$, either $a_i \in D$ or $b_i \in D$. So to each $i, D \cap A_i$ is either $\{a_i, e_i\}$ or $\{b_i, e_i\}$. Hence D can be represented by a vector $(z_1, z_2, \ldots, z_n)$ where

$$z_i = \begin{cases} 0 \text{ if } a_i \in D \\ 1 \text{ if } b_i \in D \end{cases}$$

For example, γ_t-sets $D_0 = \{a_i, e_i | 1 \leq i \leq n\}$ and $D_1 = \{b_i, e_i | 1 \leq i \leq n\}$ are represented by the vectors $(0, 0, \ldots, 0)$ and $(1, 1, ..., 1)$ respectively.
For all other γ_t-sets D of H, both 0 and 1 will appear in the vector representation of D. We claim that in the vector representation of D, $0's$ will appear together and 1's will appear together.(i.e), if D is a γ_t-set of H, and $(z_1, z_2, \ldots, z_n)$ is the vector representation of D, then for some $i \in \{1, 2, ..., n\}, z_j = z_i$ for all $j \leq i$ and $z_j = z_{i+1}$ for all $j \geq i + 1$. Let $D \neq D_0$ and D_1. Assume that $z_1 = 0$. (Similar if $z_1 = 1$). As $D \neq D_0$ and $D_1, z_j = 1$ for some j. Let $i + 1$ be the first positive integer in $\{1, 2, \ldots, n\}$ suchthat $z_{i+1} = 1$. i.e $z_1 = z_2 = ... = z_i = 0$ and $z_{i+1} = 1$. We claim that $z_j = 1$ for all $j \geq i + 1$. As $z_{i+1} = 1$ and $z_1 = z_2 = \cdots = z_i = 0$. We have $\{a_1, a_2, ..., a_i, b_{i+1}\} \subseteq D$. As $D \cap A_{i+2} = \{a_{i+2}, e_i\}$ or $\{b_{i+2}, e_i\}$ and $a_{i+1} \notin D$, in order to dominate the vertices $x_{j,i+1} (j \leq i), b_{i+2} \in D$. In the same way in order to dominate the vertices $x_{j,i+2} (j \leq i + 1), b_{i+3} \in D$, we get $b_k \in D$ for all $k \geq i+1$. So D is $(0, 0, \ldots, 0, 1, 1, \ldots, 1)$. Thus $z_j = 0$ for all $j \leq i$ and $z_j = 1$ for all $j > i$. conversely for every i, the vector $(z_1, z_2, ..., z_i, z_{i+1}, \ldots, z_n)$ where $z_j = z_i$ for all $j \leq i$ and $z_j \neq z_i$ for all $j > i$ represents a γ_t-set D other than D_0 and D_1. We can now list all the γ_t-sets of H as follows.

$D_{0,i} = (z_1, \ldots, z_n)$ where $z_j = 0 \; \forall \; j \leq i$ and $z_j = 1 \; \forall \; j > i$.
$D_{1,i} = (z_1, \ldots, z_n)$ where $z_j = 1 \; \forall j \leq i$ and $z_j = 0 \; \forall \; j > i$ for all $1 \leq i \leq n$.

It is clear that $\gamma_t \cdot H$ is a cycle $C_{2n} : D_0 = D_{0,n}, D_{1,1}, D_{1,2}, \ldots, D_{1,n} = D_1, D_{0,1}, D_{0,2}, \ldots, D_{0,n} = D_0$.

Remark. By the Theorems 2 and 3, we obtain the following result.

Theorem 4. *Every unicyclic graph is a γ_t-graph.*

References

[1] R. Balakrishnan and K. Ranganathan, *A text book of graph theory*, Springer, (2000).

[2] M.A. Henning and T.W. Haynes, Total domination excellent tree, *Discrete Applied Math.*, **263** (2003), pp. 93–104.

[3] M. Yamuna, *Excellent-Just excellent-Very excellent graphs*, Ph.D thesis, Alagappa University (2003).

Proc. of ICDM 2006, Pages 263–270

Some Combinatorial Results on Pcodes

K. Sasikala
Department of Mathematics, St. Joseph's College of Engineering, Jeppiaar Nagar, Chennai - 600 119, India
Email: sasikalaveerabadran@yahoo.co.in
V.R. Dare and D.G. Thomas
Department of Mathematics, Madras Christian College, Chennai - 600 059, India

Abstract. Partial words have been recently introduced by Berstel and Boasson [4] in the context of gene (or protein) comparison. Alignment of two genes (or two proteins) can be viewed as a construction of two partial words that are said to be compatible. Quite recently, Blanchet-Sadri [7] has made a first step towards investigating languages of partial words by introducing the concept of pcodes which are sets of partial words preserving the uniqueness of factorization of partial words. This study of pcodes is closely related to the theory of codes which draws its origin from the theory of information devised by Shannon. In this paper, we provide some characterizations of pcodes of partial words in terms of a measure of partial languages. Motivated by the work in [1, 3, 5, 8–11], we introduce unambiguous partial automata and prove the equivalence between pcodes and unambiguous partial automata.

Keywords: Partial words, partial finite automata, unambiguous partial automata, pcodes.

1 Introduction

Partial words are strings of symbols from a finite alphabet that may have a number of "do not know" symbols. Partial word were recently introduced by Berstel and Boasson [4] in the context of gene (or protein) comparison. Alignment of two genes (or two proteins) can be viewed as a construction of two partial words that are said to be compatible. The main motivation for the introduction of partial words came from molecular biology of nucleic acids. An aim of the study is to determine properties of the DNA or RNA sequences encountered in nature.

A partial word is very much like a conventional word, only at some positions we do not know, which letter it has. Similar to the definition of a word as a total function from $\{0, \ldots, |w| - 1\}$ to Σ, a partial word w is defined as a partial function from $\{0, \ldots, |w| - 1\}$ to Σ. The positions, where $w(n)$ (the nth letter of w) is not defined for $n < |w|$ are called the word's holes. If $D(w)$ stands for the domain of w, then the set of holes of w denoted by $h(w)$ is the set $\{0, \ldots, |w| - 1\} \backslash D(w)$.

Recently [6] Blanchet-Sadri introduced pcodes that play a role in the study of combinatorics on partial words. Pcodes are defined in terms of the compatibility relation which considers two words over the same alphabet that are equal except for a number of insertion and / or deletions of symbols. While a code of words X does not allow two distinct decipherings of some word in X^+, a code of partial words Y does not allow two distinct compatible decipherings in Y^+. In [6] the authors have described two element pcodes, complete pcodes, maximal pcodes and circular pcodes. In particular, many pcodes can be obtained as antichains with respect to certain partial orderings.

Motivated by these studies, in section 3, some characteristics of pcodes of partial words are discussed. Unambiguous automata are considered by Berstel and Perrin [5] in connection with codes. The main idea is to replace computations on words by computation on paths labeled by words. In section 4 we introduce partial finite automata and provide a characterization of pcodes in terms of unambiguous partial finite automata.

2 Preliminaries

In this section, we deal with some basic concepts on words, partial words and pcodes [7].

Let Σ be a finite non-empty set of symbols called an alphabet. Symbols in Σ are called letters and any finite string over Σ is called a word over Σ. The empty word is denoted by λ. The set of all words over Σ is denoted by Σ^*.

A word over Σ is a partial word over Σ with an empty set of holes (we sometimes refer to words as full words). For any partial word u over Σ, $|u|$ denotes its length. In particular $|\lambda| = 0$.

We denote by W_0 the set Σ^*, and for every integer $i \geq 1$, by W_i the set of partial words over Σ with exactly i holes. $W = \bigcup_{i \geq 0} W_i$, is the set of all partial words over Σ with an arbitrary number of holes.

If u is a partial word of length n over Σ, then the companion of u (denoted by $u_\diamond$) is the total function $u_\diamond : \{0, 1, \ldots, n-1\} \to \Sigma \cup \{\diamond\}$ defined by

$$u_\diamond(i) = \begin{cases} u(i) & \text{if } i \in D(u) \\ \diamond & \text{otherwise} \end{cases}$$

When it is more convenient, we will also refer to the companion as a partial word to simplify the syntax of our sentences. Thus we will say for example the partial word $\diamond a \diamond b$ instead of the partial word with companion $\diamond a \diamond b$.

For a subset X of W, we denote by X^* the submonoid of W generated by X. It consists of all partial words which are obtained by concatenating of elements of X.

A partial word u is a factor of the partial word v if there exist partial words x, y such that $v = xuy$. The factor u is called proper if $u \neq v$. The partial word u is a prefix (respectively suffix) of v if $x = \lambda$ (respectively $y = \lambda$). For a subset X of W, we denote by $F(X)$ the set of factors of elements of X.

Definition 2.1. *If u and v are two partial words of equal length then u is said to be contained in v, denoted by $u \subset v$, if $D(u) \subset D(v)$ and $u(i) = v(i)$ for all $i \in D(u)$. The partial words u and v are compatible, denoted by $u \uparrow v$ if there exists a partial word w such that $u \subset w$ and $v \subset w$.*

For a subset X of W, we denote by $C(X)$ the set of all partial words compatible with elements of X.

Definition 2.2. *A partial language over Σ is a subset of all partial words over Σ.*

Example 2.3. *$a \diamond b(ab)^*$ is a partial language.*

Definition 2.5 ([5]). *Let X be a non-empty subset of $W_0 \backslash \{\lambda\}$. Then X is called a code over Σ if for all integers $m \geq 1, n \geq 1$ and words $u_1, u_2, \ldots, u_m, v_1, v_2, \ldots, v_n \in X$, the condition $u_1 u_2 \ldots u_m = v_1 v_2 \ldots v_n$ implies $m = n$ and $u_i = v_i$ for $i = 1, 2, \ldots, m$.*

In the case of partial words a pcode is defined as follows:

Definition 2.6. *Let X be a non-empty subset of W. Then X is called a pcode over Σ if for all integers $m, n \geq 1$ and partial words $u_1, u_2, \ldots, u_m, v_1, v_2, \ldots, v_n \in X$, the compatibility relation $u_1 u_2 \ldots u_m \uparrow v_1 v_2 \ldots v_n$ implies $m = n$ and $u_i = v_i$ for $i = 1, 2, \ldots, m$.*

Remark 2.7.

1. A non-empty set X satisfying $X \subset \{a, \diamond\}^*$ is a pcode over $\{a\}$ if and only if X is a singleton set distinct from $\{\lambda\}$.
2. The set $Y = \{a\diamond b, a\diamond\}$ is a pcode over $\{a, b\}$.
3. The set $Z = \{u_1, u_2, u_3, u_4\}$ where $u_1 = a\diamond b, u_2 = aa\diamond bba, u_3 = \diamond b$ and $u_4 = ba$ is not a pcode over $\{a, b\}$ since $u_1 u_3 u_3 u_4 u_3 \uparrow u_2 u_3 u_1$ is a non-trivial compatibility relation over Z.
4. If X is a pcode over Σ, then X is a code over $\Sigma \cup \{\diamond\}$. But the converse does not hold. Consider, for instance, the code $X = \{a\diamond a, \diamond a\diamond\}$ over $\{a, \diamond\}$. Then X is not a pcode over $\{a\}$ since its elements are compatible.

In [7], an algorithm to test whether or not a finite set is a pcode is given. Let X be a subset of $W \backslash \{\lambda\}$. Let

$$U_1 = \{x | x \in W \backslash \{\lambda\} \text{ and there exists } u \in C(X) \text{ such that } ux \in X\}$$

and for $i \geq 1$, let

$$U_{i+1} = \{x | x \in W \text{ and there exists } u \in C(X) \text{ such that } ux \in U_i\} \cup$$

$$\{x | x \in W \text{ and there exists } y \in U_i, y' \in W \text{ such that } y \uparrow y' \text{ and } y'x \in X\}$$

Theorem 2.7 ([7], Theorem 2). *Let $X \subset W \backslash \{\lambda\}$ be pairwise non-compatible. The set X is a pcode if and only if none of the sets U_n contains the empty word.*

3 Characterization of Pcodes of Partial Words

In this section we shall prove some characteristics of pcodes of partial words.

Definition 3.1. *For any subset X of W, let $\alpha(X) = \sum_{x \in X} |\Sigma|^{-(|x| - |h(x)|)}$, where $|\Sigma|$ is the cardinality of Σ and $|h(x)|$ is the number of holes in x.*

Lemma 3.2. *For any $n \geq 0$, (i) $\alpha(\Sigma^n) = 1$ (ii) $\alpha((\Sigma \cup \{\diamond\})^n) = 2^n$ if $|\Sigma| = 2$.*

Proof. (i) Let $\Sigma = \{a, b\}$.
$\Sigma^n = \{\text{words of length } n \text{ over } \Sigma\}$.
Since there are 2^n words of length n over Σ,
$\alpha(\Sigma^n) = 2^n \times \frac{1}{2^n} = 1$.
In general $\alpha(\Sigma^i) = 1$ for all $i \geq 0$.
(ii) $(\Sigma \cup \{\diamond\})^n = \bigcup_{i \geq 0} W_{i,n}$,

where $W_{i,n}$ denotes the set of all partial words of length n having exactly $'i'$ number of holes.

$$\begin{aligned}\alpha\left((\Sigma \cup \{\diamond\})^n\right) &= \sum_i \alpha(W_{i,n}) \\ &= \alpha(W_{0,n}) + \alpha(W_{1,n}) + \alpha(W_{2,n}) + \cdots + \alpha(W_{n,n}) \\ \alpha(W_{i,n}) &= nC_i 2^{n-i} \times \frac{1}{2^{n-i}} = nC_i, \ \forall\ i, \text{ since partial words of length } n \\ &\quad \text{having exactly } i \text{ holes can occur in } nC_i \text{ ways} \\ \alpha\left((\Sigma \cup \{\diamond\})^n\right) &= 1 + nC_1 + nC_2 + \cdots + nC_n = (1+1)^n = 2^n.\end{aligned}$$

□

Lemma 3.3. *Let X be a finite subset of W*

1. *X is a pcode if and only if $\alpha(X^n) = (\alpha(X))^n$ for all $n \geq 1$.*
2. *For all $n \geq 1$, $\alpha(X^n) \leq (\alpha(X))^n$*

Proof. For any positive integer n,

$$(\alpha(X))^n = \left(\sum_{x \in X} |\Sigma|^{-(|x|-|h(x)|)}\right)^n \quad \text{and} \quad \alpha(X^n) = \sum_{u \in X^n} |\Sigma|^{-(|u|-|h(u)|)}$$

If X is not a pcode, then there exists a partial word u of X^* of minimal length having two different factorizations in terms of elements of X. If X is a pcode, then X is a code. Hence any element of X^n can be uniquely factorized by elements of X. Therefore $\alpha(X^n) = (\alpha(X))^n$.

i.e., $u = x_1x_2 \ldots x_h = y_1y_2 \ldots y_k$ with $x_1x_2 \ldots x_h \uparrow y_1y_2 \ldots y_k$, $x_1, x_2, \ldots, x_h, y_1, y_2, \ldots y_k \in X$ and $x_i \neq y_i$ for some i.
Now $u.u = x_1 \ldots x_h y_1 \ldots y_k \uparrow y_1 \ldots y_k x_1 \ldots x_h$, with $u.u \in X^{h+k}$
Thus $\alpha(X^{h+k}) < (\alpha(X))^{h+k}$. □

Example 3.4. Let $X = \{a\diamond b, a\diamond\}$. X is a pcode.

$$\begin{aligned}\alpha(X) &= \frac{1}{2} + \frac{1}{4} = \frac{3}{4} \\ X^2 &= \{a\diamond ba\diamond b, a\diamond a\diamond, a\diamond ba\diamond, a\diamond a\diamond b\} \\ \alpha(X^2) &= \frac{1}{2^4} + \frac{1}{2^2} + \frac{2}{2^3} = \frac{9}{16} = \left(\frac{3}{4}\right)^2 \\ (\alpha(X))^3 &= \frac{1}{2^6} + \frac{1}{2^3} + \frac{3}{2^5} + \frac{3}{2^4} = \frac{27}{2^6} = \left(\frac{3}{4}\right)^3\end{aligned}$$

Similarly it can be proved that $\alpha(X^n) = (\alpha(X))^n$ for all $n \geq 1$.

Lemma 3.5. *If $X \subset W$ is a finite pcode, then $(\alpha(X))^n \leq 2(2^{n\ell} - 1)$ where ℓ is the maximal length of partial words in X.*

Proof. If X is a pcode, then by Lemma 3.3, $\alpha(X^n) = (\alpha(X))^n$. Let ℓ be the maximal length of partial words in X. Then $X^n \subseteq \bigcup_{i=1}^{nl} \Sigma_1^i$ where $\Sigma_1 = \Sigma \cup \{\diamond\}$. By Lemma 3.2, $\alpha(\Sigma_1^i) = (\alpha(\Sigma_1))^i = 2^i$ and so,

$$\alpha(X^n) \leq \alpha(\Sigma_1) + \alpha(\Sigma_1^2) + \cdots + \alpha(\Sigma_1^{nl}) = 2 + 2^2 + \cdots + 2^{nl} = 2.\frac{2^{nl}-1}{1} = 2(2^{nl}-1)$$

Therefore $(\alpha(X))^n \leq 2(2^{nl} - 1)$. □

4 Unambiguous Partial Finite Automata

In this section we introduce partial finite automata and prove a characterization of pcodes in terms of unambiguous partial finite automata.

Definition 4.1. *A finite deterministic partial automaton is a 7-tuple* $PA = (Q, Q_h, \Sigma, \Gamma, E, I, F)$, *where* Q *is a finite set of states,* Q_h *is a finite set of hole states,* Σ *is a finite input alphabet,* $\Gamma = \{\Diamond\}$, $\Diamond$ *is the don't care symbol not in* Σ *with* $\Sigma \cap \Gamma = \phi$ *and* $Q \cap Q_h = \phi$, $I \subseteq Q$ *is the set of initial states,* $F \subseteq Q$ *is the set of final states and* E *is a subset of* $(Q \cup Q_h) \times (\Sigma \cup \Gamma) \times (Q \cup Q_h)$, *called the set of edges. That is*
(i) $(p, a, q) \in E$ *if* $p \in Q \cup Q_h, q \in Q, a \in \Sigma$,
(ii) $(q_h, \Diamond, q_h) \in E$ *if* $q_h \in Q_h, \Diamond \in \Gamma$,
(iii) $(p, \Diamond, q_h) \in E$ *if* $p \in Q, q_h \in Q_h, \Diamond \in \Gamma$.

A partial word w *is said to be accepted by a finite deterministic partial automaton* $PA = (Q, Q_h, \Sigma, \Gamma, E, q_0, F)$ *if* $(q_0, w, p) \in E$ *for some* p *in* F. *The partial language accepted by* PA*, denoted* $L(PA)$ *is* $\{w \in (\Sigma \cup \Gamma)^* / (q_0, w, p) \in E$ *for some* $q_0 \in I$ *and* $p \in F\}$.

Thus, an edge in E *is either*
$f_1 = (p, a, q)$ *denoted by* $f_1 : p \xrightarrow{a} q$ *where* $p \in Q \cup Q_h$, $a \in \Sigma$, $q \in Q$ *or*
$f_2 = (p, \Diamond, q_h)$ *denoted by* $f_2 : p \xrightarrow{\Diamond} q_h$, *where* $p \in Q, q_h \in Q_h, \Diamond \in \Gamma$ *or*
$f_3 = (q_h, \Diamond, q_h)$ *denoted by* $f_3 : q_h \xrightarrow{\Diamond} q_h$, *where* $q_h \in Q_h, \Diamond \in \Gamma$
A path in PA *is a finite sequence* $c = c_1 c_2 \ldots c_n$ *of consecutive edges* c_i, *each* c_i *is any one of* f_1, f_2 *or* f_3. *i.e.,* $c_i : p_i \xrightarrow{a_i} q_i (q_i = p_{i+1}$ *for* $1 \leq i \leq n-1)$
where $a_i \in \Sigma \cup \Gamma$, $q_i \in Q \cup Q_h$ *and* $p_i \in Q \cup Q_h$
We denote by $c : p \xrightarrow{w} q$, *a finite path* c *of label* w *which begins at* $p \in Q \cup Q_h$ *and terminates at* $q \in Q \cup Q_h$ *where* w *is a partial word over* Σ.

Example 4.2. The following deterministic finite partial automaton accepts the partial language $(a\Diamond b)^n, n \geq 1$.
$PA_1 = (Q, Q_h, \Sigma, \Gamma, E, I, F)$ where $\Sigma = \{a, b\}$, $Q = \{q_0, q_1, q_2, q_3\}$, $Q_h = \{q_h\}$, $\Gamma = \{\Diamond\}$, $I = \{q_0\}$ and $F = \{q_3\}$. The transition diagram is shown in Figure 4.1.

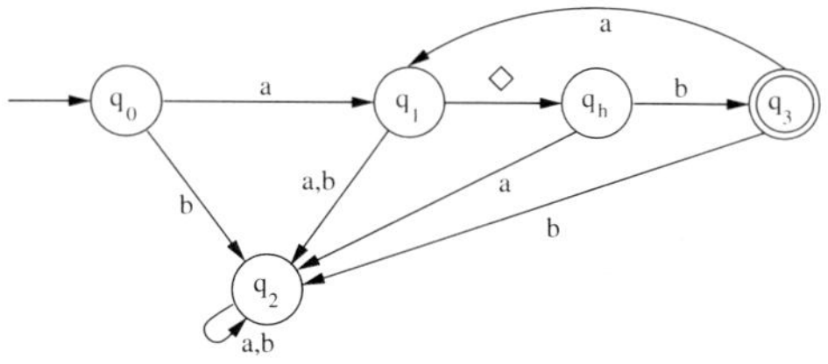

Figure 4.1: DPFA accepting $(a\Diamond b)^n, n \geq 1$

Let $\mathbb{N}$ be the set of all non-negative integers including 0. Let $\mathcal{N} = \mathbb{N} \cup \{\infty\}$. A $\mathcal{N}$ relation over $Q \cup Q_h$ is a mapping m from $(Q \cup Q_h) \times (Q \cup Q_h) \to \mathcal{N}$. Let $\mathcal{N}^{(Q \cup Q_h) \times (Q \cup Q_h)}$ be the set of all $\mathcal{N}$ relations over $Q \cup Q_h$.

Given a partial finite automaton $PA = (Q, Q_h, \Sigma, \Gamma, E, I, F)$ over Σ, define a mapping $\phi_{PA} : \Sigma \cup \Gamma \to \mathcal{N}^{(Q \cup Q_h) \times (Q \cup Q_h)}$ so that $(p, \phi_{PA}(a), q) = \begin{cases} 1 & \text{if } (p, a, q) \in E \\ 0 & \text{otherwise} \end{cases}$

$$(p, \phi_{PA}(\Diamond), q_h) = \begin{cases} 1 & \text{if } (p, \Diamond, q) \in E \\ 0 & \text{otherwise} \end{cases} \quad (q_h, \Diamond, q_h) = \begin{cases} 1 & \text{if } (q_h, \Diamond, q_h) \in E \\ 0 & \text{otherwise} \end{cases}$$

ϕ_{PA} can be extended into a morphism from $(\Sigma \cup \Gamma)^*$ into $\mathcal{N}^{(Q\cup Q_h)\times(Q\cup Q_h)}$ as follows: In particular $\phi(1) = I_Q$ and
$\phi_{PA} : (\Sigma \cup \Gamma)^* \rightarrow N^{(Q\cup Q_h)\times(Q\cup Q_h)}$ is such that
$\phi_{PA}(\lambda) = I_Q \cup Q_h$ where $I_Q \cup Q_h = \{i_p | i_p : p \rightarrow p, p \in Q \cup Q_h\}$
$u_0, v_0 \in (\Sigma \cup \Gamma)^*$
$(p, \phi_{PA}(u_0 v_0), q) = \sum_{r \in Q\cup Q_h} (p, \phi(u_0), r)(r, \phi(v_0), q).$

Proposition 4.3. *Let $PA = (Q, Q_h, \Sigma, \Gamma, E, I, F)$ be a partial finite automaton over Σ. For all $p, q \in Q \cup Q_h$ and $w_0 \in (\Sigma \cup \Gamma)^*$, $(p, \phi_{PA}(w_0), q)$ is the number of paths from p to q with label w_0.*

Definition 4.4. *A partial automaton $PA = (Q, Q_h, \Sigma, \Gamma, E, I, F)$ over Σ is unambiguous if for all $p, q \in Q \cup Q_h$ and $w_0 \in (\Sigma \cup \Gamma)^*$, $(p, \phi_{PA}(w_0), q) \in \{0, 1\}$.*

Thus a partial automaton is unambiguous if and only if for all $p, q \in Q \cup Q_h$ and $w_0 \in (\Sigma \cup \Gamma)^*$, there is atmost one path from p to q with label w_0.

Definition 4.5. *A path $c : i \rightarrow t$ is called successful if $i \in I$ and $t \in T$. The behaviour of the partial automaton $PA = (Q, Q_h, \Sigma, \Gamma, E, I, F)$ is the formal power series (characteristic series) $|PA|$ defined by*

$$(|PA|, w_0) = \sum_{\substack{i\in I\\ t\in T}} (i, \phi_{PA}(w_0), t) \tag{1}$$

The set recognized by PA is the support of $|PA|$. It is just the set of all labels of successful paths. It is denoted by $L(PA)$.

Proposition 4.6. *Let $PA = (Q, Q_h, \Sigma, \Gamma, E, I, F)$ be a partial automaton over Σ. For all $w_0 \in \Sigma \cup \Gamma$, $(|PA|, w_0)$ is the number of successful paths labeled by w_0.*

Remark 4.7. *A more compact writing of (1) consists in $(|PA|, w_0) = I\ \phi_{PA}(w_0)\ T$.*

Here the element $I \in \mathcal{N}^{Q\cup Q_h}$ is considered as a row vector and $T \in \mathcal{N}^{Q\cup Q_h}$ as a column vector.

Definition 4.8. *Let $PA = (Q, Q_h, \Sigma, \Gamma, E, I, F)$ be a partial automaton. A state $q \in Q \cup Q_h$ is accessible (respectively co-accessible) if there exists a path $c : i \rightarrow q$ with $i \in I$ (respectively a path $c : q \rightarrow t$ with $t \in T$). A partial automaton is trim if all states are both accessible and co-accessible.*

The trim part of $PA = (Q, Q_h, \Sigma, \Gamma, E, I, F)$ is the partial automaton

$$PA^\circ = (P, Q_h, \Sigma, \Gamma, E, I \cap P, F \cap P)$$

which is the restriction of PA to the set P of states of PA which are both accessible and co-accessible. It is easily seen that $|PA^\circ| = |PA|$.

Proposition 4.9. *Let $PA = (Q, Q_h, \Sigma, \Gamma, E, I, F)$ be a trim partial automaton with a unique initial and a unique final state. Then PA is unambiguous if and only if $|PA|$ is a characteristic series.*

Proof. If PA is unambiguous, then clearly $|PA|$ is a characteristic series.

Conversely, if $(p, \phi_{PA}(w_0), q) \geq 2$ for some $p, q \in Q \cup Q_h$ and $w_0 \in (\Sigma \cup \Gamma)^*$, then choosing paths $i \overset{u_0}{\rightarrow} p$ and $q \overset{v_0}{\rightarrow} t$, we have $(|PA|, u_0 w_0 v_0) \geq 2$.

If PA is an unambiguous automaton, we usually identify the series $|PA|$ with the set $L(PA)$ recognized by PA.

To each partial automaton $PA = (Q, Q_h, \Sigma, \Gamma, E, q_0, F)$, we associate a partial automaton PA^* by a canonical construction consisting of the two following steps. Let $\omega \notin (Q \cup Q_h)$ be a new state (which is not a hole state) and let

$$PB = (Q \cup \{\omega\}, Q_h, \Sigma, \Gamma, E, \omega, \omega) \tag{2}$$

be the partial automaton with edges $G = E \cup I \cup J \cup V$ where E is the set of edges of PA and

$$\begin{aligned} I &= \{(\omega, a, q)/\exists i \in I : (i, a, q) \in E\} \\ &\quad \cup\{(\omega, \diamond, q_h)/\exists i \in I : (i, \diamond, q_h) \in E\} \\ J &= \{(q, a, \omega)/\exists t \in T : (q, a, t) \in E\} \\ &\quad \cup\{(q_h, \diamond, \omega)/\exists t \in T : (q_h, \diamond, t) \in E\} \\ V &= \{(\omega, a, \omega)/\exists i \in I, t \in T : (i, a, t) \in E\} \end{aligned} \tag{3}$$

By definition, the automaton PA^* is the trim part of PB. □

Definition 4.10. *To every partial language X, we associate a characteristic series $\underline{X}$ defined by*

$$(\underline{X}, w_0) = \begin{cases} 1 & \text{if } w_0 \in X \\ 0 & \text{if } w_0 \notin X \end{cases}$$

Proposition 4.11. *Let $X \subset (\Sigma \cup \Gamma)^+$, and let PB be a partial automaton constructed as explained by (2) from the trim partial automaton PA with a unique initial and a unique final state such that $|PA| = \underline{X}$. Then $|PA^*| = (\underline{X})^*$.*

Proof. Since PA^* is the trim part of the automaton PB defined by (2), it suffices to show that $|PB| = |PA|^*$.

A path $c : \omega \longrightarrow \omega$ in PB is called simple if it is not the null path and if no interior node is equal to ω. Each path $c : \omega \longrightarrow \omega$ is a product in a unique manner of simple paths from ω to ω.

Let S be the power series defined as follows: For all $w_0 \in (\Sigma \cup \Gamma)^*$, (S, w_0) is the number of simple paths from w to w labeled with w_0. By the preceding remarks, we have $|B| = S^*$.

Thus it remains to prove that $S = \underline{X}$.

Let $w_0 \in (\Sigma \cup \Gamma)^*$. If $w_0 = \lambda$ then $(S, \lambda) = (\underline{X}, \lambda) = 0$, since a simple path is not null.

If $w_0 = a \in \Sigma$, then $(S, a) = 1$ if and only if $a \in X$, according to (3).

Assume now $|w_0| \geq 2$. Set

(i) $w_0 = au_0b$ with $a, b \in \Sigma$ and $u_0 \in (\Sigma \cup \Gamma)^*$. Each simple path $c : \omega \xrightarrow{w_0} \omega$ factorizes uniquely into $c : \omega \xrightarrow{a} p \xrightarrow{u_0} q \xrightarrow{b} \omega$ for some $p \in Q, q \in Q \cup Q_h$.

(ii) $w_0 = \diamond u_0 b$. $c : \omega \xrightarrow{w_0} \omega$. There exist at least one successful path

$$i \xrightarrow{a} p \xrightarrow{u_0} q \xrightarrow{b} t, p \in Q, q \in Q \cup Q_h$$

in PA. This path is unique because the behaviours of PA is a characteristic series. If there is another simple path $c' : \omega \xrightarrow{w_0} \omega$ in PB, then there is also another successful path labeled w_0 in PA, this is impossible. Thus there is at most one simple path $c : \omega \xrightarrow{w_0} \omega$ in PB and such a path exists if and only if $w_0 \in X$. Consequently $S = \underline{X}$ which was to be proved. □

Proposition 4.12. *Let X be a subset of $(\Sigma \cup \Gamma)^+$, and let $M = X^*$ be the submonoid generated by X. Then X is a pcode if and only if $\underline{M} = (\underline{X})^*$.*

Proof. By [8, 11] the coefficient $((\underline{X})^*, w_0)$ of a partial word w_0 in $(\underline{X})^*$ is equal to the number of distinct factorizations of w_0 in partial words in X. By definition, X is a pcode if and only if this number takes any one of the values 0 and 1 for any partial word in $(\Sigma \cup \Gamma)^*$. But this is equivalent to saying that $(\underline{X})^*$ is the characteristic series of its support. That is, $(\underline{X})^* = \underline{M}$. □

Proposition 4.13. *Let $X \subset (\Sigma \cup \Gamma)^+$ and let PA be a partial automaton such that $|PA| = \underline{X}$. Then X is a pcode if and only if PA^* is unambiguous.*

Proof. According to proposition 4.11, we have $|PA^*| = (\underline{X})^*$. Since PA is trim, proposition 4.9 shows that PA^* is unambiguous if and only if $|PA^*|$ is a characteristic series. Since $L(PA) = X^*$, this means that PA^* is unambiguous if and only if $\underline{X}^* = (\underline{X})^*$. The result follows from proposition 4.12. □

Conclusion

In this paper, we provide some characterization of pcodes of partial words in terms of a measure of a partial languages. We also introduce unambiguous partial automata and prove the equivalence between pcodes and unambiguous partial automata.

References

[1] Aldo de Luca, Antonio Restivo, Minimal Complete sets of words, *Theoretical Computer Science*, **12** (1980) pp. 325–332.

[2] Aldo de Luca, Some combinatorial results on Bernoulli sets and codes, *Theoretical Computer Science*, **273** (2002) pp. 143–165.

[3] Arturo Carpi, Aldo de Luca, Completions in measure of languages and related combinatorial problems, *Theoretical Computer Science*, **330** (2005) pp. 35–57.

[4] J. Berstel and L. Boasson, Partial words and a theorm of Fine and Wilf, *Theoretical Computer Science*, **218** (1999) pp. 135–141.

[5] J. Berstel and D. Perrin, *Theory of codes*, Academic Press, New York (1985).

[6] F. Blanchet-Sadri, Codes, orderings and partial words, *Theoretical Computer Science*, **329** (2004) pp. 177–202.

[7] F. Blanchet-Sadri and M. Moorefield, Pcodes of partial words, Manuscript, 2005 (see `http://www.uncg.edu/mat/pcode/`).

[8] Do Long Van, Codes infinitaries of automata non-ambiguous, *G.R. Acad. Sci. Paris, T. 302, Series*, **I** (1986), pp. 693–696.

[9] Peter Leupold, Partial words: Results and Perspectives, preprint (2003).

[10] G. Rozenberg and A. Salomaa, *Handbook of Formal Languages*, vol. 1–3, Springer-Verlag, Berlin (1997).

[11] D.G. Thomas, V.R. Dare and K.G. Subramanian, Bi-infinitary codes and formal power series, in *Proceedings of the 7th International Conference on Formal Power Series and Algebraic Combinatorics, Paris, France, May 29 – June 2* (1995), pp. 535–548.

Proc. of ICDM 2006, Pages 271–277

Characterizing Defect 1-Extendable Bipartite Graphs and an Algorithm

XUELIAN WEN
Faculty of computer, Guangdong University of Technology, Guangzhou 510006, P.R. China
DINGJUN LOU
Department of computer science, Sun Yat-sen University, Guangzhou 510275, P.R.China

Abstract. A near perfect matching is a matching saturating all but one vertex in a graph. If every edge in G is contained in a near perfect matching of G, then G is defect 1-extendable. Let $G=(U,W)$ be a bipartite graph with $|W|=|U|+1$ and $|U|\geq 2$ where $U\cup W$ is a bipartition of G. First, it is proved that G is defect 1-extendable if and only if G' is 1-extendable where G' is a graph obtained from G by adding a new vertex u and joining u to all vertices in W. Second, it is shown that G being defect 1-extendable is equal to the property that G has exactly one minimum point cover U or the property that for any non-empty subset X of U, $|\Gamma_G(X)|\geq|X|+1$. Third, it is proved that if M is a near perfect matching of G and w is the M-unsaturated vertex, then G is defect 1-extendable if and only if for any $v\in U$, there is an M-alternating path between w and v. Finally, an $O(|E(G)|)$ time algorithm is designed to determine whether G is a defect 1-extendable bipartite graph if a maximum matching of G is known.

Keywords: near perfect matching, defect 1-extendable, 1-extendable, maximum matching

1 Terminology and Introduction

All graphs considered in this paper are undirected, finite and simple.

A **perfect matching** is a matching covering all vertices in a graph. A **near perfect matching** is a matching covering all but one vertex in a graph. Let G be a connected graph and $n\leq(|V(G)|-2)/2$ be a positive integer. If any n independent edges in G are contained in a perfect matching, then G is **n-extendable**. If any n independent edges in G are contained in a near perfect matching, then G is **defect n-extendable**. If M is a matching and P is a path in G such that the edges on P appear in M and $E(G)\backslash M$ alternately, then P is an **M-alternating path**.

Plummer[7] introduced n-extendable graphs in 1980. Since then, extensive research has been done on this topic (see [8, 9]). However, n-extendable graphs are all of even order since they have a perfect matching. To naturally extend the property of n-extendablity to graphs of odd order, Lou and Wen [5] introduced defect n-extendable graphs as defined above.

In fact, a few results on defect n-extendable graphs have been established previously. In [2], Grant, Holton and Little characterized defect 1-extendable graphs which were called 1-covered graphs in their paper. In [4], to combine the concept of n-extendable graphs and k-critical graphs, Liu and Yu introduced (k,n,d)-graphs such that (0,n,1)-graphs are the same as defect n-extendable graphs. They gave a Tutte style characterization of (k,n,d)-graphs.

In this paper, we give three characterizations of defect 1-extendable bipartite graphs in Section 3. Then in Section 4, an $O(|E(G)|)$ time algorithm is designed to determine whether G is a defect 1-extendable bipartite graph if a maximum matching of G is known.

2 Notation and Preliminary Results

We use $G=(U,W)$ to denote a bipartite graph G with bipartite sets U and W. Let A and B be two sets. Then $A\triangle B$ denotes the symmetric difference of A and B. Let E_1 be a set of edges not in graph G_1 and each edge in E_1 has both ends in $V(G_1)$. Then $G_1\cup E_1$ denotes a graph by adding edges in E_1 to G_1. Let G_1 and G_2 be two graphs such that $V(G_1)\cap V(G_2)=\phi$. Then $G_1\cup G_2$ denotes a graph which only consists of G_1 and G_2. Let G be a graph and $S\subseteq V(G)$, then $\Gamma_G(S)$ denotes the set of all vertices in G which are adjacent to at least one vertex in S.

For the other notation not defined in this paper, the reader is referred to [1].

We now give some known results which will help to prove our main theorems.

Lemma 2.1 (Hall [3]). *Let $G=(X,Y)$ be a bipartite graph. Then G has a matching of X into Y if and only if $|\Gamma_G(S)|\geq|S|$ for all $S\subseteq X$.*

Lemma 2.2 (Grant, Holton and Little [2]). *A connected bipartite graph $G=(U,W)$ with $|W|=|U|+1$ is defect 1-extendable if and only if for any $S\subset W$ and $S\neq\phi$, $|\Gamma_G(S)|\geq|S|$.*

Lemma 2.3 (Plummer [7]). *Let n be a positive integer. If G is n-extendable, then $\kappa(G)\geq n+1$.*

3 Characterizing Defect 1-Extendable Bipartite Graphs

In this section, we give three characterizations of defect 1-extendable bipartite graphs.

Theorem 3.1. *Let $G=(U,W)$ be a bipartite graph with $|W|=|U|+1$. Let u be a vertex not in $V(G)$ and $G'=G\cup\{u\}\cup\{$edges joining u to all vertices in $W\}$. Then G is defect 1-extendable if and only if G' is 1-extendable.*

Proof. First, we prove the sufficiency. Since G' is 1-extendable, Lemma 2.3 implies $\kappa(G')\geq 2$. So $\kappa(G)=\kappa(G'-u)\geq\kappa(G')-1\geq 1$ and hence G is connected.

Choose any edge e in G. Then $e\in E(G')$ and hence there is a perfect matching M in G' containing e. Clearly, there exists a vertex $w\in V(G')$ such that $uw\in M$. Since $e\in E(G)$ and $u\notin V(G)$, $e\neq uw$ and hence $M\backslash\{uw\}$ is a near perfect matching in G containing e. So G is defect 1-extendable.

We now prove the necessity. Since G is connected and u joins all vertices in W, G' is also a connected bipartite graph. Clearly, $U'=U\cup\{u\}$ and $W'=W$ are two bipartite sets of G' with $|W'|=|U'|$.

Choose any edge f in G'. Assume $f=xy$ where $x\in U'$ and $y\in W'$. It suffices to prove that there is a perfect matching in G' containing f.

Suppose $x = u$. Consider the graph $G - y$, since G is a defect 1-extendable bipartite graph, Lemma 2.2 implies that for any nonempty set $S \subseteq W\backslash\{y\}$, $|\Gamma_G(S)| \geq |S|$. Then $|\Gamma_{G-y}(S)| = |\Gamma_G(S)| \geq |S|$. So by Lemma 2.1, $G - y$ contains a matching M' of $W\backslash\{y\}$ into U which is a perfect matching of $G - y$. So $M' \cup \{xy\}$ is a perfect matching in G' containing f.

Suppose $x \neq u$. Then $f \in E(G)$ and hence there is a near perfect matching M'' in G containing f. There is an M''-unsaturated vertex v in G and $v \in W$. By the construction of G', $uv \in E(G')$. So $M'' \cup \{uv\}$ is a perfect matching in G' containing f. □

Theorem 3.2. *Let $G = (U,W)$ be a bipartite graph with $|W| = |U|+1$ and $|U| \geq 2$. Then the following three statements are equivalent:*

(1) *G is defect 1-extendable.*

(2) *G has exactly one minimum point cover U.*

(3) *For any non-empty subset X of U, $|\Gamma_G(X)| \geq |X| + 1$.*

Proof.
(1)⇒(2). Select any minimum point cover C in G. Suppose $C_U = C \cap U \neq \phi$ and $C_W = C \cap W \neq \phi$. Then in G, vertices in $U - C_U$ can only be joined to vertices in C_W and vertices in $W - C_W$ can only be joined to vertices in C_U. Further, $\kappa(G) \geq 1$, so there is an edge e between C_U and C_W. Since G is defect 1-extendable, there is a near perfect matching M in G containing e. Observe that M matches $U - C_U$ to C_W. So $|C_W| > |U - C_U|$ and hence $|U| < |C_W| + |C_U| = |C|$. But U is a point cover of G since G is a bipartite graph and this contradicts the choice of C.

So $C \subseteq U$ or $C \subseteq W$. Since G is connected, exactly one of $C = U$ or $C = W$ holds. Further, $|W| > |U|$ and C is a minimum point cover, so $C = U$.

(2)⇒(3). Suppose to the contrary there is a non-empty subset X of U such that $|\Gamma_G(X)| \leq |X|$. Clearly, $\Gamma_G(X) \cup (U\backslash X)$ is a point cover in G. Further, $|\Gamma_G(X) \cup (U\backslash X)| \leq |X| + |U| - |X| = |U|$. Since U is the only minimum point cover by (2), $\Gamma_G(X) \cup (U\backslash X) = U$ and hence $X = \phi$ and this contradicts the assumption that X is non-empty. So (3) holds.

(3)⇒(1). Suppose G is disconnected. Then there exists a component $H = (U_H,W_H)$ in G where $U_H \subseteq U$ and $W_H \subseteq W$ such that $|W_H| \leq |U_H|$ and $U_H \neq \phi$. So $|\Gamma_G(U_H)| \leq |W_H| \leq |U_H|$, contradicting (3). So G is connected. Now it suffices to prove that any edge e in G is contained in a near perfect matching. Assume $e = xy$ where $x \in U$ and $y \in W$. Then for any $S \subseteq U\backslash\{x\}$ and $S \neq \phi$, $|\Gamma_{G-x-y}(S)| \geq |\Gamma_G(S)| - 1 \geq |S|$ as $|\Gamma_G(S)| \geq |S|+1$ by (3). So Lemma 2.1 implies there is a matching M in $G - x - y$ covering all vertices in $U\backslash\{x\}$. Obviously $M \cup \{xy\}$ is a near perfect matching in G containing e. □

Theorem 3.3. *Let $G = (U,W)$ be a bipartite graph with $|W| = |U|+1$ and $|U| \geq 2$, M be a near perfect matching in G and w be the M-unsaturated vertex. Then G is defect 1-extendable if and only if for all $v \in U$, there is an M-alternating path between w and v.*

Proof. First, we prove the necessity. Assume $x_0, x_1, \ldots, x_r$ are all the vertices in U that can be reached by an M-alternating path beginning with w. Let $A = \{x_i\colon 0 \leq i \leq r\}$. It suffices to prove that $A = U$.

Suppose to the contrary $A \neq U$. Since M is a near perfect matching and w is the only M-unsaturated vertex, there exists a vertex y_i such that $x_i y_i \in M$ for any

$0 \leq i \leq r$. Let $B = U \backslash A$, $C = \{y_i$: $0 \leq i \leq r\}$ and $D = W \backslash (C \cup \{w\})$. Then $|A| = |C|$ and $|B| = |D| \neq 0$ since $A \neq U$.

Note that no vertex in B is joined to any vertex in C. Otherwise, suppose there is a vertex $v \in B$ that is joined to a vertex y_k, $0 \leq k \leq r$. Then $x_k \in A$ and hence there is an M-alternating path P in G beginning with w and ending with x_k. Suppose $v \in V(P)$. Then in P, the subpath from w to v is an M-alternating path and hence $v \in A$, contradicting the assumption that $v \in B$. So $v \notin V(P)$. Clearly, both end edges of P are in $E(G) \backslash M$. Further, $x_k y_k \in M$ and $y_k v \in E(G) \backslash M$, then $P \cup \{x_k y_k, y_k v\}$ is an M-alternating beginning with w and ending with v and hence $v \in A$, contradicting the assumption that $v \in B$.

Similar to the proof above, we obtain that no vertex in B joins to w.

So vertices in B can only be joined to vertices in $W \backslash (C \cup \{w\}) = D$. Since $\kappa(G) \geq 1$, so there is at least an edge between D and $U \backslash B = A$. Assume ux_j is such an edge where $u \in D$ and $x_j \in A$. Since G is defect 1-extendable, ux_j can be extended to a near perfect matching of G which matches vertices in $B \cup \{x_j\}$ to D. Hence $|D| \geq |B \cup \{x_j\}| = |B|+1$, this contradicts $|D| = |B|$. So $A = U$.

Now we prove sufficiency. Since there is an M-alternating path between w and v for all $v \in U$, G is connected. Choose any edge e in G. Assume $e = xy$ where $x \in U$ and $y \in W$. It suffices to prove that there is a near perfect matching in G containing e.

Since M is a near perfect matching and $x \in U$, there is a vertex y' in G such that $xy' \in M$.

Suppose $y = y'$. Then $e \in M$ and hence M is a near perfect matching in G containing e.

Suppose $y = w$. Then clearly $(M \cup \{e\}) \backslash \{xy'\}$ is a near perfect matching in G containing e.

Suppose $y \in W \backslash \{y', w\}$. Since w is the only M-unsaturated vertex, there is a vertex x' in U such that $yx' \in M$. Then $x' \in U$ and hence there is an M-alternating path P' beginning with w and ending with x'. Observe that P' begins and ends with an edge in $E(G) \backslash M$, so $y \notin V(P')$.

If $x \in V(P')$, let P'' be the subpath of P' from x to x', then $C = P'' \cup \{x'y, xy\}$ is an M-alternating cycle and hence $M \triangle E(C)$ is a near perfect matching in G containing e.

If $x \notin V(P')$, then $y' \notin V(P')$ and hence $P = P' \cup \{x'y, yx, xy'\}$ is an M-alternating path beginning with w and ending with y'. Clearly, P begins with an edge in $E(G) \backslash M$ and ends with an edge in M. So it is easy to see that $M \triangle E(P)$ is a near perfect matching in G containing e. Thus G is defect 1-extendable. □

Corollary 3.4. *Let $G = (U, W)$ be a bipartite graph with $|W| = |U|+1$ and $|U| \geq 2$, M be a near perfect matching in G and w be the M-unsaturated vertex. Then G is defect 1-extendable if and only if for any $v \in W \backslash \{w\}$, there is an M-alternating path between w and v.*

Proof. The proof is immediate from Theorem 3.3. □

4 The Algorithm

In this section, using Theorem 3.3, we design an algorithm to determine whether a graph is a defect 1-extendable bipartite graph if one of its maximum matchings is given. It consists of two functions as follows.

Function Is_defect_1_ext(G,M)
Input: A graph G and its maximum matching M
(1) If G isn't a connected bipartite graph or $|V(G)<5$, then Return *FALSE;*
(2) Else
(3) If M is not a near perfect matching, then Return FALSE;
(4) Else
(5) Begin
(6) w = M-unsaturated vertex in G;
(7) Result=All_Reachable(G,M,w);
(8) Return (Result);
(9) End

Function All_Reachable(G,M,w) uses a BFS (Breadth First Search) strategy to build an M-alternating tree T rooted at w so as to find all vertices that can be reached by an M-alternating path from w.

Function All_Reachable(G,M,w)
Input: A bipartite graph G, a near perfect matching M in G and the M-unsaturated vertex w.
(1) Let tree $T := \{w\}$;
(2) Put w into and empty queue Q, mark w even;
(3) Repeatedly take the first vertex u from Q, do steps (4) until Q is empty;
(4) For each edge uv in G, *do the following step (5)*
(5) If v isn't marked
(6) Begin
(7) Let $vv' \in M$;
(8) $T := T \cup \{v, v'\} \cup \{uv, vv'\}$;
(9) Mark v odd and v' even;
(10) Put v' into Q;
(11) End
(12) If the number of vertices marked odd is equal to (|V|−1)/2, return (TRUE);
(13) Else return FALSE;

5 Correctness of the Algorithm

Now we prove the correctness of the Algorithm.

Theorem 5.1. *Let $G = (U,W)$ be a bipartite graph with $|W| = |U| + 1$, M be a near perfect matching in G and w be the M-unsaturated vertex. Then the function All_Reachable(G,M,w) returns TRUE if and only if for any $v \in U$, there is an M-alternating path between w and v.*

Proof. First we prove the following.

Claim 1. In function All_Reachable(G,M,w), a vertex v in G is marked odd if and only if $v \in U$ and there is an M-alternating path P from w to v.
By the construction of the M-alternating tree T in the function All-Reachable (G,M,w), it is easy to obtain that for any $u \in V(G)$, there is an M-alternating path in T between u and w if and only if in G, there is an M-alternating path between u and w. So the proof of necessity is immediate. We now prove the sufficiency.

Suppose $v \in U$ and in G, there exists an M-alternating path P from w to v. Then P begins and ends with edges in $E(G)\backslash M$. Clearly, in the M-alternating tree T constructed by the function, there is also an M-alternating path from w to v and the path begins and ends with edges in $E(G)\backslash M$. By the construction of tree T, all vertices that can be reached by an M-alternating path from w such that the path begins and ends with an edge in $E(G)\backslash M$ are marked odd, so v is marked odd and this completes the proof of the Claim.

We now return to the proof of the Theorem. Let S be the set of vertices marked odd in function All_Reachable(G,M,w). Note that All_Reachable(G,M,w) returns $TRUE$ if and only if $|S| = (|V|-1)/2$, that is $S = U$. Since S is the set of all vertices in U that can be reached by an M-alternating path beginning with w by Claim 1, function All_Reachable(G,M,w) returns $TRUE$ if and only if for any $v \in U$, there is an M-alternating path P between w and v. □

Theorem 5.2. *Let G be a graph and M be a maximum matching in G. Function Is_defect_1_ext(G,M) returns $TRUE$ if and only if G is a defect 1-extendable bipartite graph.*

Proof. Note that function Is_defect_1_ext(G,M) returns $TRUE$ if and only if G is a connected bipartite graph with $|V(G)| \geq 5$, M is a near perfect matching in G and function All_Reachable(G,M,w) returns $TRUE$ where w is the M-unsaturated vertex in G. By Theorem 3.3 and 5.1, it is not difficult to see that G is a connected bipartite graph with $|V(G)| \geq 5$, M is a near perfect matching in G and function All_Reachable(G,M,w) returns $TRUE$ if and only if G is a defect 1-extendable bipartite graph. So Is_defect_1_ext(G,M) returns $TRUE$ if and only if G is a defect 1-extendable bipartite graph. □

6 Complexity

Now we come to the time complexity of the algorithm for graph G. For simplicity, we use $|E|$ to denote $|E(G)|$ and $|V|$ to denote $|V(G)|$ in the following.

Function All_Reachable takes $O(|E|)$ time since it only executes one BFS to build the M-alternating tree T.

In function Is_defect_1_ext, step (1) takes $O(|E|)$ time because it only needs to execute one BFS to check whether G contains an odd cycle or contains more than one component to identify whether G is a connected bipartite graph, steps (2) to (9) take $O(|E| + |V|)$ time obviously. Clearly, if G is disconnected, only step (1) is executed which takes $O(|E|)$ time and if G is connected, it takes $O(|E| + |V|)$ $= O(|E|)$ time. So function Is_defect_1_ext takes $O(|E|)$ time in all.

By [6], it takes $O(|V|^{1/2}|E|)$ time to find a maximum matching M in G. So to determine whether a graph G is a defect 1-extendable bipartite graph spends the main time in finding a maximum matching in G.

Note that if we use the definition of defect 1-extendable bipartite graphs to design an algorithm, then we should delete every edge e in G and then try to identify whether $G - V(e)$ contains a near perfect matching. In this way, the algorithm needs $O(|V|^{1/2}|E|^2)$ time. So our algorithm has higher efficiency.

References

[1] J.A. Bondy and U.S.R Murty, *Graph theory with applications*, Macmillan Press, London (1976).

[2] D.D. Grant, D.A. Holton and C.H.C. Little, On defect-d matchings in graphs, *Discrete Math.*, **13** (1975), pp. 41–54.

[3] P. Hall, On representatives of subsets, *J. London Math. Soc.*, **10** (1935), pp. 26–30.

[4] G.J. Liu and Q.L. Yu, Generalization of matching extensions in graphs, *Discrete Math.*, **231** (2001), pp. 311–320.

[5] D.J. Lou and X.L. Wen, Characterizing defect n-extendable bipartite graphs with different connectivities, Accepted.

[6] S. Micali and V. V. Vazirani, An $O(|V|^{1/2}|E|)$ algorithm for finding maximum matching in general graphs, in *The 21st Annual Symposium on Foundations of Computer Science, Syracuse, NY* (1980), pp. 17–27.

[7] M.D. Plummer, On n-extendable graphs, *Discrete Math.*, **31** (1980), pp. 201–210.

[8] M.D. Plummer, Extending matchings in graphs: a survey, *Discrete Math.*, **127** (1994), pp. 277–292.

[9] M.D. Plummer, Extending matching in graphs: an update, *Cong. Numer.*, **116** (1996), pp. 3–32.

Author Index

Made in United States
North Haven, CT
05 November 2026

21764887R00163